THE SEMANTIC WEB

THE SEMANTIC WEB

CRAFTING INFRASTRUCTURE FOR AGENCY

Bo Leuf

Technology Analyst, Sweden

John Wiley & Sons, Ltd

This publication is designed to provide accurate and authoritative information in regard to the subject matter covered. It is sold on the understanding that the Publisher is not engaged in rendering professional services. If professional advice or other expert assistance is required, the services of a competent professional should be sought.

Other Wiley Editorial Offices

John Wiley & Sons Inc., 111 River Street, Hoboken, NJ 07030, USA

Jossey-Bass, 989 Market Street, San Francisco, CA 94103-1741, USA

Wiley-VCH Verlag GmbH, Boschstr. 12, D-69469 Weinheim, Germany

John Wiley & Sons Australia Ltd, 42 McDougall Street, Milton, Queensland 4064, Australia

John Wiley & Sons (Asia) Pte Ltd, 2 Clementi Loop #02-01, Jin Xing Distripark, Singapore 129809

John Wiley & Sons Canada Ltd, 22 Worcester Road, Etobicoke, Ontario, Canada M9W 1L1

Wiley also publishes its books in a variety of electronic formats. Some content that appears in print may not be available in electronic books.

Library of Congress Cataloging-in-Publication Data

Leuf, Bo, Technology Analyst, Sweden.
The Semantic Web: crafting infrastructure for agency/Bo Leuf.
      p. cm.
  Includes bibliographical references and index.
  ISBN 0-470-01522-5
  1.   Semantic Web. I. Title.
  TK5105.88815.L48 2005
  025.04–dc22   2005024855

British Library Cataloguing in Publication Data

A catalogue record for this book is available from the British Library

ISBN-13 978-0-470-01522-3 (HB)
ISBN-10 0-470-01522-5 (HB)

Typeset in 10/12pt Times Roman by Thomson Press (India) Limited, New Delhi
Printed and bound in Great Britain by Antony Rowe, Chippenham, Wilts
This book is printed on acid-free paper responsibly manufactured from sustainable forestry in which at least two trees are planted for each one used for paper production.

I dedicate this book to the many visionaries, architects, and programmers, who all made the Internet an interesting place to 'live' and work in.

And especially to Therese.

Contents

Foreword

As an individual, as a technologist, as a business person, and as a civic participant, you should be concerned with how the Semantic Web is going to change the way our knowledge-based society functions.

An encouraging sign is that one of the first large-scale community-based data pools, the Wikipedia, has grown to well over half a million articles. It is an ominous indicator that one of the first large-scale governmental metadata assignment projects is going on in China for the purpose restricting personal access to political information.

The Semantic Web is not a single technology; rather it is a cluster of technologies, techniques, protocols, and processes. As computational power becomes more powerful and more ubiquitous, the amount of control that information technology will hold over people's lives will become more pervasive, and the individual's personal control ever less.

At the same time, the employment of anonymous intelligent agents may buy individuals a new measure of privacy. The Semantic Web is the arena in which these struggles will be played out.

The World Wide Web profoundly transformed the way people gain access to information; the Semantic Web will equally profoundly change the way machines access information. This change will transform yet again our own roles as creators, consumers and manipulators of knowledge.

—Mitchel Ahren, Director of Marketing Operations,
AdTools | Digital Marketing Concepts, Inc.

Preface

This is a book that could fall into several different reader categories – popular, academic, technical – with perhaps delusional ambitions of being both an overview and a detailed study of emerging technology. The idea of writing *The Semantic Web* grew out of my previous two books, *The Wiki Way* (Addison-Wesley, 2001) and *Peer to Peer* (Addison-Wesley, 2002). It seemed a natural progression, going from open co-authoring on the Web to open peer-sharing and communication, and then onto the next version of the Web, involving peer-collaboration between both software agents and human users.

Started in 2002, this book had a longer and far more difficult development process than the previous two. Quite honestly, there were moments when I despaired of its completion and publication. The delay of publication until 2005, however, did bring some advantages, mainly in being able to incorporate much revised material that otherwise would have been waiting for a second edition.

The broader 'Semantic Web' as a subject still remains more of a grand vision than an established reality. Technology developments in the field are both rapid and unpredictable, subject to many whims of fate and fickle budget allocations.

The field is also 'messy' with many diverging views on what it encompasses. I often felt like the intrepid explorer of previous centuries, swinging a machete to carve a path through thick unruly undergrowth, pencil-marking a rough map of where I thought I was in relation to distant shorelines and major landmarks.

My overriding concern when tackling the subject of bleeding-edge technologies can be summed up as providing answers to two fundamental reader questions:

- *What does this mean?*
- *Why should I care?*

To anticipate the detailed exposition of the latter answer, my general answer is that you – *we* – *should* care, because these technologies not only can, but most assuredly *will*, affect us more than we can possibly imagine today.

Purpose and Target

The threefold purpose of the book is rather simple, perhaps even simplistic:

- Introduce an arcane subject comprehensively to the uninitiated.
- Provide a solid treatment of the current 'state of the art' for the technologies involved.
- Outline the overall 'vision of the future' for the new Web.

My guiding ambition was to provide a single volume replete with historical background, state of the art, and vision. Intended to be both informative and entertaining, the approach melds practical information and hints with in-depth analysis.

The mix includes conceptual overviews, philosophical reflection, and contextual material from professionals in the field – in short, all things interesting. It includes the broad strokes for introducing casual readers to ontologies and automated processing of semantics (not a unique approach, to be sure), but also covers a sampling of the real-world implementations and works-in-progress.

However, the subject matter did not easily lend itself to simple outlines or linear progressions, so I fear the result may be perceived as somewhat rambling. Well, that's part of the journey at this stage. Yet with the help of astute technical reviewers and the extended period of preparation, I was able to refine the map and sharpen the focus considerably. I could thus better triangulate the book's position on the conceptual maps of both the experts and the interested professionals.

The technologies described in this book will define the next generation Internet and Web. They may conceivably define much of your future life and lifestyle as well, just as the present day Web has become central to the daily activities of many – the author included. Therefore, it seems fitting also to contemplate the broader implications of the technologies, both for personal convenience and as instigator or mediator of social change.

These technologies can affect us not only by the decisions to implement and deploy them, but sometimes even more in the event of a decision *not* to use them. Either way, the decision taken must be an informed one, ideally anchored in a broad public awareness and understanding of what the decision is about and with some insight into what the consequences might be. Even a formal go-ahead decision is not sufficient in itself. The end result is shaped significantly by general social acceptance and expectations, and it may even be rejected by the intended users.

Some readers might find the outlined prospects more alarming than enticing – that is perhaps as it should be. As with many new technologies, the end result depends significantly on social and political decisions that for perspective require not just a clear vision but perhaps also a glimpse of some dangers lurking along the way.

We can distinguish several categories of presumptive readers:

- The casual reader looking for an introduction and overview, who can glean enough information to set the major technologies in their proper relationships and thus catch a reflection of the vision.
- The 'senior management' types looking for buzzwords and 'the next big thing' explained in sufficient detail to grasp, yet not in such unrelenting technical depth as to daze.
- The industry professional, such as a manager or the person responsible for technology, who needs to get up to speed on what is happening in the field. Typically, the professional wants both general technology overviews and implementation guides in order to make informed decisions.
- The student in academic settings who studies the design and implementation of the core technologies and the related tools.

The overall style and structure of the book is held mainly at a moderate level of technical difficulty. On the other hand, core chapters are technical and detailed enough to be used as a course textbook. All *techno-jargon* terms are explained early on.

In *The Semantic Web,* therefore, you are invited to a guided journey through the often arcane realms of Web technology. The narrative starts with the big picture, a landscape as if seen from a soaring plane. We then circle areas of specialist study, thermal-generating 'hot-spots', subjects that until recently were known mainly from articles in technical journals with limited coverage of a much broader field, or from the Web sites of the institutions involved in the research and development. Only in the past few years has the subject received wider public notice with the publication of several overview books, as noted in Appendix B.

Book Structure

The book is organized into three fairly independent parts, each approaching the subject from a different direction. There is some overlap, but you should find that each part is complementary. Therefore, linear cover-to-cover reading might not be the optimal approach. For some readers, the visions and critiques in Part III might be a better starting point than the abstract issues of Part I, or the technicalities in Part II.

Part I sets the conceptual foundations for the later discussions. These first four chapters present mostly high-level overviews intended to be appropriate for almost anyone.

- The first chapter starts with the basics and defines what Web technology is all about. It also introduces the issues that led to the formulation of the Semantic Web initiative.
- Chapter 2 introduces the architectural models relevant to a discussion of Web technologies and defines important terminology.
- Chapter 3 discusses general issues around creating and managing the content and metadata structures that form the underpinnings of the Semantic Web.
- Finally, Chapter 4 looks at online collaboration processes, which constitute an important motivating application area for Semantic Web activities.

Part II focuses on the technologies behind the Semantic Web initiative. These core chapters also explore representative implementations for chosen implementation areas, providing an in-depth mix of both well-known and lesser-known solutions that illustrate different ways of achieving Semantic Web functionality. The material is detailed enough for Computer Studies courses and as a guide for more technical users actually wanting to implement and augment parts of the Semantic Web.

- Chapter 5 provides layered analysis of the core protocol technologies that define Web functionality. The main focus is on the structures and metadata assertions used to describe and manage published data.
- Chapter 6 is an in-depth study of ontologies, the special structures used to represent term definitions and meaningful relationships.
- Chapter 7 introduces the main organizations active in defining specifications and protocols that are central to the Semantic Web.
- Chapter 8 examines application areas of the Semantic Web where prototype tools are already implemented and available.

- Chapter 9 expands on the previous chapters by examining application areas where some aspect of the technology is deployed and usable today.

Part III elevates the discussion into the misty realms of analysis and speculation.

- Chapter 10 provides an 'insights' section that considers the immediate future potential for Semantic Web solutions, and the implications for users.
- Chapter 11 explores some directions in which future Web functionality might develop in the longer term, such as ubiquitous connectivity and the grander theme of managing human knowledge management.

Finally, the appendices supplement the main body of the book with a terminological glossary, references, and resources – providing additional detail that while valuable did not easily fit into the flow of the main text.

Navigation

This book is undeniably filled with a plethora of facts and explanations, and it is written more in the style of a narrative rather than of reference-volume itemization. Despite the narrative ambitions, texts like this require multiple entry points and quick ways to locate specific details.

As a complement to the detailed table of contents and index, each chapter's 'at a glance' page provides a quick overview of the main topics covered in that chapter. Technical terms in **bold** are often provided with short explanations in the Appendix A glossary.

Scattered throughout the text you will find the occasional numbered 'Bit' where some special insight or factoid is singled out and highlighted. Calling the element a 'Bit' seemed to convey about the right level of unpretentious emphasis – they are often just my two-bits worth of comment. Bits serve the additional purpose of providing visual content cues for the reader and are therefore given their own List of Bits in Appendix C.

When referencing Web resources, I use the convention of omitting the 'http://' prefix because modern Web browsers accept addresses typed in without it. Although almost ubiquitous, the 'www.' prefix is not always required, and in cases where a cited Web address properly lacks it, I tried to catch instances where it might have been added incorrectly in the copyedit process.

The Author

As an independent consultant in Sweden for 30 years, I have been responsible for software development, localization projects, and design-team training. A special interest was adapting an immersive teaching methodology developed for human languages to technical-training fields.

Extensive experience in technical communication and teaching is coupled with a deep understanding of cross-platform software product design, user interfaces, and usability analysis. All provide a unique perspective to writing technical books and articles.

Monthly featured contributions to a major Swedish computer magazine make up the bulk of my technical analyst writing at present. Coverage of, and occasional speaking engagements at, select technology conferences have allowed me to meet important developers.

I also maintain several professional and recreational Internet Web sites, providing commercial Web hosting and Wiki services for others.

Collaborative Efforts

A great many people helped make this book possible by contributing their enthusiasm, time, and effort – all in the spirit of the collaborative peer community that both the Web and book authoring encourage. Knowledgeable professionals and colleagues offered valuable time in several rounds of technical review to help make this book a better one and I express my profound gratitude for their efforts. I hope they enjoy the published version.

My special thanks go to the many reviewers who participated in the development work. The Web's own creator, and now director of the *World Wide Web Consortium*, Sir Tim Berners-Lee, also honoured me with personal feedback.

Thanks are also due to the editors and production staff at John Wiley & Sons, Ltd.

Personal thanks go to supportive family members for enduring long months of seemingly endless research and typing, reading, and editing – and for suffering the general mental absentness of the author grappling with obscure issues.

Errata and Omissions

Any published book is neither 'finished' nor perfect, just hopefully the best that could be done within the constraints at hand. The hardest mistakes to catch are the things we think we know. Some unquestioned truths can simply be wrong, can have changed since we learned them, or may have more complex answers than we at first realized.

Swedish has the perceptive word *hemmablind*, literally blind-at-home, which means that we tend not to see the creeping state of disarray in our immediate surroundings – think of how unnoticed dust 'bunnies' can collect in corners and how papers can stack up on all horizontal surfaces. The concept is equally applicable to not always noticing changes to our particular fields of knowledge until someone points them out.

Omissions are generally due to the fact that an author must draw the line somewhere in terms of scope and detail. This problem gets worse in ambitious works such as this one that attempt to cover a large topic. I have tried in the text to indicate where this line is drawn and why.

Alternatively, I might sometimes make overly simplified statements that someone, somewhere, will be able to point to and say 'Not so!'. My excuse is that not everything can be fully verified, and sometimes the simple answer is good enough for the focus at hand.

A book is also a snapshot. During the course of writing, things changed! Constantly! Rapidly! In the interval between final submission and the printed book, not to mention by the time you read this, they have likely changed even more. Not only does the existing software continue to evolve, or sometimes disappear altogether, but new implementations can suddenly appear from nowhere and change the entire landscape overnight.

Throughout the development process, therefore, book material was under constant update and revision. A big headache involved online resource links; 'link-rot' is deplorable but inevitable. Web sites change, move, or disappear. Some resources mentioned in the text might therefore not be found and others not mentioned might be perceived as better.

The bottom line in any computer-related field is that any attempt to make a definitive statement about such a rapidly moving target is doomed to failure. But we have to try.

Book Support and Contacting the Author

The Internet has made up-to-date reader support a far easier task than it used to be, and the possibilities continue to amaze and stimulate me.

Reader feedback is always appreciated. Your comments and factual corrections will be used to improve future editions of the book, and to update the support Web site. You may e-mail me at *bo@leuf.com*, but to get past the junk filters, please use a meaningful subject line and clearly reference the book. You may also write to me c/o the publisher.

Authors tend to get a lot of correspondence in connection with a published book. Please be patient if you write and do not get an immediate response – it might not be possible. I do try to at least acknowledge received reader mail within a reasonable time.

However, I suggest first visiting the collaborative wiki farm (follow links from *www.leuf.com/ TheSemanticWeb*), where you can meet an entire community of readers, find updates and errata, and participate in discussions about the book. The main attraction of book-related Web resources is the contacts you can form with other readers. Collectively, the readers of such a site always have more answers and wisdom than any number of individual authors.

Thank you for joining me in this journey.

Bo Leuf
Technology Analyst, Sweden
(Gothenburg, Sweden, 2003–2005)

Part I
Content Concepts

1

Enhancing the Web

Although most of this book can be seen as an attempt to navigate through a landscape of potential and opportunity for a future World Wide Web, it is prudent, as in any navigational exercise, to start by determining one's present location. To this end, the first chapter is a descriptive walkabout in the current technology of the Web – its concepts and protocols. It sets out first principles relevant to the following exploration, and it explains the terms encountered.

In addition, a brief Web history is provided, embedded in the technical descriptions. Much more than we think, current and future technology is designed and implemented in ways that critically depend on the technology that came before. A successor technology is usually a reaction, a complement, or an extension to previous technology – rarely a simple plug-in replacement out of the blue. New technologies invariably carry a legacy, sometimes inheriting features and conceptual aspects that are less appropriate in the new setting.

Technically savvy readers may recognize much material in this chapter, but I suspect many will still learn some surprising things about how the Web works. It is a measure of the success of Web technology that the average user does not need to know much of anything technical to surf the Web. Most of the technical detail is well-hidden behind the graphical user interfaces – it is essentially click-and-go. It is also a measure of success that fundamental enhancements (that is, to the basic Web protocol, not features that rely on proprietary plug-in components) have already been widely deployed in ways that are essentially transparent to the user, at least if the client software is regularly updated.

Chapter 1 at a Glance

Chapter 1 is an overview chapter designed to give a background in broad strokes on Web technology in general, and on the main issues that lead to the formulation of the Semantic Web. A clear explanation of relevant terms and concepts prepare the reader for the more technical material in the rest of the book.

There and Back Again sets the theme by suggesting that the chapter is a walkabout in the technology fields relevant to the later discussions that chapter by chapter revisit the main concepts, but in far greater detail.

The Semantic Web: Crafting Infrastructure for Agency   Bo Leuf
© 2006 John Wiley & Sons, Ltd

- *Resource Identifiers* defines fundamental identity concepts, protocol basics, and how content can at all be located in the current Web by the user's client software.
- *Extending Web Functionality* examines proposed ways to enhance the basic Web transport protocol, as well as protocol-independent methods.
- *From Flat Hyperlink Model* describes the current navigational model of the Web, especially the hyperlink, and highlights the areas where it is lacking. After a wishlist of Web functionality, *To Richer Informational Structures* explores strategies for extending the hyperlink model with background information about the content.
- *The Collaboration Aspect* explores one of the driving forces for a new Web, after which *Extending the Content Model* shows why a unified way to handle content is important in any extension.
- *Mapping the Infosphere* discusses ways that have been tried to map what is on the Web so that users can find what they are looking for. *Well-Defined Semantic Models* introduces why current lexical mappings are insufficient for the task, especially if searching and processing is to be automated.

There and Back Again

The World Wide Web was conceived and designed as an open information space defined by the *hyperlink* mechanism that linked documents together. The technology enabled anyone to link to any other document from hyperlinks on a published Web page – a page anyone could see, and link to in turn. The whole system could self-grow and self-organize.

No permissions were required to set up such links; people just linked to whatever other published resources they found interesting and useful. The only requirements to participate were a simple Internet connection and a place to put a Web page. This open nature is fundamental to many of the early design decisions and protocol implementations, sometimes in ways that were later obscured or marginalized.

Bit 1.1   The Web is an open universe of network-accessible information
This definition of the Web, formulated by Tim Berners-Lee, provides in all its simplicity the most fundamental description of the Web's potential.

Open network access enables a potentially infinite resource, for people both to contribute to and use. The explosive growth of the Web and the content it mediates is in many ways a direct consequence of this design. It has given rise to a remarkable plethora of both content and functionality, sometimes unexpected.

I am very happy at the incredible richness of material on the Web, and in the diversity of ways in which it is being used. There are many parts of the original dream which are not yet implemented. For example, very few people have an easy, intuitive tool for putting their thoughts into hypertext. And many of the reasons for, and meaning of, links on the web is lost. But these can and I think will change.

Tim Berners-Lee (*www.w3.org/People/Berners-Lee/FAQ.html*),
'inventor' of the Web and director of the W3C.

In addition, the original design had the goal that it should not only be useful for human-to-human communication but also support rich human–machine and machine–machine interactions. In other words, the intent was that machines would be able to participate fully and help in the access and manipulation of this information space – as automated agents, for example.

Bit 1.2   The Web had the twin goals of interactive interoperability and creating an evolvable technology

The core values in Web design are expressed in the principle of universality of access – irrespective of hardware or software platform, network infrastructure, language, culture, geographical location, or physical or mental impairment.

Before going further into the nature of such interactions and the infrastructure that is to support them, we need to explore the fundamental issues of resource identity, and how naming schemes relate to the protocols used to access the resources.

Resource Identifiers

The full interoperability and open-ended nature of the Web was intended to be independent of language, as evident in the way the design specified the universality of referencing resources by identity through the **Universal Resource Identifier** (or **URI**).

The principle that absolutely anything 'on the Web' can be identified distinctly and uniquely by abstract pointers is central to the intended universality. It allows things written in one language to refer to things defined in another language.

Properties of naming and addressing schemes are thus defined separately, associated through the **dereferencing protocol**, allowing many forms of identity, persistence and equivalence to refer to well-defined resources on the Web. When the URI architecture is defined and at least one dereferencing protocol implemented, the minimum requirement for an interoperable global hypertext system is just a common format for the content of a resource (or Web object).

Anyone can create a URI to designate a particular Web resource (or anything, actually). This flexibility is at the same time both the system's greatest strength and a potential problem. Any URI is just an identifier (a 'name' often opaque to humans), so simple inspection of it in isolation does not allow one to determine with certainty exactly what it means. In fact, two different URIs might refer to the same resource – something we often also run across in naming schemes in the 'everyday' world.

The concept of unique identifiers finds expression in many fields, and is crucial to 'finding' and handling things in a useful way.

Any identifier scheme assuredly defines a useful **namespace**, but not all schemes provide any useful dereferencing protocol. Some examples from the latter category are the **MIME** content identifier (*cid*) or message identifier (*mid*) spaces, the MD5 **hash code** with verifiable pure identity (often used as secure verification of file identity), and the pseudo-random Universally Unique Identifier (*uuid*). They all identify but cannot locate.

The ability to utilize such namespace schemes in the URI context provides valuable functionality—as is partially illustrated by some peer-to-peer (p2p) networks, such as *Freenet* (described in the previous book, *Peer to Peer*).

Even a simple persistent identity concept for connection-oriented technologies, for which no other addressable content exists, can prove more useful than might be expected. Witness the ubiquitous mailbox namespace defined by the *'mailto:'* protocol – unfortunately named, however, since URIs are functionally nouns not verbs. The resulting URIs define connection endpoints in what is arguably the most well-known public URI space, persistent virtual locations for stores of e-mail messages.

Understanding HTTP Space

The other most well-known public URI space is the **HTTP** namespace – commonly called 'the Web'. It is characterized by a flexible notion of identity and supports a richness of information about resources and relating resources.

HTTP was originally designed as a protocol for remote operations on objects, while making the exact physical location of these objects transparent. It has a dereferencing algorithm, currently defined by HTTP 1.1, but augmented by caching, proxying and mirroring schemes. Dereferencing may therefore in practice take place even without HTTP being invoked directly.

The HTTP space consists of two parts:

- **Domain Name**, a hierarchically delegated component, for which the Domain Name System (DNS) is used. This component is a centrally registered top level domain (**TLD**): generic (**gTLD**, such as *example.com*) or national (**ccTLD**, *example.se*).
- **Relative Locator**, an opaque string whose significance is locally defined by the authority owning the domain name. This is often but need not (indeed should rarely) be a representation of a local directory tree path (relative to some arbitrary 'root' directory) and a file name (example: */some/location/resource*).

A given HTTP URI (or resource object identity) is commonly written as a **URL**, a single string representing both identity and a Web location. URL notation concatenates the parts and prefixes the protocol (as *http://*). As a rule, client software transparently maps any 'illegal' characters in the URL into protocol-acceptable representations, and may make reasonable assumptions to complete the abbreviated URL entry.

In practice, the domain component is augmented by other locally defined (and optional) prefix components. Although still formally DNS arbitrated, the prefix is determined by the local authority. It is resolved by the DNS and Web (or other service) servers in concert. Most common is '*www.*' but it may also be a server name, a so-called vanity domain, or any other local extension to the domain addressing scheme.

A related important feature is to dereference the physically assigned IP number bound to a particular machine. Domain names therefore improve **URI persistence**, for example when resources might move to other machines, or access providers reallocate. However, persistence of locators in HTTP space is in practice not realized fully on the Web.

Bit 1.3   URLs are less persistent overall than might reasonably be expected

Improving persistence involves issues of tool maturity, user education, and maturity of the Web community. At a minimum, administrators must be discouraged from restructuring (or 'moving') resources in ways that needlessly change Web addresses.

The design of HTTP and the DNS makes addressing more human-readable and enables almost complete decentralization of resources. Governance is freely delegated to local authorities, or to endpoint server machines. Implementing a hierarchical rather than flat namespace for hosts thus minimizes the cost of name allocation and management.

Only the domain-name component requires any form of formal centralization and hierarchical structure – or rather, only as currently implemented does it require centralized registrars and domain-name databases for each TLD.

The domain name is, strictly speaking, optional in HTTP. It is possible, if not always convenient (especially with the trend to share IP in virtual hosting), to specify HTTP resource addresses using the physical IP number locally assigned by an access provider. Other protocol spaces, such as *Freenet*, in fact dispense with domains altogether and rely instead on unique key identifiers and node searches.

As a feature common and mandatory to the entire HTTP Web, as it is currently used, the **DNS root** is a critical resource whose impartial and fair administration is essential for the world as a whole. Ownership and governance of the root of the DNS tree and *gTLD* subtree databases has in recent years been subject to considerable debate. The situation is only partially and nominally resolved under international arbitration by **ICANN**.

The Semantics of Domain Names

Another issue concerning *gTLD* allocation, with relevance to the concept of 'meaning' (or semantics) of the URI, is the design intent that the different domain categories say something about the owners.

The original international *gTLD*s and their intended application were clear, for example:

- '.com' for commercial organizations with a true international presence (those with presence just in a single national region were recommended to use instead country-code domains).
- '.org' for non-profit organizations with an international presence.
- '.net' for international providers of network services, for example Web hosts or service providers.

Such a division provides basic meta-information embedded in the URL in a way that is easy to see. However, so many U.S. companies registered under *.com* that the common perception became that *.com* designated a U.S. business. This skew was due mainly to the original strict rules concerning the allocation of *.us* domains by state, rather than allowing a company with a multi-state presence to use a single national domain.

The *.com* domains became in fact the most popular on the Web, a status symbol no matter what the purpose. It also gave rise to the pervasive '*dotcom*' moniker for any Internet-related business venture. In a similar vein, though less so, the popular but mistaken association arose that *.org* and *.net* were U.S. domains as well. Administration by a U.S. central registrar only strengthened this false association.

- These three *gTLDs* are considered *open domains*, because there are no formal restrictions on who may register names within them. Anyone, anywhere, can therefore have a *dotcom*, *dotorg*, or *dotnet* domain – business, individual, whatever.

Further confusion in the role of these *gTLDs* arose when domain names became traded commodities. Public perception tends to equate brand name with domain name, regardless of the TLD. Attractive names thus became a limited commodity and the effective namespace smaller. Most brand owners are likely to register all three at once.

Bit 1.4   The TLD level of domain names currently lacks consistent application
The TLD system defines the root of the URLs that, as subsets of URIs, are supposed to provide an unambiguous and stable basis for resource mapping by resource owners. Reality is a bit more complex, also due to hidden agendas by the involved registrars.

Therefore, over time, the original semantic division was substantially diluted, and essentially it is today lost from public awareness. However, the intent of at least one of the original seven *gTLD* was preserved:

- '.int' is used only for registering organizations established by international treaties between governments, or for Internet infrastructure databases.

Since the early Internet was mostly implemented in the U.S. and the administration of the *gTLD* names was then under U.S. governance, the other three original categories were quickly reserved for U.S. bodies:

- '.edu' for universities and corresponding institutions of higher education that qualified, became in practice the domain for U.S.-based institutions.
- '.gov' became reserved exclusively for the United States Government and its federal institutions.
- '.mil' became reserved exclusively for the United States Military.

Other countries must use qualified *ccTLDs* for the same purpose, such as *.gov.uk*. A further special TLD, '.arpa' is provided for technical infrastructure purposes.

In an apparent attempt to reclaim some inherent TLD meaning, new usage-restricted *gTLDs* were proposed in 2000. First implemented were *.biz*, *.info*, *.pro*, and *.name*, while *.coop*, *.aero*, and *.museum* are still pending (see FAQ at *www.internic.net/faqs/new-tlds.html*). Since 2004, ICANN has added *.asia*, *.cat*, *jobs*, *.mail*, *.mobi*, *.post*, *.tel*, *.travel*, and *.xxx* to the proposal list.

Unfortunately, we already see semantic erosion in some of the newly implemented TLDs. Also, the addition of more TLD namespaces not only risks further confusion in public perception (*.biz* or *.com*?), but can be detrimental to the original concept of reducing the cost of namespace management. Brand owners may feel compelled redundantly to add further TLDs to protect against perceived misappropriation.

Bit 1.5   Conflicting interests are fragmenting the Web at the TLD-level

Issues of ownership and intent involve complex and changing policy and politics, not easily pinned down in any lasting way, and often at odds with the Web's underlying concept of universality. The resulting confusion, market pressures, and market conflict subtly distort the way we can usefully map the Web.

In 2004, the W3C was, in fact, moved to state that the influx of new TLD subtrees was harmful to the Web infrastructure, or at the very least incurred considerable cost for little benefit. Detailed critique is given for some of these 'special-interest' proposals (see reasoning at *www.w3.org/DesignIssues/TLD* and *www.w3.org/2004/03/28-tld*), for example:

- Implementing *.mobi* would seem to partition the HTTP information space into parts designed for access from mobile devices and parts not so designed. Such a scheme destroys the essential Web property of device independence.
- The domain name is perhaps the worst possible way of communicating information about the device. A reasonable requirement for adaptive handling of mobile devices is that it be transparent, by way of stylesheets and content negotiation.
- Content-filtering through TLD (as in *.xxx*) is unlikely to be effective, even assuming best-case consensus, applicability, and binding international agreements on appropriate domain owners and suitable content in the respective TLD. We may also assume that many companies would merely redirect new special-interest domains (such as *.travel*) to existing ones (an established *.com*, for instance).

As it happens, even the superficially clear geographic relationship of the *ccTLDs* to the respective countries has been considerably diluted in recent years. The increased popularity of arbitrary businesses, organizations, or individuals registering attractive small-country domains as alternatives to the traditional *dotcom* ones causes more TLD confusion.

In part, this practice reflects a preference for country codes that impart some 'useful' association, which can become popular in the intended contexts – for example, Tongan '*.to*' as in '*http://come.to*', or Samoan '*.ws*' recast as meaning 'website' or 'worldsite'. In part, it is a result of the increasing scarcity of desired or relevant *dotcom* names. This 'outsourced' registration, usually U.S.-based through some licensing agreement, generates significant foreign income (in U.S. dollars) for many small Pacific island nations, but assuredly it confuses the previously clear knowledge of *ccTLD* ownership.

Pervasive Centralization

Despite the decentralized properties inherent to the HTTP space design, the subsequent evolution of the Web went for a time in a different direction.

A needless fragmentation situation arose, with different protocols used to transfer essentially the same kind of text (or message) objects in different contexts – for example, e-mail, Web page content, newsgroup postings, chat and instant messaging. Such fragmentation, which leads to multiple client implementations, is confusing to the user. The confusion is only recently in part resolved by advanced multiprotocol clients.

Web implementations also ignored to a great extent the interactive and interoperative design goals – much to the disappointment of the early visionaries who had hoped for something more open and flexible. This deficiency in the Web is part of what prompted the development of other clients in other protocols, to implement functionality that might otherwise have been a natural part of the HTTP Web.

The pervasive paradigm of the Web instead became one of centralized content-providing sites designed to serve unilaterally a mass of content-consuming clients. Such sites constrain user interaction to just following the provided navigational links.

Web pages thus became increasingly 'designer' imprinted, stylistic exercises 'enhanced' with attention-grabbing devices. Technology advances unfortunately tended to focus on this eyeball functionality alone. In fact, the main visitor metric, which tellingly was used to motivate Web advertising, became 'page hits' or 'eyeball click-through counts' rather than any meaningful interaction.

Most Web-browser 'improvements' since the original *Mosaic* client (on which *MS Internet Explorer*, the long dominant browser, is based) have thus dealt more with presentational features than any real navigational or user-interaction aspects. Worse, much of this development tended towards proprietary rather than open standards, needlessly limiting the reach of information formatted using these features.

Revival of Core Concepts

Despite the lackluster Web client implementations with respect to user functionality, the potential for interactive management of information on the Web remained an open possibility – and a realm in recent years extended by alternative peer architectures.

The way the Internet as a whole functions means that nothing stops anyone from deploying arbitrary new functionality on top of the existing infrastructure. It happens all the time. In fact, Internet evolution is usually a matter of some new open technology being independently adopted by such a broad user base that it becomes a *de facto* new standard – it becomes ever more widely supported, attracting even more users.

Bit 1.6   Internet design philosophy is founded in consensus and independent efforts contributing to a cohesive whole

Principles such as simplicity and modularity, basic to good software engineering, are paralleled by decentralization and tolerance – the life and breath of Internet.

Several open technologies have in recent years provided a form of revival in the field, with a focus on content collaboration. One such technology, explored in an earlier book, *The Wiki Way*, is a simple technology that has in only a few years transformed large parts of the visible Web and redefined user interaction with Web-published content.

- Wiki technology relies on the stock Web browser and server combination to make collaborative co-authoring a largely transparent process. It leverages the existing client-server architecture into a more peer-based collaboration between users.
- Prominent examples of large-scale deployment are *Wikipedia* (*www.wikipedia.com*) and related *Wikimedia* projects, and many open-source support sites (*sourceforge.net*).

Other extending solutions tend to be more complex, or be dependent on special extensions to the basic Web protocol and client-server software (such as plug-in components, or special clients). Yet they too have their validity in particular contexts.

This tension between peer and hierarchical models was further explored and analyzed in considerable detail in the book *Peer to Peer*, along with the concept of **agency**. Although these peer technologies lie outside the immediate scope of the current book, some aspects drawn from these discussions are taken up in relevant contexts in later chapters.

Extending Web Functionality

With the enormous growth of the Web and its content since its inception, it is clear that new implementations must build on existing content to gain widespread usage, or at least allow a relatively simple and transparent retrofit. This development can be more or less easy, depending on the intentions and at what level the change comes.

So far, we have seen such changes mainly at the application level. Long sought is a change at a more profound level, such as a major extension to the underlying Web protocol, HTTP. Such a change could harmonize many functionality extensions back into a common and uniform framework on which future applications can build.

Current HTTP does in fact combine the basic transport protocol with formats for limited varieties of **metadata** – information about the payload of information. However, because it is descended from the world of e-mail transport (and an old protocol), HTTP metadata support as currently implemented remains a crude architectural feature that should be replaced with something better.

Bit 1.7   The Web needs a clearer distinction between basic HTTP functionality and the richer world of metadata functionality

A more formalized extension of HTTP, more rigorous in its definitions, can provide such a needed distinction and thus bring the Semantic Web closer to reality.

Extending HTTP

HTTP was designed as part of a larger effort by a relatively small group of people within the IETF HTTP Working Group, but Henrik Frystyk Nielsen (the specification author) claims that this group did not actually control the protocol. HTTP was instead considered a 'common good' technology, openly available to all.

In this vein of freedom, HTTP was extended locally as well as globally in ways that few could predict. Current extension efforts span an enormous range of applications, including distributed authoring, collaboration, printing, and remote procedure call mechanisms.

However, the lack of a standard framework for defining extensions and separating concerns means that protocol extensions have not been coordinated. Extensions are often applied in an *ad hoc* manner which promotes neither reusability nor interoperability.

For example, in the variant **HTTPS** space, a protocol distinction is made needlessly visible in the URI. Although HTTPS merely implies the use of HTTP through an encrypted *Secure Socket Layer* (**SSL**) tunnel, users are forced to treat secure and insecure forms of the same document as completely separate Web objects. These instances should properly be seen as transparent negotiation cases in the process of dereferencing a single URI

Therefore, the *HTTP Extension Framework* was devised as a simple yet powerful mechanism for extending HTTP. It describes which extensions are introduced, information about who the recipient is, and how the recipient should deal with them.

The framework allows parameters to be added to method headers in a way that makes them visible to the HTTP protocol handler (unlike **CGI** parameters, for example, that must remain opaque until dealt with by the handler script on the target server). A specification of the HTTP Extension Framework is found in RFC 2774. (Among other sources, see user-friendly Web site *www.freesoft.org/CIE/RFC/r* to search and read **RFC** documents.)

Otherwise, the most ubiquitous and transparent functionality change in the Web in recent years was an incremental step in the basic Web protocol from HTTP 1.0 to HTTP 1.1. Few users noticed this upgrade directly, but a number of added benefits quickly became mainstream as new versions of server and client software adapted.

- Examples of new features include *virtual hosting* (to conserve the IP-number space), *form-based upload* (for browser management of Web sites), and *MIME extensions* (for better multimedia support).

Many would instead like to see something more like a step to 'HTTP-NG' (Next Generation) that could implement a whole new range of interoperable and interactive features within the base protocol.

- *WebDAV* is one such initiative, styled as completing the original intent of the Web as an open collaborative environment, and it is discussed in Chapter 4.

Consider, after all, how *little* HTTP 1.x actually gives in the form of 'exposed interfaces', otherwise often termed 'methods' and known by their simple names. These methods are in effect the 'action verbs' (one of many possible such sets) applied to the 'identifier nouns' (URI) of the protocol.

The main method is to access resources:

- **GET** is the core request for information in the Web. It takes the form of HTTP header specifying the desired information object as either the unique location (URL) or the process (CGI) to produce it. (A variant, HEAD, might support the return of header information but without data body.)

GET is actually the only HTTP method that is required always to be supported. It has a special status in HTTP, in that it implements the necessary dereferencing operation on identifiers – it defines the namespace. As such, GET must never be used in contexts that have side-effects. Conversely, no other method should be used to perform only URI dereferencing, which would violate universality by defining a new namespace.

Many *ad hoc* extensions and p2p-application protocols are based solely on GET.

The GET-queried resource is expected to return the requested information, a redirection URI, or possibly a status or error message. It should never change state. In the request response, an appropriate *representation* of the URI-specified object is transferred to the client – *not*, as is commonly assumed, the stored literal data.

Representations, encoding, and languages acceptable may be specified in the GET-header request fields, along with any specific client-side information. These and other factors affect both what is returned and the chosen format or encoding.

Client-side input may be handled by two other methods:

- **PUT** is the method to store information at a particular Web location specified by a valid URL. It is structurally similar to GET, except that the header is also associated with a body containing the data to be stored. Data in the body comprise opaque bitstreams to HTTP agents.
- **POST** is in effect an *indirect* method to pass an information object to a Web server. It is entirely up to the server to determine both when and how to deal with it (process, store, defer, ignore, or whatever). The header URI specifies the receiving server agent process (such as a named script), possibly including a suggested URI for the object.

Other methods sometimes seen are essentially only extensions of these basic three. It is optional for the responding HTTP agent (the Web server) to implement an appropriate process for any of these input methods.

The following are some formally registered examples:

- CHECKOUT and CHECKIN are version-control variants corresponding to GET and PUT, but with the added functionality of locking and unlocking, respectively, the data object for access/change by other users.
- TEXTSEARCH and SPACEJUMP are extended forms of GET applied to search and map coordinate positioning.
- LINK and UNLINK are variants of POST to add and remove meta information (object header information) to an object, without touching the object's content.

In fact, it is relatively rare to see PUT used these days. Most servers are configured to deny such direct, external publishing requests, except perhaps in specific, well-authenticated contexts. Instead, POST is used to pass data from client to server in a more open-ended way, by requesting that a server-defined link be created to the passed object. Although, traditionally, POST is used to create, annotate, and extend server-stored information, it was successfully MTME-extended in v1.1 to serve as a generic data upload and download mechanism for modern browsers.

The use of POST allows tighter server control of any received information. Perhaps more relevant to security, it gives server control over processing, storage location, and subsequent access. The client may suggest a storage URI, but the server is never obliged to use it. Even if a POST is accepted by the server, the intended effect may be delayed or overruled by subsequent processing, human moderation, or batch processing. In fact, the creation of a valid link may not only be significantly delayed, it may never occur at all.

Overall, the guiding design thought is that HTTP requests may be cached. From the perspective of the client/user, simple requests must be *stateless*, repeatable, and free of side-effects.

Bit 1.8   HTTP is, like most basic Internet protocols, stateless

State tracking is, however, desired in many situations, but must then be implemented using message-based mechanisms external to the base protocol. Information stored client-side in Web-browser 'cookies' are but one example of such workaround measures.

- It must be noted that POST is strictly speaking neither repeatable nor free from side-effect, and thus not a proper replacement for PUT (which is both). In fact, POST requests are by definition declared *noncachable*, even though it might prove possible in some situations to cache them with no side-effects.
- Modern browsers correctly warn the user if a POST request is about to be reissued. The reason is of course that a POST often alters information state at the server, for example to generate content change, or to commit a unique credit-card transaction.

State considerations place serious constraints on possible functionality in the context of the current HTTP Web.

Extending in Other Layers

Extending the transport protocol (HTTP) is by no means the only proposed solution to extending Web functionality. The growing adoption of **XML**, not just in preference to **HTML** as markup but crucially as a real language for defining other languages, provides an alternative extension framework.

XML is more broadly application-managed than HTML, and its functionality definitions are independent of whether the underlying HTTP is extended or not. Implementations include message-based protocols to extend Web functionality by embedding the extension methods inside the message content passed by the base protocol.

The use of XML for inter-company remote operations became prevalent in 2001, mainly because of its ability to encapsulate custom functionality and convey specialized content meaning. This kind of functionality extension is what led to the development of **Web Services**, a term implying standard functionality distributed across the Web.

Even the markup change alone has direct potential benefits on existing Web functionality – for example, search. Most search engines read the format languages, usually HTML tags that may often be applied inappropriately from the point of view of logical structure. Consequently, the search results tend to reflect the formatting tags rather than actual page content as expressed in natural language. XML markup can express semantic meaning and thus greatly improve search-result relevancy.

There are both pros and cons in the XML approach to extending the Web framework. Some of both aspects hinge on the fact that the implemented message passing often relies on the HTTP POST method. Embedding messages in GET headers imposes unacceptable message constraints.

Extended headers, for example, might be passed as POST content and thus be opaque to the HTTP agent. When referring to a particular resource addresses, this embedded reference cannot be cached or stored in the way a GET-request URI can.

- A typical workaround is a syntax extraction procedure to let the agent recover a workable URI from the POST header response to a GET request. The solution allows HTTP agents not to consider content-embedded extensions when responding to arbitrary requests.

The XML extension approach shares header opaqueness, seen from the point of view of HTTP agents (such as servers, proxies, and clients), with more proprietary extensions anchored in specific applications. Nonetheless, XML extension at least has the virtue of being an open standard that can be adopted by any implementation and be freely adapted according to context. It is ongoing work.

Functionality extension is not simply about extending the protocol; other issues must also be considered.

Content Considerations

A major obstacle facing anyone who wishes to extend the usefulness of the Web information space to automated agents is that most information on the Web is designed explicitly for human consumption.

Some people question what they see as a 'geek' focus on extending and automating the Web with new gee-wizardry. Instead, they suggest that most people seem happy with the Web as designed for human browsing, and do not necessarily have a problem with that. Most people, they say, are finding more or less what they want, and are finding other things of interest as they browse around looking for what they want.

Just because current Web content is designed supposedly for humans does not mean that it is a 'successful' design, however. For example, published Web content commonly relies on any number of implied visual conventions and assumptions to convey intended structure and meaning. Many of these often complex stylistic elements are neither easily nor unambiguously interpreted by 'most people' even with explicit instruction, and much less is it possible to decode them by machines. Furthermore, such style conventions can vary significantly between different authors, sites, and contexts.

Embedded metadata describing the intent behind arbitrary visual convention would not just aid machines, but could provide unambiguous cues to the human reader – perhaps as consistent client-side styling according to reader preference, or as cue pop-up.

- The advantage of such explicit guidance is most obvious in the case of browsers for the visually impaired, a group of human readers who often have significant problems dealing with existing content supposedly designed for them.

Even in cases where the information is derived from a database, with reasonably well-defined meanings for at least some terms in its tabled layout, the structure of the presented data is not necessarily evident to a robot browsing the site. Almost any such presentation has implicit assumptions about how to interpret and connect the data. Humans readers familiar with the material can deal with such assumptions, but unaided machines usually cannot.

For that matter, the material might not even be possible to browse for a machine following embedded hyperlinks. Other, visual navigational conventions might have been implemented (such as image maps, form buttons, downloaded *ActiveX* controls, scripted mouse-state actions, flash-animated widgets, and so on).

Bit 1.9   Machine complexities that depend on human interpretation can signal that the problem statement is incorrectly formulated

A fundamental insight for any attempt to solve a technological problem is that tools and technologies should be used in appropriate ways. In cases where it becomes increasingly difficult to apply the technology, then it is likely that significant advances require fundamentally changing the nature of the problem.

Instead of tackling the perhaps insurmountable artificial-intelligence problem of training machines to behave like people ('greater AI'), the Semantic Web approach is to develop languages and protocols for expressing information in a form that can be processed by machines.

- This 'lesser AI' approach is transparent to the human users, yet brings information-space usability considerably closer to the original and inclusive vision.

Another content issue is the proliferation of different content formats on the Web, which usually require explicit negotiation by the user. This barrier not only complicates and hides the essential nature of the information on the Web, but also makes the content more difficult to handle by software agents.

Granted, the initial design of HTTP did include the capability of negotiating common formats between client and server. Its inclusion was based on the correct assumption that the ideal of a single content format would remain a vision obstructed by the wild proliferation of proprietary data formats. However, this negotiation feature has never really been used to advantage. The rapid adoption of HTML as the common formatting language of the Web did make the need less pressing, but the real reason is that the ever-larger list of possible formats made automatic negotiation impractical.

Instead content format became classified (and clarified, compared to traditional extension-based advertising) by importing the MIME-type concept from the realm of e-mail. Reading the MIME declaration tells the handler/client how to handle an otherwise opaque data stream, usually deferring interpretation to a specified client-side application.

- Although MIME declarations now formally refer to a central registry kept by **IANA**, there is no reason why the Web itself cannot be used as a distributed repository for new types.
- A transition plan would allow such a migration. Unqualified MIME types are in this scheme interpreted as relative URIs within a standard reference URI, in an online MIME registry.

Services on the Web

The issues of distributed services (**DS**) and remote procedure calls (**RPC**) constitute a kind of parallel development to the previously considered Web extensions, and are also considered a future part of the Semantic Web. At present, they coexist somewhat uneasily with the Web, since they operate fundamentally outside Web address space yet fulfill some of the same functionality as the proposed **Web Services** (**WS**).

DS and RPC applications often use proprietary protocols, are highly platform dependent, and are tied to known endpoints. WS implementations function within the Web address space, using Web protocols and arbitrary endpoints. They also differ from DS and RPC remote operation work in that WS transactions are less frequent, slower, and occur between non-trusted parties. Issues such as 'proof of delivery' become important, and various message techniques can become part of the relevant WS protocols.

Further discussion of these issues is deferred to later chapters. Instead the next few sections trace the conceptual developments that underpin the Semantic Web.

From Flat Hyperlink Model

In the Beginning was the Hyperlink...   The click-to-browse hyperlink, that core convention of text navigation, and ultimately of Web usability, is in reality a *key-data pair* interpreted in a particular way by the client software. It reads as a visible anchor associated with a hidden representation of a Web destination (usually the URL form of a URI).

The anchor can be any structural element, but is commonly a selection of text or a small graphic. By convention, it is rendered visually as underlined or framed, at least by default, though many other rendering options are possible.

Bit 1.10   Good user interface design imbues simple metaphors with the ability to hide complex processing

The clicked text hyperlink in its basic underlined style might not have been visually very pleasing, yet it was easily implemented even in feature-poor environments. Therefore, it quickly became the ubiquitous symbol representing 'more information here'.

Functionally, the hyperlink is an *embedded pointer* that causes the client to request the implied information when it is activated by the appropriate user action.

The pragmatics of this mechanism, and the reason it defined the whole experience of the Web, is that it makes Web addressing *transparent to the user.* A simple sequence of mouse clicks allows the user freely to browse content with little regard for its location.

The concept is simple and the implementation ingenious. The consequences were far-reaching. Figure 1.1 illustrates this concept.

This 'invention of the Web' has generally been attributed to Tim Berners-Lee (knighted for his achievements in 2004, thus now Sir Tim), who also founded the World Wide Web Consortium (W3C, *www.w3.org*) in 1994. A brief retrospective can summarize the early development history.

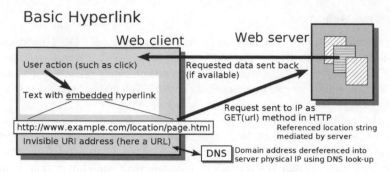

Figure 1.1 Conceptual view of hyperlink functionality as it is currently implemented to form the World Wide Web. The interlinking hyperlinks in Web content provide a navigational framework for users browsing it

In the late 1980s, Tim led an effort with Robert Cailliau at the CERN nuclear research center in Switzerland to write the underlying protocols (including HTTP) for what later came to be known as the World Wide Web. The protocols and technologies were disseminated freely with no thought of licensing requirements.

The early work on the Web was based on, among other things, earlier work carried out by Ted Nelson, another computer and network visionary who is generally acknowledged to have coined the term 'hypertext' in 1963 and used it in his 1965 book, *Literary Machines*. Hypertext linking subsequently turned up in several contexts, such as in online helpfiles and for CD-content navigation, but really only became a major technology with the growth of the public Web.

The matter of public availability deserves further discussion. Although it may seem like a digression, the issues raised here do in fact have profound implications for the Web as it stands now, and even more so for any future Semantic Web.

Patents on the Infrastructure

Since hyperlink functionality is seen as a technology, like any other innovation, it is subject to potential issues of ownership and control.

Even though the W3C group published Web protocols and the hyperlink-in-HTML concept as open and free technology (expressed as 'for the greater good'), it was only a matter of time before somebody tried to get paid for the use of such technologies when the economic incentive became too tempting.

The issue of patent licensing for the use of common media formats, client implementations, and even aspects of Internet/Web infrastructure is problematic. While the legal outcome of individual cases pursued so far can seem arbitrary, taken together they suggest a serious threat to the Web as we know it. Commercialization of basic functionality and infrastructure according to the models proposed by various technology stakeholders would be very restrictive and thus unfortunate for the usability of the Web.

Yet such attempts to claim core technologies seem to crop up more often. In some cases, the patent claims may well be legitimate according to current interpretations – for example, proprietary compression or encryption algorithms. But in the context of the Web's status as a global infrastructure, restrictive licensing claims are usually damaging.

Further complications arise from the fact that patents are issued by respective countries, each with its own variant patent laws. Trying to apply diverging country-specific patents to a global network that casually disregards national borders – in fact, appears innately to transcend such artificial boundaries – seems impossible.

It beomes especially difficult if one country's patented technology is another's public-domain technology (public-key cryptography was but one early example of this quandary). In practice, U.S. patent and copyright interpretations tend to be enforced on the Web, yet this interim state of affairs is unacceptable in the long term because it suggests that the global Internet is owned by U.S. interests.

If a particular technology is critical for a widely deployed functionality on the Web, allowing arbitrary commercial restrictions easily leads to unreasonable consequences. Many corporations, realizing this problem, are in fact open to free public licensing for such use (usually formally with the W3C). Sadly, not all business ventures with a self-perceived stake in Web technology are as amenable.

Bit 1.11   Commercialization (thus restriction) of access cripples functionality

This view, while not popular among companies that wish to stake out claims to potential pay-by-use markets, seems valid for information-bearing networks. The natural tendency is for access and transaction costs to rapidly go towards zero.

Unrestricted (free, or at least 'microcost') access benefits the network in that the increasing number of potential connections and relations exponentially increase its perceived value. Basic p2p network theory states that this value increases as a power relationship of the number of nodes – as 2^n.

Ownership issues should also be seen in relation to the general assault we can see on 'free' content. It is not easy to summarize the overall situation in this arena. Commercialization goals are pursued by some decidedly heavyweight interests, and often even long-established individual rights get trampled in the process. Legislation, interpretations, and application seem to shift almost daily.

At least for hyperlink technology, the legal verdict appears to be a recognition of its importance for the common good. For now, therefore, it remains free.

On the other hand, a failing perhaps of the current Web implementation is that it has no easy solutions to accommodate commercial interests in a user-friendly and unobtrusive way – the lack of a viable micropayment infrastructure comes to mind.

Bit 1.12   Issues of ownership compensation on the Web remain unresolved

Without transparent and ubiquitous mechanisms for tracking and on-demand compensation of content and resource usage, it seems increasingly likely that the Web will end up an impoverished and fragmented backwater of residual information.

Related to this issue is the lack of a clear demarcation of what exactly constitutes the 'public good' arena, one that despite commercial interests should remain free. Table 1.1

Table 1.1 The issue of 'free' usage and development or creation type

Type	Open	Proprietary
Free to use freely	free stuff	public good
Free to use, limited (licensed, non-commercial use)	community	promotional
Pay to use (buy or license)	rare, low cost	usual case

outlines one way of looking at the dimensions of 'free' in this context. Not specifically located in it is the 'free as in beer' stance – that is, affordable for anyone.

Hyperlink Usability

What, then, are the characteristics of usability of the hyperlink? Of particular interest to later discussions on extensibility is the question: In what way is the usability lacking in terms of how we would like to use the Web?

The current conceptual view of the hyperlink, as it has been implemented, has several implications for usability that are not always evident to the casual user. Current usage is deeply ingrained, and the functionality is usually taken for granted with little awareness of either original design intentions or potential enhancements.

The major benefit is of hiding the details of URI and URL addressing from the user, behind a simple user action: clicking on a rendered hyperlink. Another benefit is the concept of 'bookmarking' Web resources for later reference, again through a simple user action in the client. These actions quickly become second nature to the user, allowing rapid navigation around the Web.

On the other hand, despite the ingenious nature of the hyperlink construct, the deficiencies are numerous. They can be summarized as follows:

- **Unidirectional linkage**. Current implementations provide no direct means to ascertain whether hyperlinks elsewhere point to particular content. Even the simple quantitative metric that X sites refer to Web resource Y can be valuable as an approximate measure of authoritativeness. Mapping and presenting parent-child-sibling structures is also valuable. *Backlink* information of this nature can be provided indirectly by some search index services, such as *Google,* but this information is both frozen and often out-of-date.
- **Unverified destination**. There is no way to determine in advance whether a particular destination URL is valid – for example, if the server domain exists or the information is stored in the address space specified. The address can have been incorrectly entered into the hyperlink, or the server's content structure reorganized (causing the dreaded 'linkrot' effect).
- **Filtered access**. Many sites disallow direct or 'deep' linking, either by denying such linked access outright or by redirecting to some generic portal page on the site. Content bookmarked in the course of sequential browsing may therefore not be accessible when following such links out of context. Other access controls might also be in effect, requiring login or allowed categories of users – and increasingly, the subscription model of content access.

- **Unverified availability**. Short of trying to access the content and receiving either content or error code, a user cannot determine beforehand if the indicated content is available – for example, the server may be offline, incapable of responding, or just applying filtered access.
- **Relevance, reputation and trust issues**. The Web is superbly egalitarian; anyone can publish anything, and all content hyperlinks have the same superficial value. In the absence of mechanisms for hyperlink feedback, the user has no direct way of determining the relevance or trustworthiness of the linked content. Visual spoofing of link destination is easy, which can lead users astray to sites they would never wish to visit, something often exploited by malicious interests on Web pages.

As can be seen, the browsing user (also agents, and even the servers) could benefit from information about what exactly a hyperlink is pointing to, and from an indication of its status, *operational* as well as *reputational*.

One-way linkage cannot provide this information. Existing tools are 'out-of-band' in the sense that they are not integrated into the browsing experience (or the client-server transactions), but instead rely on external, human-interpreted aids such as search engines or separate applications (say hyperlink checkers).

Wouldn 't it be nice if . . .

Some features users might find useful or convenient when browsing the Web can be listed as a sampler of what the current infrastructure and applications either do not do, or do only with difficulty or limitations in special cases/applications.

At this point, we are not concerned about the 'how' of implementation, only the user experience for some fairly simple and straightforward enhancements. Some of these features are hard to describe adequately in the absence of the 'back-end' data and automation, and they will depend greatly on commercial, political, and social decisions along the way as to what is desired or allowed.

- **Context awareness**, which would allow 'intelligent' client behavior. Ideally perhaps, semantic parsing of document context could reduce the need for user decision and simply present 'most-probable' options as distinctive link elements at the rendered location. Alternatively, the client might provide an on-the-fly generated sidebar of 'related links', including information about the site owner. Several proprietary variations of the theme are found in *adware*-related generation of 'spurious' hyperlinks in displayed content, which however is a more intrusive and annoying implementation in the absence of other cues.

  A simple example of a context feature for content is found in newer browser clients, where the user can highlight text and right-click to access Web search or other useful actions, as illustrated in Figure 1.2.
- **Persistent and shareable annotations**, where users can record private or public comments about Web content automatically attached to a given document (URI) and ideally internal location, in context. Public comments would then be shareable among an entire community of users – perhaps most usefully in goal-oriented groups.

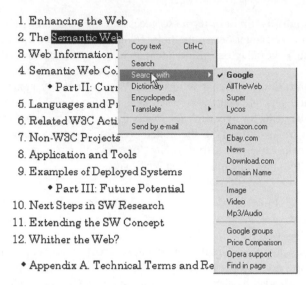

1. Enhancing the Web
2. The Semantic Web
3. Web Information :
4. Semantic Web Co:
    ◆ Part II: Curr
5. Languages and P
6. Related W3C Acti
7. Non-W3C Projects
8. Application and Tools
9. Examples of Deployed Systems
    ◆ Part III: Future Potential
10. Next Steps in SW Research
11. Extending the SW Concept
12. Whither the Web?

    ◆ Appendix A. Technical Terms and Re

Figure 1.2   A simple example of context awareness that allows a browser user to initiate a number of operations (such as Web search, Dictionary or Encyclopaedia lookup, or translation) based on a highlighted block of text and the associated context menu

- **Persistent and shareable ratings**, which is a complement to annotations, providing some compiled rating based on individual user votes. A mouse-over of a link might then pop up a tooltip box showing, for example, that 67% of the voting users found the target document worthwhile. Of course, not everyone finds such ratings useful, yet such ratings (and their quality) might be improved by a consistent infrastructure to support them.
- **More realtime data**, which would result from the ability of Web clients and services to compile Web pages of data, or embed extended information, culled from a variety of sources according to the preferences of the user.
- **Persistent and shareable categorizations**, which would be a collaborative way to sort Web documents into searchable categories.

The common theme in most wish lists and methods of addressing the deficiencies of the current Web model is that the issues are based on adding information about the content. Then the user client can access, process, and display (or use it) in ways transparent but useful to the user – ideally with some measure of user control.

To Richer Informational Structures

The further information about the content that the previous shortcomings and wish list highlight include the metadata and relational aspects, some of which might be implemented as distributed services on the Web.

One idealized extension of the originally unidirectional information flow could look as in Figure 1.3.

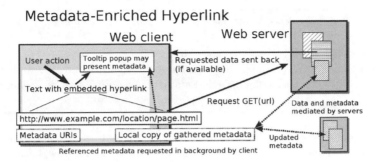

Figure 1.3 Conceptual view of hyperlink metadata functionality as it might be implemented to form an enhanced Web. Metadata and relational data are gathered and presented to the user for each embedded link

The point in the enriched hyperlink model is that further information about each hyperlink destination is automatically gathered and processed in the background. This behavior is an extension of how some caching solutions already now parse the currently loaded content and pre-fetch all hyperlink-referenced content into a local cache, ready for the next user selection.

In the new model, the client software would also be busy compiling information about the referenced content, ready to present this to the user even before a next destination is selected. Link metadata might be presented to the user as tooltip text, for example, or in ancillary windows on a mouse-over of the link.

A very limited and static simulation of this feature can be seen in the use of the 'title' attribute in a hyperlink markup (as shown in Figure 1.4) – the difference is that this kind of metainformation indicator must be precoded by the content author as an attribute in the hyperlink tag itself.

The Collaboration Aspect

The way that we have viewed the Web has generally been one of computer users browsing a web of content; independent readers of static information, or possibly also sole publishers of personal Web sites. This view was never the whole truth.

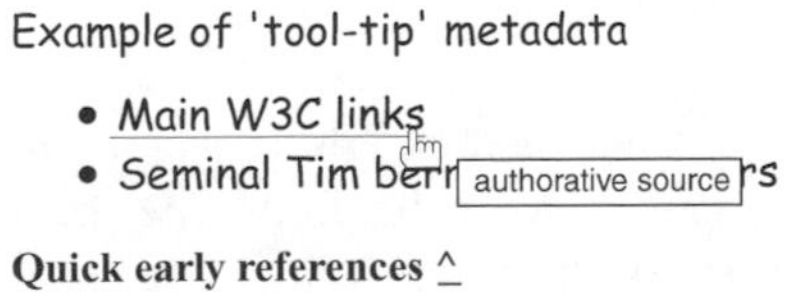

Example of 'tool-tip' metadata

- Main W3C links
- Seminal Tim berr authorative source rs

Quick early references ∧

Figure 1.4 Static meta-information precoded by content author into the hyperlink tag of the content itself, using the 'title' attribute. When rendering the page, the browser pops up a 'tooltip' box to show the hidden text when the cursor is placed over the link

Bit 1.13   In the early days of the Web, considerable collaboration was the rule

What is interesting about the early collaborative effort is that it occurred in a smaller and more open community, with implicit trust relationships and shared values. Among those who maintained Web pages, compiling and publishing resource lists became a major part of the effort.

To find information on the Web, a user would visit one or another of these lists of hyperlinks and use it as a launchpad, long before the idea of 'portal sites' became popular on the Web. The implicit metadata in such a reference represents the value assessment of the person creating the list. Such personal lists became more specialized and interlinked as their maintainers cross-referenced and recommended each other, but they also became vastly more difficult to maintain as the Web grew larger.

Some directory lists, almost obsessively maintained, eventually grew into large indexing projects that attempted to span the entire Web – even then known to be an ultimately hopeless goal in the face of the Web's relentless and exponential growth.

One such set of lists, started as a student hobby by David Filo and Jerry Yang in 1994, quickly became a mainstay for many early Internet users, acquiring the name 'Yet Another Hierarchical Officious Oracle' (YAHOO). By 1996, Yahoo! Inc. (*www.yahoo.com*) was a viable IPO business, rapidly diversifying and attaining global and localized reach. *Yahoo!* quickly became the leader, in both concept and size, with a large staff to update and expand manually the categories from a mix of user submissions, dedicated browsing, and eventually search-engine database.

Other search engines went the other way: first developing the engine, then adding directories to the Web interface. A more recent collaborative venture of this kind is the *Open Directory project* (*www. dmoz. org*).

An offshoot to cross-linked personal lists and a complement to the ever-larger directories was the *'WebRing'* (*www.webring.org*) concept. Sites with a common theme join an indexed list (a 'ring') maintained by a theme ringmaster. A master meta-list of categories tracks all member rings.

- To appreciate the size of this effort, *WebRing* statistics were in July 2002 given as 62,000 rings and 1.08 million active sites, easily capable of satisfying any user's special interests indefinitely.
- However, the *WebRing* phenomenon appears to have peaked some years ago. presumably shrinking in the face of the ubiquitous search-engine query. The figures show a clear trend: 52,250 rings with 941,000 active sites in August 2003, and 46,100 rings with 527,000 sites in February 2005

The individual user of today tends increasingly to use the more impersonal and dynamic lists generated by search engines. An estimated 75% of user attempts to seek Web information first go through one of the major search engines.

Although search-engine automation has reduced the relative importance of individual collaboration in maintaining the older forms of resource lists, it is still not the panacea that most users assume. Issues include incomplete indexing of individual documents, query-relevancy ranking, and handling of non-HTML and proprietary formats.

Search engines also face the same problem as other approaches of scaling in the ever-growing Web. They are unable to index more than a small fraction of it. Estimates vary, but the cited percentage has been shrinking steadily – recently 10%, perhaps even less.

The bottom line is that centralized indexing in the traditional sense (whether manual or automatic) is increasingly inadequate to map the distributed and exponential growth of the Web – growth in size, variety, updates, etc.

Bit 1.14   New requirements and usage patterns change the nature of the Web

The adaptation of any network is driven by requirements, technology adoption, resource allocation, and usage – if only because new and upgraded resources usually target areas of greatest loading in order to maintain acceptable levels of service. *Peer to Peer* (a previous book) is replete with examples.

Returning to collaboration, it has become more important to support new and more flexible forms – between individuals and groups, between people and software, and between different software agents.

Loose parallels could be drawn between this adaptive situation for the Web and the changing real-world requirements on telephone networks:

- As first, subscriber conversation habits changed slowly over the years, and group conferencing became more common for business.
- Later, cellular usage became ubiquitous, requiring massive capacity to switch calls between line and mobile subscribers from many diverse providers.
- As dial-up Internet access became more common, individual subscriber lines and exchanges became increasingly tied up for longer intervals.
- Digitizing the network and applying packet switching to all traffic helped solve some congestion problems by letting separate connections coexist on the same circuit.
- In all cases, telecom operators had to adapt the network, along with associated technologies, to maintain an acceptable service in the face of shifting requirements.

The Web is a virtual network. The underlying Internet infrastructure (and the physical access and transport connectivity under it) must constantly adapt to the overall usage patterns as expressed at each level. New collaboration technology has generally just deployed new application protocols on top of the existing Internet and Web ones, forming new virtual networks between the affected clients.

Better collaboration support is achieved by instead enhancing the basic Web layer protocol. Other, proprietary solutions are possible (and some do exist), but making these features a fundamental part of the protocol ensures maximum interoperability. Just as we take for granted that any Web browser can display any standard-compliant Web page, so should clients be able to access collaboration functionality. Providing mechanisms for dealing with content meaning that will automatically provide intelligent mechanisms for promoting collaboration in all its forms is also one of the aims of the Semantic Web.

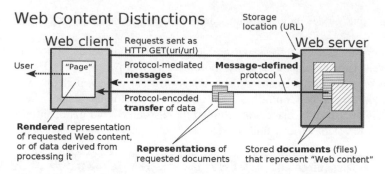

Figure 1.5 The traditional distinction between "document" and "message" is illustrated in the context of the client-server model, along with some terms relevant to requesting and viewing Web content

Extending the Content Model

Discussions of 'content' without qualifications or explanations might be misleading for several reasons, and the term is easily misunderstood; this is why the terms 'information' and 'information object' crop up in these discussions.

It is a common assumption that 'Web content' refers to the text or other media objects provided by Web pages and their related links – static information physically stored in well-defined server locations. In addition, historically, there has always been a distinction maintained between 'messages' on the one hand, and 'documents' (or files, or content) on the other (see Figure 1.5); that protocols are defined on top of messages in order to transport documents, and that further protocols are defined in terms of messages exchanged by previously defined protocols, and so on.

Such recursive definitions, while sometimes confusing, occur all the time and can prove very useful. As implied earlier, p2p protocols that define separate virtual networks are defined by the exchange of underlying HTTP messages.

However, especially in the context of XML, it is often better to view *content as the side-effect of an ongoing exchange of messages between various endpoints.* It matters not whether these endpoints are servers of static Web pages, database servers, Web services, other users, or any form of software agent. Also, it does not matter whether the information is actually stored somewhere, compiled from several sources, or generated totally on the fly.

Higher-level protocols are generally implemented as *message-based transactions* that can encapsulate information and pointers to information in many complex ways.

Recall that requesting information on the Web, even in a basic protocol such as HTTP, only returns a representation of the information, encoded in some way acceptable at both endpoints – only a message, in effect, from the process providing it and thus inextricably bound to the protocol.

Bit 1.15   Messages pass representations of information

The issue of information representation by proxy and subsequent interpretation is far more fundamental than it might seem. The concept has deep philosophical roots, and far-reaching implications about how we know anything at all and communicate.

This generalized view, that all content is the same exchange of messages expressed in a single protocol, amounts to a useful simplification. It can result in a merging of the current fragmentation of different protocols for e-mail, file transfer, newsgroup postings, instant messaging, chat, Web pages, and other 'content'.

The same client, based on such a harmonized protocol, could transparently handle arbitrary information objects in a consistent way from the user perspective.

Mapping the Infosphere

Navigating the wealth of information on the Web is not a trivial task, even assuming that it is all openly accessible. Various methods have gradually increased or decreased in relative importance for users over the years. In roughly chronological order, these navigational methods are summarized in the following list:

- **Resource lists** present hyperlinks that particular users have collected and recommended on their own Web sites. An outgrowth of this resource was the 'meta list', or list of hyperlinks to sites that maintained such lists. The quality and comprehensiveness of such lists could vary immensely.
- **Meta directories** are large hyperlink lists compiled by dedicated staff into some kind of hierarchical category trees, often with the aim of being a comprehensive guide to all content. Examples are the *Yahoo! Web Site Directory* (*www.yahoo.com*), *Google directory* (*directory.google.com*) and the *Open Directory Project* (*dmoz.org*). A problem with hierarchical listings is classification, which can be arbitrary, and may or may not coincide with the user's perception of an appropriate category.
- **Search engines** provide *ad hoc* lists of hyperlinks to Web pages (or other Internet resources) culled from indexed databases collected and updated by automated 'spiders' that systematically traverse the Web hyperlinks on each Web site they can reach. Search results are often ranked according to some opaque metric of 'relevancy', but depend critically on the user's ability to formulate 'optimal' search criteria based on suitable terms, along with the quality of the indexing process.
- **Special interest compilations** lie somewhere between idiosyncratic lists compiled by individuals, and the comprehensive Web directories. They aggregate self-nominated, like-themed sites into interlinked Web rings. Each site in a ring can directly or indirectly refer to its neighbors through a ring index, each ring organized in a category hierarchy in some meta ring.
- **Web services** can automate and manage resource discovery and content retrieval (including compilation and recasting) on behalf of a user. Such services might incorporate 'learning routines' to adapt to each user's history of requests to determine scope and relevancy criteria, for example.

Currently, Web information is largely 'atomistic', with relatively few and arbitrary relations defined between the different information objects – see also the previous discussion about hyperlink usability. This deficiency and the constraints of the current hyperlink model clearly have a severe impact on user navigation. The lack also severely hampers the development of automated agents for such purposes.

Structure and Relationship

Traditionally, maps of Internet resources in general, and the Web in particular, have by necessity been very simplistic and one-dimensional:

- **Human-compiled hierarchy directories** have the virtue of including some, albeit variable, amount of metadata based on human assessment of the link target and classification into categories. Expressed relational metainformation is an external construct, however, independent of any explicit hyperlinks between sites.
- **Automated link-walk compilations** result in maps that are in this case purely structural – a 'snapshot' of Web hyperlinks. While subsequent indexing of at least portions of content from each node does allow for term searches to create simple maps that reflect literal term usage (and thereby indicate term-related linking), the content semantics are largely ignored.

We can illuminate this distinction and some of the issues involved by looking to some well-established conventions from paper publishing.

Paper-published content is inherently linear, because it has its origins in oral narrative. Unlike transient speech, however, published sentences are frozen in a medium that the reader can to some extent perceive outside of time, freely choosing what to read at any time. This characteristic enables the concept of multiple entry points based on some overall mapping.

Paper publishing thus developed structural and semantic aids to the reader to provide such entry points into the content:

- **Chapters and sections**. As a structural device, the division of a text into chapters and sections imparts a layer of semantics in that the logical division helps order content and implies relationships between the parts.
- **Table of contents**. Given an expressed logical structure, it is possible to provide a list of headings and corresponding numbered pages. The reader can use the table of contents to gain an overview of the content structure and quickly find specific content by following the logical divisions.
- **Abstracts and summaries**. Older novels commonly provided short abstracts for each chapter, and although this practice was later dropped in fiction, the abstract lives on in academic publications. Summaries (especially executive summaries) are a similar device. Both variants are commonly set apart from the main text by special styling (such as extra-wide margins, vertical whitespace, and italic font). A listing of keywords may be considered as an extremely abbreviated summary.
- **Index**. From the other end (and it is not coincidence that the index is found at the back of a book), a list of terms and their page-location in the text constitutes an index. Creating a useful index is a highly developed art; it is not simply a compilation of occurrences of all the words. Instead It critically relies on the compiler's understanding of relevancy and importance – of which words or phrases to include, which occurrences to index in terms of the semantic context, which occurrences to ignore, which relationships to highlight using categories and multi-level listings, just to name a few considerations.

Other semantic-mapping contructs also exist in publishing, but those listed here suffice for the purpose of comparison with the Web. For example, comparing with the previous examples of Web maps, we can see a rough correspondence between the human-compiled category list and the table of contents.

- In passing, we may note that despite well-developed mechanisms for multiple entry points for the reader, not to mention advanced non-linear reading possibilities in electronic form, most written content maintains the fiction of linear narrative, from beginning to end.

The search-engine list of links derived from automated indexing would perhaps approximate the index, except for the fact that such a list compilation is currently an index created solely by listing all occurrences of the search terms. It lacks the semantic order and weighting that the expert indexer imparts, and critically, the distinction between different usages and meanings – for example, between the noun 'book' and the verb 'to book' (as in make a reservation).

Natural language is replete with ambiguities of usage that only information about semantic context can even begin to resolve. With clever programming, Web searches can at least be made tolerant of some common typographical mistakes and regional spelling variations, but these searches remain semantically ignorant.

Proper logical markup of text structure in Web content can somewhat offset this analytical lack, but at the cost of reliance on external conventions that may or may not be correctly applied in particular instances. Text markup still does not convey inherent semantic information, only overlain syntactical.

Although *Google* is commonly cited as 'the best' search engine, mainly due to the way it leverages backlinks to provide relevant results ranking, it is not hard to find consistently 'bad' or skewed results in simple searches.

- Such frustrations are often due to the engine's strictly lexical basis combined with the skew effect of some kinds of linked content. For search words that also relate to product names, for example, *Google* ranks online shopping pages higher than any other results, probably because of the wealth of product links both on-site and from affiliate sites. Advanced syntax may therefore be needed to find information sites not devoted to selling products.

Links, Relationship, and Trust

The fact that object **A** unidirectionally links to object **B** says little about any real relationship between them. Any metainformation about this relationship is largely inferred by the user. For example, we are willing to believe in a high degree of relevance of a hyperlink on a page that refers to another page on the same site, even in the absence of any explicit declarations about it. Such trust may prove unfounded, however, especially if the link is 'broken' (that is, the URL is temporarily or permanently invalid).

But the point is that the hyperlink itself contains no metadata whatsoever about the relationship or target status. At best, the contextual text around the link might explain something about the target page; at worst, it might be totally misleading.

The situation is even less reliable when the link targets an object outside the site, because then the page/link author does not even have nominal control over the target object. Target site contact might have changed drastically since the link was created, for example, or the target site might employ filtering or redirection that makes the direct reference invalid and the target hard or impossible to track down manually.

Trust is also at issue when a link (URL) points to a Web page that defines an association between URL and URI – perhaps the easiest way to create and publish a URI Anyone can do it, and anyone can on the face of it misrepresent an association to a particular URI.

Trust issues of this nature are probably best resolved by embedding currently accepted mechanisms based on content hashes and public key signing of the published assertion. The assertion can then be independently verified as unaltered and the publisher identity verified through some established chain of trust – in the worst case, just identified as the same publisher of other similarly-signed material.

Well-Defined Semantic Models

When considering strategies to map information on the Web, one can begin by outlining some of the features that the mapped content, commonly expressed as 'documents' in these contexts, should exhibit or promote:

- It must be possible to define precisely a language (as a set of tokens, grammar, and semantics) in which to deal with 'first class objects' (which means document objects that are URI-referenced). Such a language is a prerequisite to processing.
- It must be possible to make documents in a mixture of languages (that is, allow language mixing). It is a fact of life that no single language can or will be used in all contexts. Extensibility is natural and alternate structures must be allowed to coexist.
- Every document should be self-defining by carrying the URI(s) of the language(s) in which it is written. This mechanism enables both humans and more importantly software to determine the appropriate methods for analyzing and processing the information.
- It must be possible to process a document while understanding only a subset of the languages (that is, partial understanding). In other words, full functionality must 'degrade gracefully' so that simpler processing can safely ignore the full set of features.

The requirements in this list form the basis for XML development, for example, and it defines a first step towards enabling the encoding of semantic meta-information.

The current Web recommendation is that XML should be used when a new language is defined. A new language (or, for that matter, new features extending an existing language) must then be defined as a new namespace within XML.

Although this kind of language defining might be seen as akin to misguided attempts to (re)define all natural languages in terms of some grand metalanguage (as in the tradition of applying Latin grammar), as an artificial construct *a priori* applied to other artificial constructs, it can be both attainable and practical.

Given the language, how do we deal with the meaning of what is expressed in that language? This issue is at the core of the Semantic Web, and a discussion around it forms the bulk of Chapter 2.

2

Defining the Semantic Web

When discussing the Semantic Web, it is prudent to note that we are in fact dealing with but one of several interrelated emerging technology domains: Web Services (providing distributed functionality), Grid Services (integrating distributed resources), and the Semantic Web (mapping and managing the resources). This book's primary focus is on the last of these.

In order to map Web resources more precisely, computational agents require machine-readable descriptions of the content and capabilities of Web accessible resources. These descriptions must be in addition to the human-readable versions of that information, complementing but not supplanting.

The requirement to be machine-readable is non-negotiable, because in future – in fact, practically already now – almost all software (and devices) will be innately Web-aware. Leveraging autonomic-computing technology enables agents and empowered devices to display independent activity and initiative, thus requiring less hands-on management by users. Already, an important expectation about software is that even 'dumb' applications can deal with basic Web connectivity in ways appropriate to their primary function.

Chapter 2 at a Glance

This chapter introduces the architectural models and fundamental terms used in the discussions of the Semantic Web. *From Model to Reality* starts with a summary of the conceptual models for the Semantic Web, before delving further into detailed terms.

- *The Semantic Web Concept* section analyzes and motivates this particular extension to the Web.
- *Representational Models* reviews document models relevant to later discussions.

The Road Map introduces some core generic concepts relevant to the Semantic Web and outlines their relationships.

- *Identity* must uniquely and persistently reference Web resources; interim solutions are already deployed.

The Semantic Web: Crafting Infrastructure for Agency   Bo Leuf
© 2006 John Wiley & Sons, Ltd

- *Markup* must enable the distinction between the content representation and the metadata that defines how to interpret and process it.
- *Relationships* between data objects must be described.
- *Reasoning* strategies for inference engines must be defined.
- *Agency on the Web* can be implemented to apply reasoning on the other layers.
- *Semantic P2P* outlines the role of peer-to-peer technology in this context.
  The Visual Tour summarizes the high-level view.
- This section highlights *The Map, The Architectural Goals* and *The Implementation Levels*, and discusses the recent W3C architecture recommendations.

From Model to Reality

The concept of a 'semantic Web' supposes a plan for achieving a set of connected applications for data on the Web in such a way as to form a consistently readable logical Web of data – the actual implementation of the Semantic Web ('**sweb**'). The steps needed to get from the model of the Semantic Web to reality are many. With our current understanding of the model, the process will require many iterations of model refinement as we design the systems that will implement any form of Semantic Web.

The Web was designed as an information space, with the goal not only that it should be useful for human–human communication, but also that machines would be able to participate and help. One of the fundamental obstacles to this vision has been the fact that most information on the Web is designed for human consumption. Even if derived from a database with well defined meanings (in at least some terms) for its columns, the structure of the data is rarely evident to a robot browsing it.

Bit 2.1   When parsing data, humans read, while machines decode
Actually, humans do both: first decoding the syntactical representation, then applying built-in semantic rules and context inferences to interpret the meaning of the data. For machines to 'read' in this way, they also need semantic rules and inference logic.

Initially leaving aside the 'greater AI' problem of training machines to behave like people (though assuredly some research along these lines is always being pursued), the 'lesser AI' and sweb immediate approach is instead to develop languages and codified relationships for expressing information in a machine-processable form. The machines must be able to do more than simply decode the data representations on the syntactical level; they must be able to perform logical processing and inference on well-defined meanings of the information tokens parsed from the data.

Before going on to discuss possible models for this kind of machine enhancement, we might reasonably ask: *What are the requirements on the content representation for meaningful reading to be feasible?* The answer depends a fair bit on whom you ask. However, there is some consensus for the following content features:

- distinguishing markup from data;
- avoiding ambiguity;
- discouraging arbitrary variation.

The linguist might at this point object that natural languages apparently thrive with a remarkably high degree of the latter two, ambiguity and variation – despite the many formal rigidities of syntax, grammar, and definition. Even the far more rigid and literal programming languages have their fair share of both characteristics.

As it happens, people not only cope with ambiguity and variation, they actively crave and enjoy it, as evidenced in various forms of literature, poetry, and humour. It would seem that we as humans are predisposed forever to balance on the linguistic cusp between coherency and incoherency, playing constant games with the way we use language. This constant play and variation is one of the hallmarks of a natural, living language.

On the other hand, computer languages (and by extension computer-logic implementations) have as a basic assumption the elimination of ambiguity, or at least they implement some formal abstraction so that the programmer need not consider it.

Bit 2.2   Current computer logic design is blind to important parts of human communication

It seems paradoxical that in this way we deliberately exclude the ability to automate the processing of a significant component of the information we produce.

A few computer-languages do exist that are arguably exceptions to this literalistic constraint, inherently allowing some idiomatic variations and implicit referencing in ways that are similar to human languages. However, most programmers find such languages 'difficult' or 'quirky'.

- While few might appreciate the ability to write poetry in code, such frivolous activities must have at least some bearing on the issue of creating programs that can deal with texts in ways that approximate human semantic processes.

One practical field that demonstrates the *necessity* of semantic processing is that of machine translation of human languages. The limitations of syntactical and lexical parsing, not to mention the profound misconceptions of how humans process language meaning, have for decades confounded the practical realization of automated translation.

At best, current implementations can achieve only tolerable results within narrow specialist fields, providing but a first approximation of meaning for a human reader.

The Semantic Web Concept

Returning to the 'plan' to extend the Web (more of a vision, actually), we can start with a sort of 'mission statement' made by three of the top names behind the Semantic Web initiative:

> *The Semantic Web is an extension of the current Web in which information is given well-defined meaning, better enabling computers and people to work in cooperation.*
>
> Tim Berners-Lee, James Hendler, Ora Lassila

A reformulated version appears later to guide the formal W3C 'initiative':

> *The goal of the Semantic Web initiative is as broad as that of the Web: to create a universal*
> *medium for the exchange of data.*   *www.w3.org/2001/sw/Activity*

To achieve this goal, existing Web content needs to be augmented with data (for example in the form of semantic markup) to facilitate its handling by computers, and with documents (or more generally, resources) that are intended solely for computers and software agents.

It is envisioned that the resulting infrastructure, transforming the Web into the Semantic Web, will spur the development of automated Web services to fulfill a wide variety of purposes. Computers will be able discern the meaning of semantic data by following hyperlinks to term definitions and rule sets. Thus armed, they can reason about content logically and make decisions that are relevant to the context and to user expectations.

Bit 2.3   Humans are the current Web's semantic component

Today, humans are required to process the information culled from Web (Internet) resources to determine ultimately its meaning and relevance (if any) for the task at hand. Sweb technology would move some of that processing into software.

Some writers want to (and often do) call such semantic processing simply AI. But after more than half a century of shifting expectations and broad popularization, the generic term 'AI' seems too loaded with old baggage and popular misconceptions to use without further qualification. For now, a better approach is to use more neutral and specific terms (such as 'business logic' or 'semantics') when discussing processing functionality in the context of the Semantic Web.

Reasoning aside, we are not as yet dealing with any sort of 'thinking' software in the popularized 'self-aware AI' sense. Part III revisits the issue, however, and does discuss the broader AI aspects from a speculative perspective.

The Near-term Goal

More useful for conceptualization might be to see the envisioned change as moving from a Web that now reads like a random spread of open books, to a Web that instead reads like a well-structured database – not a centrally dictated order, though some might advocate such a vision, just a locally applied, machine-decipherable structure.

In other words, the near-term goal is for a Web that is better suited to answering queries about the information published there. At its simplest level, such a change would mean that, for example, search engine results could be pre-filtered for some measure of 'correctness' in relation to the query's semantic context. Such filtering is impossible in lexical searches. But perhaps more profoundly, it can mean the contextual structure to make actual user-invoked 'searching' less relevant.

When asked what the 'killer application' of the Semantic Web would be, the clued-in proponents reply: *The Semantic Web itself!* They justify this reply by noting that the killer application of the current Web is precisely the Web itself – the Semantic Web is in this view just another expression of new innate functionality of the same magnitude. The capabilities of the whole are therefore too general to be thought about in terms of solving one particular problem or creating one essential application. The Semantic Web will have many undreamed-of uses and should not be characterized in terms of any single known one.

Bit 2.4   The 'Metadata Web' must grow as the original Web did
As was the case with the growth of the original hyperlinked Web of content, it can be argued that until anyone can easily create metadata about any page and share that metadata with everyone, no true 'semantic web' will arise.

The Semantic Web is most likely to develop first as expanding 'islands' of new functionality that over time become more generally adopted as various service providers realize the utility (and derived benefits or profits) of doing so. Such 'semantic intrawebs' are a reasonable first step to test the new approaches and tools, at a smaller yet fully functional scale, and therefore improve them before they are unleashed into the Web at large.

To provide some context for this evolution, consider the following:

- The original Internet was designed to *pass messages* between systems.
- The Web as we know it is largely about *finding information.*
- Web Services *implement distributed functionality* on the Web.
- The Grid intends to *integrate functionality* provided by the parts.
- The Semantic Web is concerned with *describing the available resources, making the data accessible*, and *providing the agency to manage them.*

Although very superficial descriptions, these items do capture something of the essence of the development. The latter three give rise to a qualitatively different dynamic nature in the Web than what we are currently used to.

Addressing some Practical Features

Let us put the abstract vision into a few simple possible contexts by recalling some of the deficiencies and desired features discussed in Chapter 1 and seeing what a Semantic Web infrastructure could add in practice to the personal user experience.

- **User-context tracking**. Imagine that your Web client could (if you want it to) track your semantic context. For example, it might pre-filter search results to show only those hits with a high correlation to the meaning implied by previous contexts.

    Imagine further that meaningfully related search terms were automatically generated to catch resources that would otherwise be missed only because a particular search word did not occur or was spelled differently in the target.

    When browsing, the client might always highlight Web links that lead to resources with a high semantic correlation to the current context, based on background metadata processing. It might even suggest links not mentioned in the displayed content, and provide pop-up comments on demand.

- **Context-defined interactivity**. Suppose the semantic context could bring to the foreground appropriate media types, or even different interactions with the data browsed. A table could be just a matrix of values, a graph could be just a bitmap picture of relations. Each could be a gateway to further information about each element/relationship. Elements could be manipulated in various ways to investigate other relationships. All would depend on the current or selected context.
- **User-preference tracking**. Anyone experiencing the ever-increasing prevalence of e-mail spam, and faced with the imponderable oddities of available junk-mail filters (such as in MS Outlook or in subscribed Web-based services), must wish for some simpler solution.
  Ideally one would simply tell a filtering agent to 'ignore further mails of this kind' and have it identify 'similar' content reliably, transparently adapting to a wide range of variation. Actually, what the user might often find more useful is the converse feature of the agent being able to highlight and give priority processing to desired mail that is semantically 'valid' and matches a current context or preference setting.

The common thread here is an *enhanced cooperation* between resource server, browsing client, and user – where detailed content analysis and classification relevant to the current user is delegated to software agents, and where even the selection of data can depend on the context. Such enhancement can only work in a decent way if the software is capable of 'seeing' the meaning of content in a defined context, not just the lexical words and structural tags.

Although this list may give the impression that we assume all such enhancements to be beneficial to and desired by the average user, this conclusion would be unfounded. At a minimum, as implied in the first item, the features should be optional and easily controlled by the user. At worst, otherwise useful features might be hijacked by content-publisher or advertising interests, reducing them to the equivalent of intrusive commercial spots in broadcast media.

Bit 2.5   An important issue: Metadata functionality on whose terms?

The ideal vision presupposes that the user is in control and is the benefactor of the outlined enhancements. Although the distributed and peer architecture of the Web is inclined to promote (or at least allow) such a development, other strong interests in government and business often follow different, less benign agendas.

Early precursors to context and tracking features have already appeared in limited ways, but they are not as yet semantically very clever. More critically, most of them are in fact *not* addressing user requirements at all, but are being pushed by business and advertising interests. Thus the user has little control over how these features work, or of when they are active. Often, the data collected is just reported back to the advertiser site, not helping the user in any way.

The best that can be said for the context-tailored user profiles we have seen so far is that they may reduce the incidence of totally irrelevant banner ads – such as U.S. mortgage rates displayed to apartment dwellers in London. The user must of course first agree to such external profiling.

What it Means for the Publisher

In the Web of today, publishing content all too often means creating one-off presentations. Data are selected, a layout chosen, and the result stored as a page on a server somewhere. Updating changing data requires new presentations to be created and uploaded to the server. (Initially, I used the term 'static' here, but found it misleading on reading, as much such content is distressingly 'dynamic' to the user visually, in the form of animated images and intrusive *shockwave-flash* elements.)

Moving to database-driven, dynamic pages (in the intended updated-content sense) is a step forward, in that it ensures presentations of the most current data. However, these pages remain locked to a particular layout. The publisher still needs to determine (or guess) the requirements of the visiting user, and design the database and page format accordingly.

Trouble is, there is really no way of knowing what the visitor is looking for, or how the data will be used. Both issues are crucial, however, to a successful presentation design.

Bit 2.6   Metadata allows 'application-neutral' publishing of data

The ideal for multi-purpose publishing is to provide 'raw data' that can be used and reused in a variety of contexts. Data descriptions can decouple the resource itself from issues of presentation, media type, and client behavior – user aspects otherwise to be determined by the collecting agent in the current context of the user.

Granted, *not everyone wants to publish raw data*, and there are many legitimate reasons for retaining strict control of presentation, media type, and purpose. Clearly, the decision must remain with the publisher. However, by restricting access to fixed formats, the publisher also constrains the possible uses of the information – and incidentally risks that another, unsupported use will simply lift a snapshot of the data (soon out of date, perhaps) to embed in another presentation.

Many useful applications of data are unforeseen (and unforeseeable) by the original publisher. The realistic choice for control thus lies somewhere between the two following extremes:

- Constrained access with complete control of presentation ('link to this page'), yet paradoxically no control at all of ultimate information use and relevancy elsewhere.
- Unfettered access, with little direct control of how the information is presented, yet with the high and desired probability that users are linking to the resource directly to pick up the most current data on demand, instead of relying on stale static snapshots of earlier data.

The latter approach does work in business, and it is gradually being adopted as companies realize the competitive benefits of allowing real-time access to the raw content data.

- For example, *Amazon.com*, with a long history of encouraged product linking from affiliate Web pages, has moved in the direction of more service-oriented support, such as for dynamic backlinks to Amazon servers. These define-once links automatically pull and embed up-to-date product images and pricing data when the affiliate page is viewed.

With similar access to manufacturer or vendor databases, resellers in general can quickly produce online catalogues as **living documents**, where product information and pricing is sourced and embedded in real time. Service providers can implement real-time inventory status, customer ordering, and support for sold services.

All such solutions become easier to implement flexibly when consistent metadata exists alongside the required information. It is a question of business realizing that open-ended formats for published data do in fact benefit the bottom line. Extending in-house information to reuse by intermediaries greatly extends the reach of both the manufacturer and the vendor.

Bit 2.7   People do not just want to read about it, they want to interact with it

Metadata descriptions and access to 'raw' data brings the virtual world closer to our real world, because we can then do more things online, on our own terms.

As an example, imagine the added value of user/agent interaction affecting the presentation of a Web-published cookbook of recipes. The concept as such is fairly simple (and very popular), but a common reader issue is about scaling the size of the recipe to fit the dinner context.

A sweb version that 'understands' content and context could, for example, recast browsed recipes to your current requirements (such as the specific number of people for dinner), offer substitutions for ingredients that you flag as unavailable (or point you to more suitable recipes), and suggest links to shops near your location that stock what the recipe requires.

Smart agents could even warn that a particular recipe will not easily scale to the required size in the context of your kitchen, and the user interaction would start to resemble a conversation as you vary parameters and preconditions to find a better solution.

Disentangling the Markup

As noted, a core part of any Semantic Web implementation must concern the concept of machine-understandable documents. The 'easy' part is the clean separation of actual content (and critically, its meaning) from the presentation aspects – that is, from how the content is to be rendered by any given client in a particular context.

Current HTML encoding practices routinely entangle structural and visual markup (see detailed discussion of markup in Chapter 5), despite the long availability of **CSS** to separate out the latter in a consistent and reusable way.

Part of the blame for this state of affairs lies with still-inconsistent support of CSS in authoring and client implementations, and part lies with the original design of tag sets to aid computer typesetting. This legacy concern with typefaces, explicit visual layout, and the need to check the consistency of the document structure lives on in HTML. In addition, HTML is unable to address self-defining semantic markup; it can only deal with a limited and externally predefined syntactic structure.

Bit 2.8   Useful semantic markup must be extensible and self-defining

We can never know (and therefore never standardize markup for) all possible contexts in advance. Usage evolves, and local yet portable extensions must be supported.

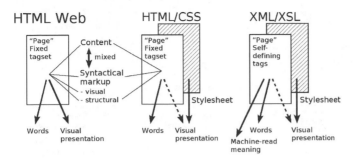

Figure 2.1 Making Web content machine-understandable requires disentangling the meaningful content from the markup. Separating structure and presentation is only part of the answer. XML goes further by allowing self-defining tags to set apart meaningful parts of the text

Figure 2.1 depicts something of this relationship, adding a comparison with XML which, with its extensible and self-defining nature, appears to be the markup language of choice.

Style sheets can be specified as CSS or **XSL**. Customized XML tags can set apart bits of text in ways that are semantically meaningful for understanding the context, and so provide a starting point for machine analysis of it. Even in XML, however, systematic reuse and re-purposing of content requires that all presentational characteristics be stripped from the data-embedded markup. Although this task would seem straightforward, along the lines of CSS but applied to XSL, it in fact proves very difficult.

One surprising result when attempting to formulate simple and consistent XSL rules for existing content is that much erstwhile 'content XML' idiom turns out to be mostly presentational in nature.

Humans seem predisposed to mix presentation and semantics – it may be just the way we perceive the world. Perhaps our human perspective is so innately skewed to presentation scaffolding that we literally cannot isolate the semantic meaning without using strict tools to resolve it.

Bit 2.9   Meaning is in the eye of the beholder

More research is needed into discovering how humans parse and classify meaning from various forms of presentation and communication.

The later sections on ontologies, however, do provide some tools to grapple with this situation, and perhaps suggest a way out of the dilemma when we try to devise structures that can be processed by machines.

Framework Constraint

Any descriptive framework contains both visible and hidden assumptions, and XML is no different. For example, XML (as a derivative of **SGML**) presumes a tree structure when used to describe data. It expects, or tries to enforce, a universal *linear representation* of hierarchical structure, with important consequences for what it can and cannot describe easily in its syntax.

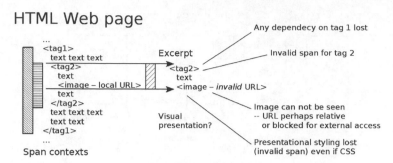

Figure 2.2 An illustration of the problem of excerpting content that is found in the context of linear markup. Missing or out-of-context tags break the syntactical sequence for correctly parsing markup

This artificial constraint can be a problem, especially when it comes to 'excerpting' content for use elsewhere, such as is easily envisioned when machines are to process information for remote presentation. In particular, the tree-based syntax makes assumptions about markup scope and span that can be undefined outside the linear context of the original text, as illustrated in Figure 2.2.

Tree-based markup under any recognized SGML/XML convention is unable to easily represent all arbitrary structures. It fails to adequately represent overlapping or concurrent structures (as in the illustrated example). This fundamental limitation does not belittle the proven usefulness of the possible representations, but any system based on such a markup convention must work under constraint and therefore will not be able to handle easily other content structures.

Another simple example is *text citation*, where a number of common conventions (such as annotations, quoter's emphasis, ellipsis, and clarifying insertions) would perhaps translate to multiple layers of overlay markup, complicated by the additional fact that the cited content is no longer within the original markup context. Note that such constraints in practice also apply to much research and many constructs that relate to markup, as is shown in the following sections.

The presumption of a tree for syntax (and hence for the structure of texts), however, is not necessary, despite its pervasiveness and seeming logic – it is only an assertion about data, not an inherent property of the data, and it is a partial description at best.

Available processing technologies in the XML realm only obscure the issue. They rely on the same tree-based syntax and thus on the consequent presumption about the structure of data. Not all solutions are amenable to such descriptions.

Embedding or Not

Separating *all* the markup from the content, not just the presentational parts, has been proposed as a solution to several current markup issues. Relational pointers into the untagged data structure would then associate specified spans with the relevant metadata information. The additional overhead of maintaining parallel data structures and relational pointers between them is offset by the vastly greater ease of processing content in a more consistent and context-independent way. (A reasoned argument on this topic is 'Embedded Markup Considered Harmful' published by Ted Nelson in 1997 – see *www.xml.com/pub/a/w3j/ s3.nelson.html.*)

Although the approach promises many advantages, it seems unlikely to be widely adopted any time soon. Interest in such a radical change of Web infrastructure has been low among content creators, so markup will continue to be embedded in the foreseeable future.

- We might see it as yet another example of how new technologies are constrained by previously chosen implementations, where identifiable advantages are offset by the perceived investment of globally adopting a revolutionary change.

Parsing, Human versus Machine

Even at the syntactical level, humans and machines parse data differently. In this context, we can speak of the 'readability' (or conversely, the 'opacity') of a document.

- Machine parsing inherently prefers binary material with no extraneous material or redundancy – a long opaque string of data that humans find hard or impossible to read. Such material is easy to process, using pattern-matching algorithms to extract tagged elements.
- Human readers prefer visual patterns, such as ample white space to group and separate semantic objects. They also prefer to have visual 'noise' (such as markup tags and other human-opaque linear elements) suppressed or transformed into typographical or layout conventions. Finally, humans experience a semantic dimension that enables, among other qualities, the appreciation of well-formed prose as an enhancement to easy reading.

What we see in XML (and in fact the entire SGML family), when applied in isolation, is unfortunately a schizophrenic approach that concentrates on machine representation (that is, a syntactical tag system) to solve human-use issues on a case-by-case basis. In particular, the lack of a generic solution approach causes significant problems when trying to create syntax rules for machine processing.

Informational Model

Syntactic details aside, the Semantic Web approach must tackle the task of how information is to be represented. It is not just about improving the capability to search for information, although this goal is important enough given the pervasive use of search engines as a starting point for browsing.

Just how important is search-engine technology?

In only a decade the dominating method for information discovery and research in general has become user-initiated search on the Web – which is on the face of it an amazing development. It is often perceived as faster to look something up using a Web browser than to walk to the bookshelf and consult a printed reference work. It is assuredly convenient . . .- *when it works*.

Precise statistics are hard to come by (and interpret properly), but yearly surveys (such as reported at *www.searchenginewatch.com*) suggest that at least a tenth of all traffic to Web sites today is due to users first clicking on links from search-engine results – individual sites may report as many as three of four hits from such sources. Eight of ten first-time visits might arise from search results.

Users looking for products, for example, are far more likely to just plug the product name into a search engine (a surveyed third do so) and explore the resulting links, than to start by browsing shopping sites or to click on ads (less than a tenth together).

On the other hand, many surveyed users follow a direct link to a site (such as a bookmark or typed-in Web address) and thus presumably already know where they are going. How did they learn of this address originally? In some cases, it is easy to surmise that casting a known brand name as a *dotcom* address might work; in other cases, it is not so evident.

Although clearly significant, it remains highly conjectural how search-engine statistics actually correlate with the way people successfully find relevant sites to revisit (and hence generate the non-search related hits). Nevertheless, the assumption is driving a fast-growing, multi-million-dollar advertising industry devoted to favorable placement of ads on search result pages. Also, it motivates considerable efforts by webmasters to 'optimize' search engine ranking for their sites.

Significant in this context is a 2001 survey that found that 97% of Fortune 100 companies had some type of site architecture problem that might give them problems being found by current search engines. The 'might' should be emphasized, however, since the most common problem related to missing or incorrect meta tags (that is, keywords to describe content) – these days the major search engines tend to ignore embedded metadata keywords in Web pages due to earlier misuse.

Bit 2.10   Lexical (that is, syntactic) search characterizes the current Web

The pervasiveness and apparent usefulness of this model can easily blind us both to its inherent failings and to the potentials of alternative models.

As the Web continues to expand, it becomes increasingly difficult for users to obtain relevant information efficiently. The common search strategies rarely scale well. The best results tend to come from engines that have useful ranking strategies such as backlink algorithms, or from engines that use a mix of sources, both human-classified directories and general indexing databases. All of these methods have known problems, however.

The ultimate filter remains human inspection and judgement, and the effects of this extra user effort can be considerable. This last point is why result pages that provide a few lines of hit context are so useful. Yet even the best search-engines can be maddeningly inefficient in some situations and then generate mostly irrelevant results. Reasons for search engine inefficiency are detailed in Chapter 1.

When search engines fail to satisfy, however, it is mainly because they are limited to reading the format tags and doing syntax analysis on a selection of published documents, sometimes not in their entirety. Despite indexing several billions of pages, such a database still represents only a small and arbitrary selection of total Web textual content – probably less than a tenth. The search situation for non-text content is far worse, though slowly improving.

The greater volume of Web content thus remains inaccessible to automated remote indexing, for one or another reason. It matters little to the frustrated user whether a search is blocked due to restrictive access control, site-local archival solutions, non-standard navigation elements, or application-specific content formats (that lack Web markup).

A new set of representational tools for metadata is required to make this immense repository of information accessible for a wide range of new applications. New interfaces are envisioned for users looking for information, and for applications that might process and compile the information for other, higher-level purposes. Requested information may not reside in any single Web location, or in a specific format, but semantic logic will enable programs to identify metadata and data elements on various sites, understand or infer relations, and piece them together reliably.

Such will be the deployed infrastructure of the Semantic Web. Over time, much of the information published on the Web will tend to be packaged in metadata as a matter of course, and some resources may even lack any immediate 'page' location as such. Instead of searching for pages of information, the new model will involve framing queries to submit to information services, locally generating 'content pages' on the fly that satisfy the initial search requirements and provide further links to in-depth material.

The result may be more like informal briefings or summaries. In fact, most retrieved information might be for agent consumption only and not reach the user directly at all.

Bit 2.11   The implied query in searching will shift from 'Where?' to 'What?'

This shift corresponds to a general shift in addressing from URL to URI. Strictly speaking, most users do not care where the information is stored, they just want useful answers and working functionality. The answer may be 'out there', but the user wants it 'here' and ready to use.

Representational Models

In this section, we briefly review the main representational models that have relevance for the later discussion of the Semantic Web and related work. Of necessity, numerous technical terms are introduced here, albeit in very abbreviated contexts. Some of these topics are later discussed in detail in Chapters 5 and 6.

Document Object Model

The Document Object Model (**DOM**) is a tree-like model of objects contained in a document or Web page, defining its elements and allowing selective access to them for reading or manipulation. The abstraction of the DOM programming interface (**API**) provides a standardized, versatile view of a document's contents that promotes interoperability.

The DOM is an established tool for dynamically manipulating document data structures, not a data structure descriptor itself. The description is derived from the document-embedded markup. Like the SGML family of markup languages, the DOM also presupposes (or imposes) a tree-like structure in the data. However, a linear structure might not always be relevant or useful.

Complementing the tree-based DOM API is an event-based Simple API for XML (**SAX**). It reports parsing events (such as the start and end of elements) directly to the application through callbacks, and does not usually build an internal tree at all. The application implements handlers to deal with the different events, much like handling events in a graphical user interface. SAX is ideal for small-resource devices that need to locate specific elements in a single pass.

As the custodian of DOM, the W3C maintains it as an open API to encourage standards compliance. See *www.w3.org/DOM/Activity* for an overview of its status.

Early on, however, Microsoft added its own proprietary extensions to the DOM in implementations of its Internet Explorer browser. Since IE still remains the dominant desktop Web browser, this divergence creates interoperability problems for anyone creating Web content or programming applications that rely on DOM.

Resource Description Framework

The Resource Description Framework (**RDF**) is a general-purpose language for representing information in the Web. It is a common specification framework to express document metadata (that is, the meaning of terms and concepts used to annotate Web resources) in a form that computers can readily process. The purpose of RDF is to provide an encoding and interpretation mechanism so that resources can be described in a way that particular software can understand it.

Bit 2.12   RDF is a core technology in the Semantic Web

In addition, most higher-level representations depend on or derive from the RDF model of expressing relationships and defining resources.

RDF commonly uses XML for its syntax and URIs to specify entities, concepts, properties, and relations. The model is based upon the idea of making statements about resources in the form of a *subject-predicate-object* expression (called a **triple**). Essentially, the subject is the resource being described, the predicate is the resource's trait or aspect that is being described, and the object is the value of that trait.

- This terminology is taken from logic and linguistics, where *subject-predicate* or *subject-predicate-object* structures have very similar but clearly distinct meanings.

Chapter 5 discusses RDF in greater detail.

Machine-usable RDF statements tend to be very verbose by human standards due to the requirement that resource descriptions be unique. By contrast, human-readable statements to convey the same meaning are typically much shorter, but prove useless to a machine trying to parse the meaning. This difference derives from the way human readers intuitively understand and use context and implicit reference. Machine parsing needs all the relations explicitly spelled out.

Schema

The term **schema** denotes a map of concepts and their relationships. Invariably, such a map is both incomplete and simplified, often with a tight focus on a particular mapping need. The role of a schema as a *representational model* in the context of Web information is to *mediate and adjudicate between human and machine semantics.*

Schemas based on RDF are described using **RDFS**, a language for defining RDF vocabularies, which specifies how to handle and label the elements.

This schema view is intentionally broader than that of operational semantics. In the latter, the meaning of a token is what we want it to be for processing software to deal appropriately with the data. The former view is more likely to represents a structuralist stance in suggesting a sign relation between the signifier and the signified.

As noted earlier, a common presumption is that the data structure is linear and tree-based. This constraint appears again in DTD and XML schema constructs. Nevertheless, XML-based schemas are successfully used to model data in many database applications, and thus viewed as 'superior' for this purpose. XML provides the self-defining relationships; no need then to supply the data in half a dozen different formats, or to create and impose a new 'standard' exchange format.

Regular tree grammar theory has established itself as the basis for structural specification in XML-based schemas. The other two constituents of a data model are the constraint specification, and the set of operators for accessing and manipulating the data. The field of constraint specification is still being studied, with different approaches such as 'path-based constraint specification' and 'type-based constraint specification'. XML operators are available as part of XPath, XSLT, and XQuery. In Part II we examine much of the mechanics involved, mainly in Chapter 5.

An alternative to constructing schemas using regular grammars is the *rules-based* approach. Its proponents often critique the otherwise uncritical adoption of grammar-based systems to define all schemas, sometimes retrofitting various 'layer' structures of pointers to handle the graph-like cases where the inherent tree-structure is inadequate.

Some purists might beg to differ with the polarization often seen, especially when 'rules-based' discussion segues into the claim of being a 'semantic' approach – to them, rule-sets are just another expression of a grammar, while semantics is about the means to choose between different rules and syntax. The pragmatic view for now is to follow current usage in this respect. However, the contrast between the two views can perhaps be illustrated by the following analogy, which is admittedly simplistic:

- **Grammar-based schemas** are similar to road *maps*; they try to locate precisely all mapped items on the grid. Implicit is the attempt to describe what is true for all of the data, all of the time. Getting from one point to another in such a map involves parsing the map and making decisions based on what one finds in the relationships (for instance, the topology and the available road connections).
- **Rules-based schemas** are like route *descriptions*; they try to express the things that make some particular information at some particular time and context different from other data of the same type. In practical terms, this operative approach is required for the data-entry and debugging phase.

To summarize: the first tries to abstract away invariants, the second tries to find abstractions to express variations. In general, XML schemas and DTDs belong to the first category because they are concerned with mapping relationships.

Bit 2.13   Road-map and route-description schemas are largely complementary

This seeming dichotomy appears in most descriptive contexts, as classic state-oriented versus process-oriented views. Inspired design can leverage optimized use of each.

The basic polarity between grammar-based and rules-based might usefully be expanded into a description of four layers to characterize schema languages, going from grammar to semantics. (If the example terms are not familiar at this time, please just ignore them for now.)

- **Structural patterns**. Processing is the binary result of matching the source against patterns expressed by a pattern language. Examples are DTD, TREX, RELAX, and the XSchema content model.
- **Datatypes**. Processing is the binary result of matching the source against datatypes expressed by a datatype language. Cited examples are XSchema datatypes and UML/XMI.
- **Structural rules**. Processing evaluates generalized expressions about structure against the source, and the results trigger actions – for example, to accept or reject the source as valid.
- **Semantics**. Processing is based on semantic rules that express meaning rather than formal structure. The rules may be constructed using RDFS, defining ontologies of meaning, such as for 'business logic' rules.

A developer discussion relating to the last layer was once seen to make the implied criticism that our understanding of the requirements for a comprehensive semantic layer is at present still so rudimentary that a strict formalized description of it is not possible. Such a nihilistic view is contested by those who in fact work at developing semantic rules, but there is truth to the supposition that we still have much to learn.

The Road Map

To explore the Semantic Web – design, implementation, and potential – we need to outline many aspects of its probable architecture, in particular what might be termed the essential components for it. In many respects, the outline describes an architectural plan untested by anything except thought experiments, more vision than substance, yet well on the way to becoming a reality.

Even though many of the following concepts were introduced in Chapter 1, and are covered in far greater detail in later chapters, it is relevant here to summarize systematically the essentials as they pertain to the Semantic Web. Therefore, we adopt the viewpoint of the sweb developer looking for the general solutions that each concept might offer, rather than considering the deficiencies that are evident in current implementations.

The following are then the 'essential components' of the Semantic Web.

Identity

Handling resources on the Web requires a strong and persistent implementation of unique identity. In local contexts, looser or transient forms of identity might be acceptable, but such constructs will not work in the global Web context.

The Uniform Resource Identifier is the generic solution, but it was not directly implemented in the early Web. The URI subset of Uniform Resource Locators (URL) was used to base global identity on an abstract 'location' instead – as protocol plus domain, plus local path or access method. However, the URL implementation has serious deficiencies in terms of persistency, among other things, in part due to its visible dependency on these arbitrarily and locally defined paths. No longer included in technical specifications, the now-deprecated term URL is only found in 'informal' usage and legacy documents.

Bit 2.14   Actual location on the Web (URL) is less useful than it may seem
At best, a given URL represents a logical (and reasonably persistent) 'location' locally dereferenced by the resource server at the specified Internet domain. It may just reflect an actual but arbitrary (and likely to change) filesystem path for server storage.

The user gains little and loses much by having the resource's actual storage location or access method exposed in the URL.

- The unpredictable mobility or availability of Internet resources at specified URLs is an inconvenience at best, often resulting in cryptic error messages from the server instead of helpful redirection, but in the expanded context of the Semantic Web, explicitly exposing location becomes a critical flaw.

For librarians (to take an example of a group on the front lines of implementing functionality in the direction of a semantic Web), it is a serious problem already. The deficiencies of the URL model compromise library services to patrons and impose unacceptably large burdens on catalog maintenance. Hence the desire for central repositories of uniquely named resource identifiers.

The existence of a named identifier must be maintained even when the referenced resource no longer exists or is unavailable. Therefore, another URI subset, **Uniform Resource Names (URN)**, has been under careful development by the IETF for some time, with the intent to provide persistent identities based on unique names within a defined **namespace** – as *urn:namespace-identifier:named-resource*. It is no coincidence that national libraries in many countries actively solicit the free registration of URN-URI addresses to online resources.

A modification to the current DNS system, for example, would resolve current issues with changing resource URLs by providing dynamic translation from URN pointers to the actual server-relative locations (as expressed by updated URLs).

Bit 2.15   Resource reference by name (URN) is better than by location (URL)
Most human-friendly locator schemes since the beginning of recorded history have been based on named references, leveraging the fundamental ability of humans to give and use names in a mapping or descriptive context. Importantly, names are portable.

In practice, the URL and URN subsets do overlap somewhat. For example, a mailbox URL address for e-mail is also a URN, in that the domain-qualified target name is sufficient and globally unique, and does not visibly depend on any server-local specifics such as path or method.

The persistence requirement of URN schemes is more an outcome of the social structures that evolve to meet a common community need than a technological issue. Such evolution can be slow – in the case of librarians, too slow.

Therefore, an interim solution was deployed: the **Persistent Uniform Resource Locator** (**PURL**) service. PURL can often be seen referenced in XML database-exchange constructions. The OCLC (*Online Computer Library Center, Inc., www.oclc.org*) maintains *PURL Org (www.purl.org)*, a global resource in the form of a database of registered unique identifiers. OCLC is a non-profit, global library cooperative serving 41,000 libraries in 82 countries and territories around the world. The PURL service is in effect a URL redirector, as shown in Figure 2.3.

A PURL URI associates a particular URL and XML term with a defined classification element, such as '*http://purl.org/dc/elements/1.1/publisher*' (which, by the way, is a *semantic* definition). Resource references are defined using PURLs and references to the PURL server. The PURL service responds to the client request with the currently valid associated URL in its database, which the client can then use to access directly the resource.

PURLs have been assigned to records cataloged in the *Internet Cataloging Project*, funded by the U.S. Department of Education to advance cataloging practices for Internet resources. This project represents the leading edge of MARC description of Internet resources, a mechanism by which computers exchange, use, and interpret bibliographic information. MARC data elements make up the foundation of most library catalogs used today.

Bit 2.16   PURL is a pragmatic hack that implements URN-like functionality

Implemented as a transparent Web service, PURL is an example of a transitional infrastructure enhancement referenced using resource-embedded pointers.

The expected syntax of URNs is clear enough to afford confidence that the syntax of transitional PURLs can be inexpensively and mechanically translated to the later URN form that is expected to be integral to the next-generation-internet infrastructure.

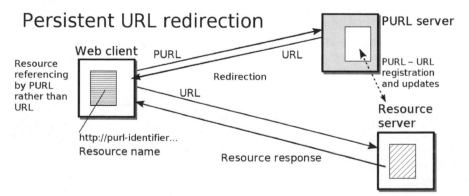

Figure 2.3 How PURL provides an intermediate step between the deficiency of impermanent URLs and the permanency of URNs. The PURL server acts as a redirector, dereferencing unique PURL identities into the currently valid URL

In the meantime, the PURL model lends itself to distribution across the Net, with servers run by organizations with a commitment to maintaining persistent naming schemes (libraries, government organizations, publishers, and others).

Markup

The syntactical component of the Semantic Web is the markup language that enables the distinction between the content representation and the metadata that defines how to interpret and process it.

For various reasons, the current choice of markup language is XML, because among other things it fulfills the dual requirements of self-defining and extensible document descriptions. XML also provides well-trodden paths for normalizing markup and upgrading existing HTML content, possibly through transitional formats such as XHTML 1.0.

Bit 2.17   XML is a fundamental building block for the Semantic Web
XML depends on URI, but in turn it is the foundation for most of the higher layers.

The visible part of this component is the markup syntax, expressed as the reserved text-pattern tags embedded in the document but invisible in human-use rendering. Primarily, markup language is for describing content structure.

- In practice, relevant parts of markup are client-interpreted and visually rendered in ways appropriate to accepted typographical and layout conventions. For non-visual rendering or other contexts, other forms of rendering would be more appropriate. When read aloud for the blind, 'emphasis', for example, would be a shift in inflection rather than an italic font style, and list or table items would be explicitly declared.

XML is well developed, and many sources and books provide both overview and in-depth studies. The principles of XML markup are simple, as are the methods for extending the language with new markup – it is after all called 'extensible' for a reason. In the context of this book, it is sufficient to note the salient concepts:

- Reserved text-pattern 'tags' that define 'containers' in which content and metadata can be placed in a structured way.
- URI references to the resources that can provide information about how the XML tags used are to be interpreted.

The basic syntax of XML is also found in other constructs, since it lends itself well to embedded definitions of more complex structures. An overview of relevant markup languages and protocols is found in Chapter 5.

Relationships

Given identities and markup, a natural step is to attempt to identify and describe relationships between data published on the Web. Much useful and helpful functionality can be achieved by combining data from several independent sources.

Since most content is published independently in a variety of formats that cannot directly be parsed and 'understood' by agents, the sweb solution is to introduce a metadata framework in which such relationships can be expressed, and thus provide a query and conversion mechanism for content elements.

Bit 2.18   Relationship expressions form the core of semantic query services
RDF and RDFS are languages well-suited to constructing relationship expressions.

The simplest such relationship is the concept of *equivalence*, so that for example corresponding fields in different databases can be recognized as describing the same property, despite arbitrary field labels. Shared fields provide ways to answer queries that a single database cannot satisfy.

Significant to the Semantic Web is the idea that such data relationships can be identified and published independently as metadata by anyone. Metadata resources enable agents to correlate data from multiple databases by identifying common fields and providing conversion rules. Actually, the application is even broader in that it enables the semantic wrapping of legacy resources and software in general. A simple representation of such a model is given in Figure 2.4.

Related to this relationship layer, and implied by the concept of conversion, we also require logical rules somewhere. In the Semantic Web, such rules are assumed to be published documents themselves, linked to relevant data documents.

Obviously, relationship rules can be added to the document stores of either database, but the openness of the concept frees us from a reliance on specific database owners to maintain such relationships. The relationship rules can be published as independent annotations, leveraging the ability of users to notice such relationships.

Relationships and conversions

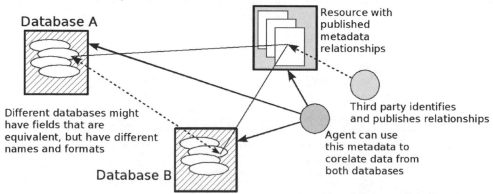

Figure 2.4   Relationships between data in independently published databases can be identified and published as a separate Web resource. Agents can thereafter use the published metadata to convert and correlate data from multiple, independent sources

A logical layer enables a large number of important functions, for example:

- *Deduction* of one type of document from a document of another type.
- *Validation* of a document against a set of rules of self-consistency.
- *Authentication* for allowed access and use of a document.
- *Inference* of valid relationships based on conversion and inference rules.
- *Deduction* of implicit knowledge from existing assertions.
- *Conversion* of a general query into more specific sub-queries.
- *Resolution* of a query by conversion from unknown terms into known ones.

Implementing such functionality is straightforward: a mechanism for quoting, simple predicate logic, and quantification expressions. It is the basis for creating a query language, and subsequently resolving the query into a result. RDF is the language framework chosen for this role in the Semantic Web.

Reasoning

Application of logic requires some form of *reasoning engine.* Implementing high-level reasoning is in effect the holy grail of designing agents. This conceptual level is as yet underdeveloped in the Semantic Web – it is in crucial parts unspecified, unlike the lower protocol, language and logic layers. At this stage in sweb development, the architects are happy to defer the problem of reasoning.

Currently, there is no definitive algorithm for *answering questions* – basically, finding proofs. Implementations rest on pragmatics. Construction of a proof is done in most applications according to some set of constrained rules; validation of a general proof is left to third parties. Thus the 'proof' of any query's answer is given as the chain of assertions and reasoning rules invoked by transactions, with pointers to supporting material in other documents. RDF and derived knowledge structures have recently added well-defined model theories with standard associated proof theories.

The design goal here is that practical queries be sufficiently constrained that the answers are computable – or are at least ensured to terminate in finite, reasonable time. The HTTP GET request will contain a proof that the client has a right to the response, while the response will be a proof that the response is what was asked for.

A second goal is a form of *adaptive forward compatibility.* This interesting idea means that a reasoning engine of a given version should be able to reason its way through a descriptive document sufficiently well to be able to deduce how to read a document written in a later version of the language. A related idea is that of *adaptive emulation*, which means that an engine should be able to read a description of a second independent implementation (for similar functionality) well enough that it can process data produced by it.

Bit 2.19   Reasoning engines are a prerequisite for adaptive Web agents

Independent software agents with initiative need to reason about things and infer appropriate responses from context. They also need to adapt for forward compatibility.

In this context we become concerned with *ontologies*, at least in the sense that computer science has redefined the term for reasons of practicality, if only because we become concerned with vocabularies and some form of meaning. In addition, perhaps more significantly, we need a mechanism to ensure a reasonable expectation of repeatable results to the same query. The full scope of investigating ontologies is examined in Chapter 6.

Agency on the Web

A diversity of algorithms exist today in search engines on the Web. Taken with the proof-finding algorithms in pre-Web logical systems, this abundance suggests that the Semantic Web will see many forms of agent evolve to provide answers to different forms of query and to perform a wide variety of tasks.

In fact, it will probably be wrong to speak of 'agent' in the singular, isolated sense that we are used to when thinking of applications. Instead, user queries and delegated tasks seem likely to be handled by collaborative services using distributed instances of functionality all over the Web.

A perceived 'agent' might therefore just be the local interface instance to transient swarms of collaborating services, each summoned for a particular task. This model is a multidimensional analog to the common Unix practice of 'piping' independent programs to process data sequentially. It may appear to be a complete application, but is only a transient combination.

It seems probable that a wide range of agents will be tailored to particular contexts, to particular query scope and span constraints, and to various forms of cooperative 'swarming' to fulfill user requests. A veritable ecology of agent software, capable of varying degrees of reasoning and initiative, could in time become an 'intelligence infrastructure' (please excuse the hype) to overlay the virtual networks of data transfer that exist today.

Bit 2.20   Adaptive Web agents enable interactive discovery of Web resources

First, as a rule, people are interested in the answers and results, not the detailed process of obtaining them. Second, people are interested in easy ways to assess and validate the results and the sources, but often it is enough to know an abstract metric of trust without necessarily and immediately seeing a detailed list of the actual sources and their credentials or reputations.

In that context, people using the new Web would no longer be concerned about the location or format of information, or precisely which services need to be found and accessed, or for that matter if the requested information is even compiled in the required way. They would delegate the queries to their immediate agent interfaces and simply await 'digest' results. Alternatively, they might engage in a 'conversation' around the request, interactively refining precision while background processing keeps up to the changing criteria.

Semantic P2P

Peer-to-peer technology is invoked implicitly when speaking of distributed collaboration and Web services. Simply providing computers on the Internet with more capabilities of processing and understanding the semantics of information will not be sufficient to bring Semantic Web technology to its full potential. Moreover, the way in which information is accessed on the Internet must undergo significant changes, away from the prevailing centralized publish-and-store-pages model.

In part, the shift in infrastructure can be seen as a belated response to the shift in computer technology, with ever more powerful desktop PCs deployed and interconnected to provide an alternative to centralized mainframe storage and processing. We are seeing a general move towards a 'ubiquitous computing' infrastructure, and in consequence, ever distributed storage and retrieval methods – a complex tapestry of p2p topologies.

Centralized storage and processing is unlikely to disappear completely – various contexts will either require or prefer such resources. However, even these traditional server architectures seem likely to connect increasingly to, and utilize, peer-based services to accomplish their tasks. After several pendulum swings between the extremes of centralized and distributed models, designs favoring hybrid solutions appear to be gaining acceptance in many application areas.

Bit 2.21   Distributed p2p can make precise publishing location irrelevant

Just as resources will be accessed mainly by location-agnostic name (URN), publishing may often be to a named service rather than any specific server location.

The key shift is that sweb resources, both data storage and processing, seem likely to be primarily distributed in adaptive, virtual networks. Who really cares where the data are stored and processor cycles are consumed, as long as it is all secure enough and available on demand? There are many advantages to encrypted and redundant storage across multiple machines on a global network, and such storage could easily become the norm for most cases. Prototype and commercial systems already exist on intranets and in the wild, and more are under development.

As an example, the *OceanStore Project* (at *oceanstore.cs.berkeley.edu*), is a well-established distributed storage solution. It is intended to scale to billions of users around the world using an infrastructure of untrusted server resources. Technically, it is called solving distributed (and guaranteed) object location in a dynamic network. A related project is *Tapestry*, an overlay location and routing infrastructure that provides location-independent routing of messages directly to the closest copy of an object or service. It uses only point-to-point links, without centralized resources, and is self-organizing, fault-resilient, and load-balancing.

OceanStore looks to be a viable model for providing one aspect of infrastructure for the Semantic Web, that of location-agnostic ubiquitous storage and processing. The interesting part of the prototyping is that anyone can download working software and set up mini-clusters on selected machines on the LAN or across the Internet. This process can form the islands that over time can merge into a complete global infrastructure, leveraging existing machine and network resources.

The Visual Tour

Putting all the preceding into a coherent picture may not be the easiest task, for we often venture into *terra incognito* – uncharted regions about which we can only speculate, since neither the finished applications, nor sometimes even the data models they are to process, exist.

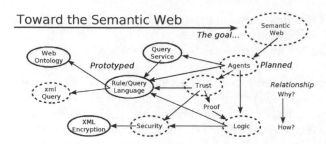

Figure 2.5   A 'road map' of sorts, based on a more detailed W3C diagram, which although dated, still reflects the relationships between different Semantic Web components examined in this book. See text for a discussion on how to read the arrows

Still, even the barest outline contours on the map serve a purpose by orienting the viewer in the relationships between prototypes, planned components, and vision. Not everyone is comfortable with simple graphs, but the subsequent discussions through this book might be augmented by the visual relationships they imply.

The Map

Figure 2.5 illustrates the lay of the land, as it were, by describing some of the relationships between now-identifiable components in the vision of the Semantic Web.

The diagram is based on a larger, more detailed road map prepared by the W3C some years ago, and describes the relationships as Why?/How? links. In other words, following any arrow answers 'How do we realize the previous component?' while going in the reverse direction answers 'Why do we need the preceding component?'.

Some of these components have been built (solid outline), at least as functional prototypes, others were planned (dashed outline). Still others remain preliminary or conjectural (no outline) until it is seen how such a system will work in practice. Some components, such as 'trust', require not just technological solutions, but also critical political decisions and social acceptance.

Although this diagram may seem unusual due to the combined relationships, depending on which direction you read the arrows, and perhaps 'backwards' to some in its overall sense, it is still useful in seeing dependency relationships.

You can read the 'flow' as in the following example, starting in an arbitrarily chosen node:

- *To realize* **Agents** (follow arrows emanating from this node), we *need* to implement Query Services, Trust mechanisms, and Logic systems.
- *Why we need* **Agents** (in this view, backtrack the arrow leading to this node) is to *reach* the goal of the Semantic Web.

You may wish to refer back to this diagram from later discussions of specific components just to refresh their particular placement in the overall scheme of things. Variations and extensions of this diagram also appear later (in Chapter 5) to place other components in this same, goal-oriented scheme.

Figure 2.6 shows an alternative 'stacking' view of layer dependencies, one of several variants popularized by Tim Berners-Lee (see *www.w3.org/DesignIssues/diagrams/sw-stack-*

Building the Semantic Web

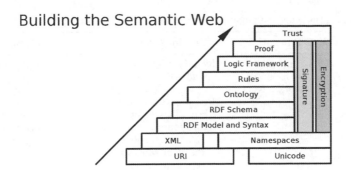

Figure 2.6 An alternative view of dependency relationships, as depicted in the 'stack layer' model popularized by Tim Berners-Lee

2002.png). It intends to show how certain implementation areas *build on* the results of others but, like most simple diagrams of this nature, it is only indicative of concept, not formally descriptive.

The Architectural Goals

It is useful to summarize the goals introduced in the earlier text:

- **Identity**, by which we understand URIs (not locator URLs).
- **Data and Structuring**, as represented by Unicode and XML.
- **Metadata and Relationships**, as defined in RDF.
- **Vocabularies**, as expressed in RDF Schema.
- **Semantic Structure**, as represented in Web Ontologies.
- **Rules, Logics, Inferencing, and Proof**, which to a great extent still remain to be designed and implemented to enable agency.
- **Trust**, as implemented through digital systems and webs of trust, and also requiring further development and acceptance.

Now in the first years of the 21st century, we begin to see a clearer contour of what is to come, the implementation of components that recently were merely dashed outlines of conjectured functionality on the conceptual chart.

Mindful of the rapid pace of sweb development in 2003 and 2004, the W3C issued a new recommendation document in late 2004, the first of several planned, that specifies more clearly the prerequisites and directions for continued development and deployment of sweb-related technologies: 'Architecture of the World Wide Web' (Vol 1, December 2004, *w3.org/TR/webarch/*).

The recommendation adopts the view that the Web builds on three fundamentals that Web Agents (which includes both human users and delegated software in the form of user agents) must deal with in a compliant way:

- **Identification**, which means URI addressing of all resources.
- **Interaction**, which means passing messages framed in standard syntax and semantics over standard protocols between different agents.

Table 2.1.   W3C Principles and Good Practice recommendations for the Web

Design Aspect	W3C Principle	W3C Good Practices	Constraints
Global Identifiers	Global naming leads to global network effects.	Identify with URIs. Avoid URI aliasing. Consistent URI usage. Reuse URI schemes. Make URIs opaque.	Assignment: URIs uniquely identify a single resource. (Agents: consistent reference.)
Formats		Reuse representation formats. (New protocols created for the Web *should* transmit representations as octet streams typed by Internet media types.)	(Transparency for agents that do not understand.)
Metadata		Representation creators *should* be allowed to control metadata association.	Agents *must not* ignore message metadata without the consent of the user.
Interaction	Safe retrieval.	(Resource state must be preserved for URIs published as simple hypertext links.)	(Agents must not incur obligations by retrieving a representation.)
Representation	Reference does not imply dereference.	A URI owner *should* provide *consistent* representations of the resource it identifies. URI persistence.	
Versioning		A data format specification *should* provide for version information.	
Namespace		An XML format specification *should* include information about change policies for XML namespaces.	
Extensibility		A specification *should* provide mechanisms that allow any party to create extensions. Also, specify agent behavior for unrecognized extensions.	(Useful agent directives: 'must ignore' and 'must understand' unrecognized content.)
Composition		Separation of content, presentation, interaction.	
Hypertext data		Link identification. Web-wide linking. Generic URIs. Hypertext links where expected.	(Usability issues.)
XML-based data		Namespace adoption. Namespace documents. *QNames* must be mapped to URIs.	*QNames* are indistinguishable from URIs.
XML Media type		XML content should not be assigned Internet media type 'text', nor specify character encoding.	(Reasons of correct agent interpretation.)
Specifications	Orthogonal specifications		
Exceptions	Error recovery based on informed user consent.	(Consent may be by policy rules, not requiring interactive human interruption for correction.)	

- **Formats**, which defines standard protocols used for representation retrieval or submittal of data and metadata, and which convey them between agents.

The document goes on to highlight a number of *Principles* and *Good Practices*, often motivated by experiences gained from problems with previous standards. Table 2.1 summarizes these items.

This table may seem terse and not fully populated, but it reflects the fact that W3C specifications are conservative by nature and attempt to regulate as little as possible. Items in parenthesis are expanded interpretations or comments included here for the purpose of this summary only.

The recommendation document goes into descriptive detail on most of these issues, including examples explaining correct usage and some common incorrect ones. It motivates clearly why the principles and guidelines were formulated in a way that can benefit even Web content authors and publishers who would not normally read technical specifications.

As with this book, the aim is to promote a better overall understanding of core functionality in order that technology implementers and content publishers achieve compliance with both current and future Web standards.

The Implementation Levels

Two distinct implementation levels are discernible when examining proposed sweb technology, not necessarily evident from the previous concept maps:

- 'Deep Semantic Web' aims to implement intelligent agents capable of performing inference. It is a long-term goal, and presupposes forms of distributed AI that have not yet been solved.
- 'Shallow Semantic Web' does not aspire as high, instead maintaining focus on the practicalities of using sweb and KR techniques for searching and integrating available data. These more short-term goals are practical with existing and near-term technology.

It is mainly in the latter category we see practical work and a certain amount of industry adoption.

The following chapters examine the major functionality areas of the Semantic Web.

3

Web Information Management

Part of the Semantic Web deals necessarily with strategies for information management on the Web. This management includes both creating appropriate structures of data-metadata and updating the resulting (usually distributed) databases when changes occur.

Dr. Karl-Erik Sveiby, to whom the origin of the term 'Knowledge Management' is attributable, once likened knowledge databases to wellsprings of water. The visible surface in the metaphor represents the explicit knowledge, the constantly renewing pool beneath is the tacit. The real value of a wellspring lies in its dynamic renewal rate, not in its static reservoir capacity. It is not enough just to set up a database of knowledge to ensure its worth – you must also ensure that the database is continually updated with fresh information, and properly managed.

This view of information management is applicable equally to the Web, perhaps the largest 'database' of knowledge yet constructed. One of the goals of the Semantic Web is to make the knowledge represented therein to be at least as accessible as a formal database, but more importantly, accessible in a meaningful way to software agents. This accessibility depends critically on the envisioned metadata infrastructure and the associated metadata processing capabilities.

With accessibility also comes the issue of readability. The Semantic Web addresses this by promoting shared standards and interoperable mapping so that all manner of readers and applications can make sense of 'the database' on the Web.

Finally, information management assumes bidirectional flows, blurring the line between server and client. We see instead an emerging importance of 'negotiation between peers' where although erstwhile servers may update clients, browsing clients may also update servers – perhaps with new links to moved resources.

Chapter 3 at a Glance

This chapter deals with the general issues around a particular implementation area of the Semantic Web, that of creating and managing the content and metadata structures that form its underpinnings. *The Personal View* examines some of the ways that the Semantic Web might affect how the individual accesses and manages information on the Web.

- *Creating and Using Content* examines the way the Semantic Web affects the different functional aspects of creating and publishing on the Web.
- *Authoring* outlines the new requirements, for example, on editing tools.
- *Publishing* notes that the act of publishing will increasingly be an integral and indistinguishable part of the authoring process.
- *Exporting Databases* discusses the complement to authoring of making existing databases accessible online.
- *Distribution* considers the shift from clearly localized sources of published data to a distributed and cached model of availability based on *what* the data are rather than *where* they are.
- *Searching and Sifting the Data* looks at the strategies for finding data and deriving useful compilations and metadata profiles from them.
- *Semantic Web Services* examines the various approaches to implementing distributed services on the Web.

Security and Trust Issues are directly involved when discussing management, relating to both access and trustworthiness.

- *XML Security* outlines the new XML-compliant infrastructure being developed to implement a common framework for Web security.
- *Trust* examines in detail the concept of trust and the different authentication models that ultimately define the identity to be trusted and the authority conferred.
- *A Commercial or Free Web* notes that although much on the Web is free, and should remain that way, commercial interests must not be neglected.

The Personal View

Managing information on the Web will for many increasingly mean managing *personal* information on the Web, so it seems appropriate in this context to provide some practical examples of what the Semantic Web can do for the individual.

Bit 3.1   Revolutionary change can start in the personal details
Early adoption of managing personal information using automation is one sneak preview of what the Semantic Web can mean to the individual.

The traditional role of computing and personal information management is often associated with intense frustration, because the potential is evident even to the newcomer. From the professional's viewpoint, Dan Connolly coined what became known as Connolly's Bane (in 'The XML Revolution', October 1998, in *Nature's Web Matters,* see *www.nature. com/nature/webmatters/xml/xml.html*):

The bane of my existence is doing things I know the computer could do for me.

Dan Connolly serves with the W3C on the Technical Architecture Group and the Web Ontology Working Group, and also on Semantic Web Development, so he is clearly not a newcomer – yet even he was often frustrated with the way human–computer interaction works. An expanded version of his lament was formulated in a 2002 talk:

The bane of my existence is doing things I know the computer could do for me . . . and getting it wrong!

These commentaries reflect the goal of having computer applications communicate with each other, with minimal or no human intervention. XML was a first step to achieving it at a syntactic level, RDF a first step to achieving it at a semantic level.

Dan's contribution to a clearer perception of Semantic Web potential for personal information management (PIM) has been to demonstrate what the technology can do for him, personally.

- For example, Dan travels a lot. He received his proposed itineraries in traditional formats: paper or e-mail. Eventually, he just could not bear the thought of yet again manually copying and pasting each field from the itinerary into his PDA calendar. The process simply had to be automated, and the sweb approach to application integration promised a solution.

The application-integration approach emphasizes data about real-world things like people, places, and events, rather than just abstract XML-based document structure. This sort of tangible information is precisely what can interest most users on the personal level.

Real-world data are increasingly available as (or at least convertible to) XML structures. However, most XML schemas are too constrained syntactically, yet not constrained enough semantically, to accomplish the envisioned integration tasks. Many of the common integration tasks Dan wanted to automate were simple enough in principle, but available PIM tools could not perform them without extensive human intervention – and tedious manual entry.

A list of typical tasks in this category follows. For each, Dan developed automated processes using the sweb approach, usually based on existing Web-published XML/RDF structures that can serve the necessary data. The published data are leveraged using envisioned Web-infrastructure technologies and open Web standards, intended to be as accessible as browsing a Web page.

- *Plot an itinerary on a map.* Airport latitude and longitude data are published in the Semantic Web, thanks to the DAML project (*www.daml.org*), and can therefore be accessed with a rules system. Other positioning databases are also available. The issue is mainly one of coding conversions of plain-text itinerary dumps from travel agencies. Applications such as *Xplanet* (*xplanet.sourceforge.net*) or *OpenMap* (*openmap.bbn.com*) can visualize the generated location and path datasets on a projected globe or map. *Google Maps* (*maps.google.com*) is a new (2005) Web solution, also able to serve up recent satellite images of the real-world locations.
- *Import travel itineraries into a desktop PIM* (based on *iCalendar* format). Given structured data, the requirements are a ruleset and *iCalendar* model expressed in RDF to handle conversion.

- *Import travel itineraries into a PDA calendar.* Again, it is mainly a question of conversion based on an appropriate data model in RDF.
- *Produce a brief summary of an itinerary suitable for distribution as plain text e-mail.* In many cases, distributing (excerpts of) an itinerary to interested parties is still best done in plain text format, at least until it can be assumed that the recipients also have sweb-capable agents.
- *Check proposed work travel itineraries against family constraints.* This aspect involves formulating rules that check against both explicit constraints input by the user, and implicit ones based on events entered into the user's PDA or desktop PIM. Also implicit in PDA/PIM handling is the requirement to coordinate and synchronize across several distributed instances (at home, at work and mobile) for each family member.
- *Notify when the travel schedule might intersect or come close to the current location of a friend or colleague.* This aspect involves extended coordination and interfacing with published calendar information for people on the user's track-location list. An extension might be to propose automatically suitable meeting timeslots.
- *Find conflicts between teleconferences and flights.* This aspect is a special case of a generalized constraints analysis.
- *Produce animated views of travel schedule or past trips.* Information views from current and previous trips can be pulled from archive to process for display in various ways.

A more technical discussion with example code is found at the Semantic Web Applications site (*www.w3.org/2000/10/swap/pim/travel*).

Most people try to juggle the analysis parts of corresponding lists in their heads, with predictably fallible results. The benefits of even modest implementations of partial goals is considerable, and the effort dovetails nicely with other efforts (with more corporate focus) to harmonize methods of managing free-busy scheduling and automatic updating from Web-published events.

Bit 3.2   Good tools let the users concentrate on what they do best

---

Many tasks that are tedious, time-consuming, or overly complex for a human remain too complex or open-ended for machines that cannot reason around the embedded meaning of the information being handled.

The availability of this kind of automated tool naturally provides the potential for even more application tasks, formulated as *ad hoc* rules added to the system.

- For example: 'If an event is in this schedule scraped from an HTML page (at this URL), but not in my calendar (at this URI), generate a file (in *ics* format) for PIM import'.

The next step in automatic integration comes when the user does not need to formulate explicitly all the rules; that is, when the system (the agent) observes and learns from previous actions and can take the initiative to collect and generate proposed items for user review based on current interests and itineraries.

- For example, the agent could propose suitable restaurants or excursions at suitable locations and times during a trip, or even suggest itineraries based on events published on the Web that the user has not yet observed.

Creating and Using Content

In the 'original release' version of the Web ('Web 1.0'), a platform was provided that enabled a convenient means to author, self-publish, and share content online with the world. It empowered the end-user in a very direct way, though it would take a few iterations of the toolsets to make such authoring truly convenient to the casual user.

In terms of ease-of-use, we have come a long way since the first version, not least in the system's global reach, and practically everyone seems to be self-publishing content these days on static sites, forums, blogs, wikis, etc. (call it 'Web 2.0'). But almost all of this Web page material is 'lexical' or 'visual' content – that is, machine-opaque text and graphics, for human consumption only.

Creating Web content in the context of the Semantic Web (the next version, 'Web 3.0') demands more than simply putting up text (or other content) and ensuring that all the hyperlink references are valid. A whole new range of metadata and markup issues come to the fore, which we hope will be adequately dealt with by the new generation of tools that are developed.

Bit 3.3   Content must be formally described in the Semantic Web

The degree of possible automation in the process of creating metadata is not yet known. Perhaps intelligent enough software can provide a 'first draft' of metadata that only needs to be tweaked, but some user input seems inevitable.

Perhaps content authors just need to become more aware of metadata issues. Regardless of the capabilities of the available tools, authors must still inspect and specify metadata when creating or modifying content. However, current experience with metadata contexts suggests that users either forget to enter it (for example, in stand-alone tools), or are unwilling to deal with the level of detail the current tools require (in other words, not enough automation).

- The problem might be resolved by a combination of changes: better tools and user interfaces for metadata management, and perhaps a new generation of users who are more comfortable with metadata.

A possible analogy in the tool environment might be the development of word-processing and paper publishing tools. Unlike the early beginnings when everything was written as plain text, and layout was an entirely different process relegated to the ranks of professional typesetters and specialized tools, we now have integrated authoring-publishing software that supports production of ready-to-print content. Such integration has many benefits to be sure. However, a significant downside is that the content author rarely has the knowledge (or even inclination) to deal adequately with this level of layout and typographical control. On the other hand, the issue of author-added metadata was perhaps of more concern in the early visions. Such content, while pervasive and highly visible on the Web, is only part of the current content published online.

Many existing applications have quietly become RDF-compliant, and together with online relational databases they might become the largest sources of sweb-compliant data – we are speaking of calendars, event streams, financial and geographical databases, news archives, and so on.

Bit 3.4   Raw online data comes increasingly pre-packaged in RDF

The traditional, human-authored Web page as a source of online data (facts as opposed to expressed 'opinion') might even become marginalized. Aggregator and harvester tools increasingly rely on online databases and summaries that do not target human readers.

Will we get a two-tier Web with diverging mainly-human-readable and mainly-machine-readable content? Possibly. If so, it is likely to be in a similar way to how we now have a two-tier Web in terms of markup: legacy-HTML and XML-compliant. Whether this becomes a problem or not depends – all things considered, we seem to be coping fairly well with the markup divide.

Authoring

As with earlier pure-HTML authoring, only a few dedicated enthusiasts would even presume to author anything but the actual visible content without appropriate tool sets to handle the complexities of markup and metadata. Chapter 8 explores the current range of available tools, but it is clear that much development work remains to be done in the area.

Minimum support for authoring tools would seem to include the following:

- XHTML/XML markup support in the basic rendering sense, and also the capability to import and transform existing and legacy document formats.
- Web annotation support, both privately stored and publicly shared.
- Metadata creation or extraction, and its management, at a level where the end-user can 'easily' do it.
- Integration of numerous types of information into a consistent view – or rather, into several optional views.

Several of these criteria are addressed in at least the proof-of-concept stage by the W3C *Amaya* combined editor/browser client (see *www.w3.org/Amaya/*).

The ideal Web 3.0 tool thus looks back at the original Web browser design, which was to have been both browsing and publishing tool. It adds, however, further functionality in a sensible way to provide users with the means to manage most of their online information management needs. In short, the user interface must be capable of visualizing a broad spectrum of now disparate local and online information items and views.

Bit 3.5   The separation of browsing and authoring tools was unfortunate

In the Semantic Web, we may expect to see a reintegration of these two functions in generalized content-management tools, presenting consistent GUIs when relevant.

The concept of browse or author, yet using the same tool or interface in different modes, should not be that strange. Accustomed as we are to a plethora of distinct dedicated 'viewers' and 'editors' for each format and media type, we do not see how awkward it really is.

Tools can be both multimedia and multimode. This assertion should not be interpreted as sanctioning 'monolithic bloatware, everything to everyone' software packages – quite the opposite.

In the Semantic Web, any application can be browser or publisher. The point is that the user should not have to deal explicitly with a multitude of different interfaces and separate applications to manage Web information. The functional components to accomplish viewing and editing might be different, but they should appear as seamless mode transitions.

Working with Information

One of the goals of the Semantic Web is that the user should be able to *work with information* rather than with programs. Such a consistent view becomes vital in the context of large volumes of complex and abstract metadata.

In order to realize the full vision of user convenience and generation of simple views, several now-manual (usually expert-entry) processes must become automatic or semi-automatic:

- creating semantic annotations;
- linking Web pages to ontologies;
- creating, evolving, and interrelating ontologies.

In addition, authoring in the Semantic Web context presumes and enables a far greater level of collaboration. Chapter 4 specifically explores collaboration space in this context.

Publishing

Some people might debate whether creation and publishing are really very separate processes any more. Even in the paper-publishing world, it is more and more common for the content author also to have significant (or sometimes complete) control of the entire process up to the final ready-to-print files. The creation/publishing roles can easily merge in electronic media, for example, when the author generates the final PDF or Web format files on the distribution server.

In the traditional content-creation view for electronic media, any perceived separation of the act of publishing is ultimately derived from the distinction between interim content products on the local file system, accessible only to a relative few, and the finished files on a public server, accessible to all. This distinction fades already in collaborative contexts over the existing Web.

Now consider the situation where the hard boundary between user system and the Web vanishes – or at least becomes arbitrary from the perspective of where referenced information resides in a constantly connected, distributed-storage system. Interim, unfinished content may be just as public as any formally 'published' document, depending only on the status of applied access control.

Web logs are just one very common example of this blurring. Open collaboration systems, such as Wiki, also demonstrate the new content view, where there is no longer any particular process, such as a discrete FTP-upload, that defines and distinguishes 'published' content from draft copy.

Bit 3.6   The base infrastructure of the Semantic Web supports versioning
With versioning support in place, managing interim versions of documents and collaboration processes online becomes fairly straightforward.

Exporting Databases

Increasingly important, and a complement to explicit authoring, are technologies for exporting existing database content to the Semantic Web. In fact, a significant frustration in the business and consumer world, and consequent disenchantment with Web technology in general, has been the lack of direct access to business databases.

Customers must often manually browse to another company's site, drill down to the sought-for published document, and in worst case, parse a downloaded and hard-to-read PDF or MS Word document for the required information. A frozen PDF rendition of a database selection of records might in addition be critically out of date.

Bit 3.7   Online business requires online database access
It is that simple, and any other offline solution, including Web-published static extracts (HTML tables or PDF) simply will not do. That more companies on the Web do not understand this issue can only be attributed to the 'printed brochure' mentality.

Data export, or rather its online access, is easiest if the database already is RDF compliant, because then the problem reduces to providing a suitable online interface for the chosen protocol. Then customers can access realtime data straight from the company database.

Distribution

In the traditional Web, it would seem that content is published to one specific location: a Web server on a specified machine. This content-storage model, however, is a serious over-simplification. Distribution was an issue even in the early Internet, and for example USENET is based on the idea of massive replication across many connected servers, motivated by the fact that overall bandwidth is saved when users connect to their nearest news server instead of to a single source.

For reasons of reliable redundancy and lessened server loading, mirror sites also became common for file repositories and certain kinds of Web content. Such a visible mirror-server model requires a user to choose explicitly between different URLs to reach the same resource. However, transparent solutions for distributed storage and request fulfilment that reference a single URL are much more common than the surfing user might imagine.

Content Caching or Edge Serving

In particular, the growth of very high-demand content, such as provided by search engines, large corporations and institutions, and by popular targets such as major news sites, *Amazon.com* and *eBay*, all rely on transparent, massively distributed content caching.

Behind the scenes, requests from different users for content from the same stem URL are actually fulfilled by distributed server clusters 'nearest' their respective locations – an approach often termed 'edge' serving. An entire infrastructure fielded by specialized companies grew up to serve the market, such as *Akamai* (*www.akamai.com*) that early on made a place for itself on the market by deploying large-scale solutions for aggressively distributed and adaptive storage.

Distribution is also about functionality, not just serving content, and about having clusters or swarms cooperatively implement it. In the Semantic Web, we might be hard pressed to separate cleanly what is 'content' and what is 'functionality'. The rendered result of a user request can be a seamless composite constructed by agents from many different kinds of Web resources.

Storage can be massively distributed in fragmented and encrypted form as well, so that it is impossible to locate a particular document to any individual physical server. In such a case, user publishing occurs transparently through the client to the network as a whole instead of a given server – or rather to a distributed publishing service that allocates distributed network storage for each document. Retrieval is indirect through locator services, using unique keys generated from the content. Retrieval can then be very efficient in terms of bandwidth usage due to parallel serving of fragments by swarms of redundant hosts.

Syndication Solutions

An area related to publishing and distribution is known as *syndication*, which in one sense, and as commonly implemented, can be seen as a user-peer form of mirroring. Syndication of content, however, differs from mirroring of content in the way the name (borrowed from news media and broadcast television) implies; it distributes only subscriber-selected content from a site, not entire subwebs or repositories.

Syndication is currently popular in peer-to-peer journalism (**p2pj**) and for tracking new content in subscribed Web logs (commonly called 'blogs'). Web syndication technology is currently based on two parallel standards, both with the same **RSS** acronym due to historical reasons of fragmented and misconstrued implementations of the original concept:

- **Really Simple Syndication** (RSS 0.9x, recently re-engineered as RSS 2.x to provide namespace-based extensions) is based on XML 1.0. It is now used primarily by news sites and Web logs to distribute headline summaries and unabridged raw content to subscribers. This standard and its development are described at Userland (see backend.userland.com/rssO91 and blogs.law.harvard.edu/tech/rss).
- **RDF Site Summary** (known as RSS 1.x) is a module-extensible standard based on RDF, in part intended to recapture the original semantic summary aspects. It includes an RDF Schema, which is interesting for Semantic Web applications. The home of this standard is found at *Resource.org* (see *web.resource.org/rss/1.0/*for details). Official modules include *Dublin Core, Syndication* and *Content.* Many more are proposed.

Somewhat conflicting perspectives exist on RSS history (see *rsswhys.weblogger.com/ RSSHistory*). In any case, we find in all about *nine* variants of RSS in the field, adhering in greater or lesser degree to one of these two development forks.

The name issue is especially confusing because RSS 0.9 was originally called *RDF Site Summary* when it was deployed in 1999 by Netscape (the company). It then provided the capability for 'snapshot' site summaries on the Netscape portal site for members wishing to advertise their Web site. Although conceived with RDF metadata summary in mind, the actual Netscape implementation dropped this aspect.

Other developers and RSS portal sites later spread the concept, leveraging the fact that RSS could be used as an XML-based lightweight syndication format for headlines.

Some advocates redefine RSS as 'just a name' to avoid the contentious acronym interpretation issue altogether. Others note the important distinction that the two parallel standards represent what they would like to call 'RSS-Classic' and 'RSS 1.0 development' paths, respectively:

- MCF (Meta Content Framework) → CDF (Channel Definition Format) → scriptingNews → RSS 0.91 → RSS-Classic.
- MCF → XML-MCF → RSS 0.90 → RSS 1.0.

Be that as it may. The formats can to some extent be converted into each other.

In any case, a plethora of terms and new applications sprang from the deployment of RSS:

- **Headline** syndication, as mentioned, carries an array of different content types in channels or feeds: news headlines, discussion forums, software announcements, and various bits of proprietary data.
- **Registries** provide classified and sorted listings of different RSS channels to make user selection easier.
- **Aggregators** decouple individual items from their sites and can present one-source archives of older versions of content from the same site, and in some cases with Scrapers, advanced searching and filtering options.
- **Scrapers** are services that collect, 'clean', and repackage content from various sources into their own RSS channels.
- **Synthesizers** are similar to Scrapers, except that the source channels are just remixed into new ones, perhaps as a user-configured selection.

Somewhere along the way, the 'summary' aspect was largely lost. Even as RSS, for a while, became known as *Rich Site Summary* (v0.91), it was morphing into full-content syndication. In this form, it became widely adopted by Web-log creators (and Web-log software programmers).

The blog community saw RSS in this non-semantic, non-summary sense as a way for users to subscribe to 'the most recent' postings in their entirely, with the added convenience of aggregating new postings from many different sources. However, nobody knows how much of this subscription traffic is in fact consumed actively.

- Many users with unmetered broadband are in fact 'passive' RSS consumers; they turn on many feed channels with frequent updates (default settings) but hardly ever read any of the syndicated content piped into their clients. It is just too time-consuming and difficult to overview, yet the user is loath to disconnect for fear of missing something.

Bit 3.8   Using RSS to syndicate entire content is a needless threat to Internet bandwidth

In 2004, RSS-generated traffic was getting out of hand, effectively becoming a form of network denial-of-service load from some sources. In particular, many clients are poorly implemented and configured, hitting servers frequently and needlessly for material that has not been updated.

Adoption of the RDF version for syndicating metadata (RSS 1.0), on the other hand, is an important asset for Semantic Web publishing since it easily enables agent processing. Additionally, it requires far less bandwidth. This kind of publishing and distribution of metadata, with pointers to traditionally stored content, is arguably more important and useful than distributing the actual content in its current format.

Syndicated metadata might yet become an important resource to find and 'mine' published content in the future Web. At present, however, much RSS traffic is less useful.

Searching and Sifting the Data

'Mining the Web' for information is by now an established term, although it might mean different things to different actors. However, we are likely to think of simply harvesting 'published data' as-is, without really reflecting on how to compile the results.

Pattern matching and lexical classification methods similar to current search engines might suffice for simple needs, but the current lack of machine-readable 'meaning' severely limits how automated or advanced such literal filtering processes can be.

Nevertheless, much current data mining is still based on simple lexical analysis, or possibly search engine output. A simple example is the way e-mail addresses are harvested from Web sites simply by identifying their distinctive syntactical pattern. The ever-mounting deluge of junk e-mail is ample testament to the method's use – and unfortunate effectiveness.

To accomplish more than just sorting key-value associations based on discrete words and phrases requires a different approach, one that the enhanced infrastructure of the Semantic Web would provide. Perhaps then too a modicum of control may limit abusive harvesting.

Usage Mining

Broader application of Web mining deals not only with Web data as such, nor even with the metadata relationships, but rather with actual *usage*.

The field of *Web Usage Mining* (**WUM**) has attracts significant commercial interest. Although commercial ventures probably account for most innovation and actual deployments in the field these days, their proprietary solutions are rarely published openly.

Generally speaking, the visible products (deliverables) in WUM are the finished analysis results and the services to produce them – for example, the *Nielsen/NetRatings* real-time research and analysis products about Internet users (see *www.nielsen-netratings.com*). Lists of available mining tools and resources are otherwise available at the *Knowledge Discovery Nuggets Directory* at *www.kdnuggets.com* – a searchable portal site.

The *WebSIFT* project (Web Site Information Filter, formally the WEBMINER project, see *www.cs.umn.edu/Research/websift/*) defines open algorithms and builds tools for WUM to provide insights into how visitors use Web sites. Other open sources for WUM theory and relevant algorithms are found in the papers presented at the annual IEEE International Conference on Data Mining (ICDM, see *www.cs.uvm.edu/~xwu/icdm.html*).

WUM processes the information provided by server logs, site hyperlink structures, and the content of the Web site pages. Usage results are often used to support site modification decisions concerning content and structure changes. As an overview, WUM efforts relate to one or more of five application areas:

- **Usage characterization** has its focus on studying browser usage, interface-design usability, and site navigation strategies. Application might involve developing techniques to predict user behavior in interactions with the Web and thus prepare appropriate responses.
- **Personalization** means tracking and applying user-specified preferences (or assumed contexts) to personalize the 'Web experience'. It is often seen as the ultimate aim of Web-based applications – and used to tailor products and services, to adapt presentations, or to recommend user actions in given contexts.
- **System improvement** means to optimize dynamically server (or network) performance or QoS better to meet actual usage requirements. Application includes tweaking policies on caching, network transmission, load balancing, or data distribution. The analysis might also detect intrusion, fraud, and attempted system break-ins.
- **Site modification** means to adapt an individual site's actual structure and behavior to better fit current usage patterns. Application might make popular pages more accessible, highlight interesting links, connect related pages, and cluster similar documents.
- **Business intelligence** includes enterprise Web-log analysis of site and sales related customer usage. For example, IBM offers a number of WUM-based suites as services to other corporations.

Most WUM applications use server-side data, and thus produce data specific to a particular site.

In cases where multi-site data are desired, some form of client or proxy component is required. Such a component can be as simple as a persistent browser cookie with a unique ID, and having participating sites use page elements to reference a common tracking server that can read it. This profiling method is commonly used with banner ads by advertisers.

By contrast, *browser toolbars* can actively collect profiled usage data from multiple sites directly from the client – for example, feedback to a search engine page ranking function.

- Many Web users are unaware of browser-deployed WUM strategies such as these. Undisclosed profiling is considered privacy invasion, but it is widely practiced.

Bit 3.9   WUM technologies have relevance to adaptive sweb agents

The main difference to current WUM is how the data are used. In sweb agents, most profiling data might never perculate beyond a small circle of agents near the user.

Technologies based on usage mining might come into play when agent software in the Semantic Web is to adapt automatically to shifting user and device contexts. In addition to explicit user-preference data, agent-centric WUM components might mine an assortment of published data and metadata resources to determine suitable configurations and to match offered device capabilities to the tasks at hand.

It is reasonable to suppose that profile disclosure to third parties be both user configurable and otherwise anonymized at the agent-Web interface.

Semantic Web Services

Query services, Web annotation and other online toolset environments provide simple forms of *Web Services* (**WS**), a catch-all term that has received extra prominence in the *.NET and Sun ONE* era of distributed resources.

The development of deployable WS is driven largely by commercial interests, with major players including *Microsoft, IBM and BEA,* in order to implement simple, firewall-friendly, industry-accepted ways to achieve interoperable services across platforms. WS advocacy should be seen in the context of earlier solutions to distributed services, such as CORBA or DCOM, which were neither easy to implement nor gained broad industry support.

The perception has also been that commercial WS development existed in opposition to the more academic Semantic Web (SW). WS developers drew criticism for not respecting the Web's existing architecture, while the SW project was critiqued for its overly visionary approach, not easily adaptable to current application needs.

However, contrary to this perception of opposition, WS and SW are inherently complementary efforts with different agendas, and as such can benefit each other:

- Web Services can be characterized as concerned with program integration, across application and organizational boundaries. Traditionally, WS handles long-running transactions and often inherits earlier RPC design.
- The Semantic Web is primarily concerned with future data integration, adding semantic structure, shareable and extensible rule sets, and trust in untrusted networks. It builds on URI relationships between objects, as uniform yet evolvable RDF models of the real world.

For example, WS discovery mechanisms are ideally placed to be implemented using sweb technology. A WS-implemented sweb protocol (SOAP) may prove ideal for RDF transfer, remote RDF query and update, and interaction between sweb business rules engines. Therefore, to gain focus we can at least identify a joint domain of *Semantic Web Services* (**SWS**).

The tersest definition of modern WS is probably 'a method call that uses XML', and it is not a bad one, as far as it goes. The W3C defines WS as 'a software application defined by a URI, whose interfaces and bindings are capable of being defined, described and discovered as XML artifacts' (as illustrated in Figure 3.1).

The current state of support suggests that SWS is an excellent applications integration technology (despite the way many weary IT professionals might cringe at the term 'integration'), but that it may not yet be ready for general use – important layers in the SWS model are not yet standard, nor available as ready-to-use products.

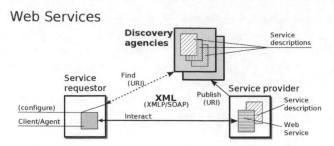

Figure 3.1 Web services can be seen as a triangular relationship of publishing, finding, and interacting with a remote service on the Web

The W3C Web Services Workshop, led by IBM and Microsoft, agreed that the WS architecture stack consists of three stack components:

- *Wire* is based on the basic protocol layers, XML overlaid by SOAP/XML, with extension layers as required.
- *Description* is based on XML Schema, and is overlain with the layers Service Description, Service Capabilities Configuration, Message Sequencing, and Business Process Orchestration.
- *Discovery* involves the use of the discovery meta-service, allowing businesses and trading partners to find, discover, and inspect one another in Internet directories.

The conceptual stack underlying the WS layer description is expressed in two dimensions. The stack layers run from transport specifications, through description aspects, up to discovery. Descriptions include both business-level and service-level agreements. Overlaying this stack are **QoS**, Security, and Management aspects.

Unfortunately, each vendor, standards organization, or business defines Web Services differently – and in the case *of Microsoft WS .NET Architecture*, even differently over time. Despite repeated calls for broad interoperability and common standards, some vendors continue to use (or return to) proprietary solutions.

The state of WS development is indicated by the general list of resources published and sorted by category at the *Directory of Web Services and Semantic Web Resources* (at *www.wsindex.org*).

Discovery and Description

Two main discovery and description mechanisms are used in WS contexts: **UDDI** (Universal Description, Discovery, and Integration) and **WSDL** (Web Services Description Language). They operate at different levels.

UDDI is a specification that enables businesses to find and transact with one another dynamically. It functions as a registry interface maintaining only business information, each record describing a business and its services. The registry facilitates discovering other businesses that offer desired services, and integrating with them. Typically, UDDI entries point to WSDL documents for actual service descriptions.

WSDL is an industry standard for describing Web-accessible services (that is, their Web interfaces). As an XML-based language, it is machine processable. However, the intended semantics of the interface elements are only accessible to human readers. In WSDL a service is seen as a collection of network endpoints which operates on messages. A WSDL document has two major parts: first, it specifies an abstract interface of the service, then, an implementation part binds the abstract interface to concrete network protocols and message formats (say SOAP, HTTP).

Purpose of WS

One way to classify Web Services is by their primary purpose. For example, a few broad application categories are: *Administrative Metadata, Agents, Architecture, Brokering, Citation, Collection, Description, E-commerce, Education, Government, Libraries,* and *Registry.* Professionals working within these application areas have increasing access to WS technologies, as these are developed and deployed, but only to the degree that they become aware of them.

An example might illustrate the issues and directions. Efforts related to the management of metadata currently tend to focus on the standards and tools required by applications within the group's main constituency. An early primary focus is e-license management for *digital libraries.*

Librarians everywhere have increasingly been struggling with the management of electronic resources, but they have not known where to go for help in an area of the profession that seemed devoid of expert advice. For example, expenditures for electronic resources have grown enormously in the past decade, accounting for upwards of half the library budget in some cases. Yet the ability of library staff to manage this growing collection of resources typically remains unchanged; most of the time is still spent managing print resources.

Commonly used integrated library systems do not provide tools to help manage electronic resources. Administrative metadata, elements about licensed resources, do not fit comfortably into most library systems. It is difficult for libraries to know what use restrictions might apply to e-journals and aggregated online texts, and how to triage and track related access problems. Most libraries simply lack the new kinds of tools required by these processes.

Only in recent years have Web-based resources became available and been published in any greater way. One such resource, useful for tracking developments in this field, is the Cornell University 'Web hub for Developing Administrative Metadata for Electronic Resource Management' (at *www.library.cornell.edu/cts/elicensestudy/home.html*). The Web Hub is especially useful in highlighting e-resource management solutions implemented at the local level by different libraries.

Such systems are typically developed by large research institutions, perhaps affiliated with a university library, but a growing number of vendors seem interested in developing generic products to meet the needs of most academic libraries.

Whither WS?

The entire field of Web Services remains in a state of flux despite several years of intense development, and not all proposed 'Web services' have very much to do with the Semantic

Web. In part, there has been a trend to call every new incarnation of e-commerce functionality an e-service, or if on the Web, a WS. In part, some confusion has been due to vacillation on the part of *Microsoft* (MS) in both its definition and deployment of *.NET* component technologies.

Confusion and caution reign. On the one hand, there is the risk that the MS approach would fragment the Web into incompatible WS segments, simply due to the massive dominance by MS in the desktop market. On the other, there has been the fear that the new Web would drift into the embrace-and-extend sphere of proprietary-modified open standards, thus defaulting into a kind of corporate hegemony control of protocol and security by a single company. It is a fact that core MS WS runs on proprietary code only on MS Windows servers.

A number of independent projects have been progressing under the guidance of several major actors. We will now look briefly at the major WS efforts, ending with *Microsoft's* *.NET*. A realistic view might be that, ultimately, a successful deployment of SWS as a new Web infrastructure will require that all the actors who are committed and cooperating around open standards should achieve the vision, or at least interoperate with it. At present, however, the major players are in critical areas still (or again) pushing proprietary solutions, which risks locking corporate environments into vendor-specific commitments difficult to migrate from later.

Fortunately, the WS paradigm is firmly message based, depending on XML, SOAP (or any other XMLP-compliant protocol), and other platform-independent specifications. This independence gives hope for bridging vendor divides.

The Sun Approach

Sun Microsystems promotes the J2EE platform for WS (*java.sun.com/j2ee/*), through the intermediary of the *open Java Community Process* (JCP, *www.jcp.org*). Both standards and architecture are open and distributed. The platform is based on Java, in theory independent of operating systems.

- As Sun invented the Java language and promotes competitor Unix-based operating systems, WS-platform debates often become slanted as a 'Sun-interests against MS-interests' confrontation. Such acrimony has often been detrimental to Web-universal solutions.

For developers working in Java, the announcement in 2002 of an official suite of API specifications for Web services from Sun was welcome news. Sun had long been criticized for inaction on this front, given the early availability of the *MS .NET SDK*. However, rather than proposing transitional hack solutions to as-yet uncharted technical areas, Sun's strategy had been instead to work with open source initiatives on the major pieces of the Web services puzzle.

Also announced was the *Java Web Services Developer Pack* (JWSDP) to provide an integrated development platform. In the JWSDP, Sun packaged some reference technology implementations into a single, free download, adding some infrastructure to ease Web service deployment. Thus, the JWSDP let Java developers create useful Web services with available components, instead of forcing them to either assemble those tools themselves or use another vendor's tools. Importantly, it is all open source, not proprietary code, and anchored in a developer community.

Other aided tasks include compiling and registering a WSDL file for the implemented Web service, so that others can find it online and use it. Supported in this context is the UDDI v2.0 registry standard. Metering and monitoring capability as of JXTA v2.1 enables automated query and inspection of traffic and memory usage on remote network peers.

Using Java makes core WS-implementations platform independent, though at the apparent cost of committing to a single language. In actuality, support for other languages does exist, though Java remains the primary language for implementations. Java is good for building complex systems but has itself been perceived as complex.

Recent technology releases have positioned developers in a better position to address the challenges of building real-world Web services.

Apache Group

The *Apache Group* collects a number of evolving WS efforts at *ws.apache.org*. Since the group is historically defined by the web server, it is natural that several WS projects evolved here.

Axis (*ws.apache.org/axis/*) is an implementation of SOAP v1.1, the proposed standard fulfilling the W3C XMLP requirements (see the protocol overview in Chapter 5). Essentially, it corresponds to what would have been Apache SOAP v3.0, a previous project, but rewritten from scratch. The project has focus on the basic building blocks for implementing WS primitives.

Since Apache is a major server workhorse on the Web, having native SOAP support goes a long way to implementing WS applications on existing Apache servers, perhaps best expressed in the *Tomcat* and *WebSphere* application server contexts. Axis builds on the Apache notion of pluggable modules and APIs to extend core server functionality with advanced message processing and request forwarding to defined services.

Other projects of interest include:

- WSIF, a simple Java API (invocation framework) for transparently invoking Web services irrespective of their location.
- WSIL, a distributed Web service discovery method (inspection language) specifying how to inspect a Web site for available Web services.

Several other projects can also be found that are considered more experimental.

DARPA DAML WS

The *DAML Services* arm (*www.daml.org/services/*) of the DAML program (*DARPA Agent Markup Language*, see the DARPA section in Chapter 7) started by developing a DAML-based Web Service Ontology (DAML-S), with early releases of specifications in 2000–2003. The DAML focus is mainly on the languages, specifications, and protocols involved in the client-WS communications, defining the common interfaces that should be shared by technologies developed by others.

In a subtle refinement of focus in November 2003, however, the renamed *Semantic Web Services* arm released the expected language implementation specifications as *OWL-S v1.0* instead. **OWL** (*Web Ontology Language*, described in Chapter 6) received official W3C recommendation in August 2003, and this move toward implementation status has naturally affected other projects.

The current OWL-S program also works on supporting tools and agent technology to enable automation of services on the Semantic Web.

The focus is on descriptions of WS at the application level, describing *what* a particular WS implementation does, rather than specifying how it does it. Other WS description languages, such as WSDL, are specifications at the communications level of message and protocol, while discovery mechanisms such as UDDI serve mainly as pointers to service addresses and ports. DAML-S/OWL-S complements the lower-level specifications prevalent in these other approaches that are mainly concerned with the core platform architecture.

DAML-S/OWL-S deals with the application-level issues of discovery, invocation, inter-operation, composition (of new services from existing selections), verification, and execution monitoring. A service implementation must in this context specify two things:

- A *service profile* to describe what it requires from agents and what it provides for them.
- A *service model* (known as a *Process*) to describe workflow and possible execution paths.

A profile would be used both by the WS implementation to advertise its availability and capabilities, and by requestors to specify their needs and expectations. Brokering services can then match requestors and providers based on their respective profiles. A third sub-ontology may specify how service details are implemented (*Grounding*).

Despite having generated a lot of interest through its promise to add semantics to WS descriptions, only a small number of these WS descriptions appear as yet to exist. Furthermore, most of these do not point to real services, representing instead constructs from the DAML-S community. A possible cause may be the perceived lack of sustained contact with and interest from the main implementers driving deployed WS-technology, Microsoft and Sun, both of whom have vested interests in promoting in-house WS languages and specifications.

A 2003 Dutch study of the situation, 'An experience report on using DAML-S' (*www.iids.org/publications/essw03.pdf*, also see *www.iids.org/pubs*), concluded that DAML-S was superior to existing WS languages as it allowed the use of formally defined domain knowledge. However, it did not yet adequately address real-life situations.

With regard to the following years of OWL-S development (v1.1 was released in November 2004), much the same concerns and conclusions appear applicable. Effort is being made to integrate WS made up of many different services into a semantic-interoper-able context for practical business use.

The DAML site references tools, security issues, example services, and a number of ontology/agent use cases, sorted by architecture and by language. Environment categories studied include B2B, Grid, Ubiquitous, Web, and Agent. The language-oriented categories include Advertising and Matchmaking, Negotiation and Contracting, Process Modelling, Process Enactment, and a catch-all Unclassified.

Recognizing that information security and privacy play an increasingly important role in a world in which the Web is central to business, industry, finance, education, and government, several OWL-S categories have been proposed and developed to resolve such issues:

- *Credential Ontology* defines the capability to specify access control restrictions of Web pages or Web services that use authentication as a requirement for authorized access.
- *Security Mechanisms Ontology* defines the capability to interface on a high level of abstraction among various security standards and notations.
- *Service Security Extensions Ontology* defines the capability to annotate the security properties of SWS.
- *Agent Security Extensions Ontology* defines the capability to annotate the security properties of agents.
- *Privacy Ontology* defines the capability to express privacy policies to protect information, and a protocol to support matching of privacy policies across different contexts.
- *Cryptographically Annotated Information Object Ontology* defines the capability to capture encrypted or signed input or output data of services.

One may see these efforts as a measure of technology maturity on the road to implementing practical services to realize the greater sweb vision.

.NET Web Services

Assessing the *.NET* ('dot-net') way of implementing an 'intelligent Web' and its related services proved difficult. Where the previous constellations of effort, even at Sun, had clear focus on technical implementation and open discussion among developers, with Microsoft one entered the realm of corporate marketing strategies and largely proprietary technology.

Whatever the expressed (and shifting) aims of the .NET effort, the dominance of MS applications on the desktop and in the corporate intranet (and the way these applications work) also ensures that any Microsoft policy to 'embrace and extend' existing open or industry standards inevitably affects deployment of sweb technologies.

Current MS overviews (for example at *www.microsoft.com/net/* and *msdn.microsoft.com/webservices/*) describe mainly *XML Web Services*. These XML-WS are somewhat loosely defined – in business terms by promised synergies, and in technology terms by selected client-server capabilities:

- Partner and client connectivity through exposure of useful functionality over an 'industry-standard' Web protocol (mainly SOAP over HTTP).
- Personal, integrated user experiences through smart devices that utilize sufficiently detailed descriptions of WS interfaces to allow client applications to talk to the services (based on metadata published as WSDL).
- Increasing revenue streams by enabling businesses to create their own WS, advertised through registry discovery so that potential users can find appropriate services easily (using UDDI).

Security aspects inherent in using open HTTP and site-dependent authentication systems are currently resolved by extending SOAP with security components, following the published WS-Security Specifications. Other, more secure transports are promised as soon as the technology matures and higher-level services are developed.

The official view is regularly restated that Web Services 'have made a lot of progress' over the last few years, and that work on the platform continues. However, the long status documents from Microsoft (see *www.microsoft.com/net/basics/*) remain remarkably vague about practical applications, even after several years.

- The reader may consider it fortunate that although Microsoft has a long-standing bad habit of sowing serious linkrot by often and gratuitously 'reorganizing' its extensive Web sites, in recent years it does implement a modicum of redirection from older URLs, at least for a time. The previous URL paths have if fact 'only' suffered changes to mixed case.

One functional service *is Microsoft Alerts* (*www.microsoft.com/net/services/alerts/*), by now fairly established. The original '.NET' moniker in the name was dropped at some point. *Alerts* is a notification service that delivers personalized news, weather, and information to people on the move – in effect, a streaming-content aggregator service using RSS. It is hard to assess how pervasively used it is, but it has been promoted heavily for several years and integrates with the typical suite of Windows productivity applications.

Other staple services offered (such as *Passport, Messenger Connect for Enterprises*, and *MapPoint*) are not 'complete' but are components requiring integration with larger business processes. The view expressed as early as 2001 remains curiously unfulfilled, despite the constant hype, that by simply subscribing to a rich selection of available 'plug and play' WS functions, developers could create a plethora of 'richer user experiences' quickly and easily.

Larger-scale deployment of .NET services appear mainly as restricted trials within Microsoft itself, as described in a series of whitepaper studies. Development focus at Microsoft has been on 'management services' such as download center, content storage and retrieval, and generic Web platforms. Less seriously, Microsoft employees developed an internal service to track left-over snacks, *SnackAlerts*, which can notify subscribing employees where the snacks are in 79 Microsoft buildings around the world.

Microsoft's WS specifications are a moving target, evolving in many places in parallel – an ongoing process where they are defined, refined, and sometimes deprecated seeming at whim. Neither are we sure of who the major players are, or will be, since it depends on the practical scope of interoperability and chosen authentication mechanisms – how open MS chooses to make them.

Therefore, it is impossible to say with any great precision what .NET is, or will be, until the respective applications are 'fully deployed' and functional. It is not enough to say it 'is a set of software technologies for connecting information, people, systems, and devices', yet many MS explanations say little more. Even the promise of 'small, reusable applications' in XML for WS seems to ignore the complications of real-world business applications. Perhaps the reason is that most of the implementations are buried in corporate systems.

Bit 3.10   Visible MS XML-WS appears mainly to be a commoditized side-effect of selling Web-aware solutions to implement core corporate business processes

This considered view is based on the fact that MS has long had only two core business models, supporting corporate solutions and selling the ubiquitous Windows desktop, combined with a proven track record of successful salesmanship based on the concept of convenience.

Some amusement has been occasioned in the press over the fact that even lead spokes-people for Microsoft have had great difficulty explaining the '.NET vision' intelligently or consistently. Developers in the .NET area might reasonably take issue with the apparent dismissal suggested by this critique, engrossed as they are in specific technical details that appear well-defined from their insider perspective, but many analysts agree that the overall MS strategy is muddled at best.

Bit 3.11   Sweb WS is mainly an architecture model rather than a market sector

The Microsoft approach is typically characterized by a product focus on component-application-suite. Therefore, management appears genuinely confused in pursuing the 'Web Services market' and by how to transition from vision to product. Stated goals slide into obscurity each year as the next great initiative is announced. Is that any way to run a vision?

One critical factor is that the design of MS software has often had lock-in effects on ensuing versions of hardware, making alternative solutions difficult. Consider how many companies are locked to the MS-Word document format(s) despite attempts to develop format-compatible third-party software. Many aspects of current .NET WS implementation require proprietary MS software at some level or another, possibly with lock-in features to specific hardware.

Of particular concern is the introduction of a proprietary 'information rights management' (IRM) mechanism in MS Office 2003 and later applications. This implementation of IRM can block general access to Web-published documents produced using them. At the very least, it may prove legally impossible to use third-party applications to view such documents. Either way, it breaks the established concept of universal access to Web-published content.

We cannot deny that there are legitimate reasons for corporate users (and even individuals) to desire effective content-control mechanisms. However, once deployed, it will be awkward and rare to produce any content without inadvertently 'locking' it for other applications and users.

Bit 3.12   Current IRM/DRM mechanisms are a threat to the Semantic Web vision

When the dust finally settles, ways may have been found to let free exchange of information on the Web coexist with economic interests. Non-intrusive micropayment solutions and less draconian copyright and licensing enforcement appear to be necessary ingredients for such a resolution.

A mitigating circumstance is the recently adopted policy of the W3C to recommend particular technologies for the Web only if they are freely offered for the public good without restrictive ties. Although this policy ensures open standards in the infrastructure and WS, it does little to stem user adoption of needlessly restrictive technologies when producing and publishing Web content.

DotGNU

The *DotGNU* project (*www.dotgnu.org*) was started in reaction to, and aims to be a complete and open-source replacement for, Microsoft's .NET initiative. It independently implements the Common Language Infrastructure (CLI) that defines '.NET' functionality.

Revisited in 2005, the project has evidently gained considerable adoption, claiming to be the industry leader and provider of Free Software solutions for desktop and server applications. Three major components of this initiative are currently being implemented:

- *DotGNU Portable .NET* suite of tools to build and execute .NET-compatible applications on any platform. The cross-platform suite includes, among other things, C/C# compilers, assembler, disassembler, and runtime engine.
- *PhpGroupWare* suite to provide multi-user collaborative Web-space, including many useful XML-RPC Web-service components ready to use in custom WS implementations.
- *DotGNU Execution Environment* (DGEE) to provide virtual server implementations and resource management support to run the WS implementations.

The project goals are to provide a reasonably .NET-compatible system and then improve on what Microsoft is offering – for example, by decentralizing it. Therefore, *DotGNU* is based on a peer-to-peer system for service discovery, avoiding the use of centralized authentication/authorization portals such as the 'Passport' system.

The strategy for *DotGNU* is not only technologically superior, it is claimed, but it also serves to ensure that no single entity would ever have the kind of monopoly power to control (or end) the free exchange of digital information. The latter is openly presented as a strong motivation:

> ... *to prevent Microsoft from achieving their stated goal 'that the era of "open computing", the free exchange of digital information that has defined the personal computer industry, is ending'. Read the quotation in context and learn how the Free Software movement works to prevent this disaster. The contribution of the DotGNU project might turn out to be important.*

In keeping with the open approach, the developer community runs *DotGNU Wiki* (*wiki. dotgnu.org*) as its forum of discussion and general Web resource.

Using WS

At this point, you might well wonder how to use Web services when the platforms are still evolving, and when there is so much flux, uncertainty, and lurking incompatibility. In fact, today's Web service tools and frameworks, irrespective of vendor, do provide enough core functionality to build interesting distributed applications. Expect a steep learning curve, however, and be prepared to develop your own solutions, or to pay for proprietary solutions developed by others.

Some higher-level services such as WS-Security are beginning to gel with the advent of support in a number of new tools (such as the Web Services Development Kit). Other services are still in their development stage as specifications are reviewed and early implementations expose areas where the specifications may need solidifying. Until then, you can rely on traditional HTTP-level mechanisms for security and other features.

From the business perspective, some make the analogy that Web Services are similar to business services. The comparison is to a great degree a valid one, but at the same time it is constrained by the products and services view. However, just as most companies outsource aspects of their business process that are not their core competencies, such as shipping and materials handling, Web services might correspond to outsourced development of certain services they use all the time but may not wish to implement in-house.

Authentication is the one such service implemented and deployed in a major way on the public Web. A substantial number of Web sites today rely on the *.NET Passport* authentication services, even though it has both suffered numerous embarrassments and received considerable criticism. The service is required when using any other .NET WS implementation.

Originally introduced with an absolute dependency on Microsoft central servers, the current .NET approach appears to allow at least a limited delegation of authentication authority to other non-MS authentication networks.

- In late 2002, this security interoperability was mediated between MS and a large coalition of independents under *Liberty Alliance* (*www.projectliberty.org*) by MS-partner and LA-founding member *RSA Security* (*www.rsasecurity.com*). RSA, with its critical focus on Secure Web Services, and a deeper understanding of the positions on both sides of the divide, clearly had a pivotal role here.

Security and Trust Issues

Reliability, availability, and security of data and services are goals that sometimes seem difficult to attain even under centralized control and governance. How then can one possibly realize them under the distributed and *ad hoc* management that characterizes in the Semantic Web, with its mix of roaming users and authority-delegated software agents?

How indeed? Part of the answer depends critically on the adopted solutions for trust, discussed later. Central to the issue is a pervasive and transparent management of strong public key cryptography, especially in the form of digital signatures, complemented by webs of trust and proofing logic.

- This same technology can be used, not only to verify reliably unique person identities, even anonymous ones (!), but also to identify uniquely and make tamper-proof particular documents, document versions, and in fact any Web resources. It is for this reason that the visual map of the Semantic Web (Figure 2.5 in Chapter 2) includes encryption, signature, and proof logic as key parallel components.

Bit 3.13   Authenticated authority must work in a roaming context

As one of the cornerstones in the Semantic Web context is that access should be automatic and transparent regardless of location or used devices, the underlying authentication model must remain valid and secure across many *ad hoc* platforms.

These same infrastructure components can also resolve some of the problems facing attempts to implement reasonable business models in a Web context.

Technologies may change, but the essential requirements remain much the same, comprising the key concepts *of Authentication, Authorization, Integrity, Signature, Confidentiality, Privacy,* and more recently, *Digital Rights Management* and *Information Rights management.* Table 3.1 summarizes the meaning of these concepts.

Another essential requirement of security standards in the Web context is that they work naturally with content created using XML (or with XML-derived languages and protocols).

Transparency is another essential characteristic, in that integrity, confidentiality, and other security benefits should apply to XML resources without preventing further processing by standard XML tools – whether at message endpoints, or in intermediate processing.

Although older security technologies provide a core set of security algorithms and technologies that can be used in XML contexts, the actual implementation of these is inappropriate for most XML applications:

- The proprietary binary formats require specialized software for interpretation and use, even just to extract portions of the security information.
- Older technologies tend to assume specific components integrated into the endpoint applications, introducing awkward dependencies.
- Older standards are not designed for use with XML and thus lack support for common XML techniques for managing content (such as URI or XPath pointers).

Table 3.1   Summary of essential security concepts

Concept	Question answered	Comments
Authentication	Who am I? (Verify asserted identity against some trusted authority.)	The later Trust section further discusses authentication, identity, and role issues.
Authority	What may I access and do?	Individual and role. See above.
Integrity	Is the information intact?	Prevent accidental or malicious change, or at least detect it.
Signature	Is the information certified?	Ties in with identity issues. Might certify an identity or authority.
Confidentiality	Is the information safe from unauthorized disclosure?	Encryption makes information unreadable even if access controls are circumvented.
Privacy	Is individual and sensitive information safe from unauthorized disclosure?	Governance issues of how to use sensitive information. Consent.
Digital Rights Management	How may I use or share this information?	Usually now a combination of access control and embedded enforcement of usage license.

A unified and open framework of new sweb-oriented standards and implementations, however, is evolving to address these issues on the Web.

Since public and corporate awareness seems dominated by a focus on Microsoft's .NET solutions, it is important to explore the subject in general terms. In particular, we need to highlight the alternatives to central authentication by proxy authorities, with their proprietary interpretations of 'trust' and 'security' as a 'product' to sell, in order to assess properly the role of these concepts in the broader Semantic Web context.

XML Security

XML Security defines a common framework with processing rules that can be shared across applications using common tools. Therefore, no extensive customizing of applications is needed to add security. While XML Security reuses the concepts, algorithms, and core technologies of the legacy security systems in the interests of interoperability, it also introduces the changes necessary to support extensible integration with XML.

The core XML Security standards comprise two Semantic Web cornerstones seen in the map and stack views (Figures 2.5 and 2.6) introduced at the end of Chapter 2:

- **integrity and signatures**, as XML Digital Signature (*XML DigSig*);
- **confidentiality**, as XML Encryption (*XML Enc*).

Persistent integrity and security is associated with the content itself, rather than with any single transport, processing, or storage method. The implementation thus provides seamless end-to-end security, irrespective of intermediate processing, and reusing existing technologies when possible.

XML Security realizes these goals by the simple expedient of transforming binary representations (such as for digital keys, certificates, or encrypted data) into plain text or rather to a well-defined subset of text (such as *Base64*) that is *guaranteed invariant* in other message processing on any platform.

- Residual issues do remain in the conflict between exact representation in signed text and digests, for example, and the relative laxness in XML that allows whitespace variations and some token substitutions (such as single and double quotes) in 'identical' messages. Such issues are generally resolved by transforming to 'canonical' form before applying a signature.

Related supporting core standards not explicitly on the simple sweb map are also needed:

- **Key Management**, as *XML Key Management Specification* (**XKMS**), defines protocols for the public-key management services and identity binding needed to implement data integrity and signatures.
- **Authentication and Authorization Assertions**, as *Security Assertion Markup Language* (**SAML**), define authentication and authorization assertions and shared framework used to establish identity and authority.
- **Authorization Rules**, as *XML Access Control Markup Language* (**XACML**), state the rules and policies used to manage access.

In addition to these standards, we must add major XML Security applications:

- **Web Services Security**, as WS-Security specifications that outline how **XML DigSig** and **XML Enc** may be used with XMLP (SOAP) messages. Also specified is how security claims (for example, identity credentials) may be included with a message.
- **Privacy Management**, as specified in the *Platform for Privacy Preferences* (**P3P**), defines the XML vocabulary to manage human-policy issues ensuring that organizations and individuals with legitimate access to information collected about an individual do not misuse it.
- **Digital Rights Management**, in the form of *extensible Rights Markup Language 2.0* (**XrML**), provides a framework, a language and processing rules to enable managed interoperability in content workflow with multiple participants and applications, and among different 'viewers'.

We might note in the context of WS-Security that *Microsoft* and *IBM* released a Web Services security architecture and roadmap in 2002 that outlined a strategy and series of specifications to bring together different security technologies. Despite the stated aim of providing a unified, flexible, and extensible security framework for WS-Security, the core group *of IBM, Microsoft, Sun*, and *VeriSign* subsequently started pushing proprietary solutions instead.

Trust

The concept of trust has both formal and informal components, but in both cases ultimately it is based on accepted knowledge, such as **an established identity**. In the more formal contexts, this knowledge requires some form of **explicit verification**; in the informal ones, it is *implicit* (typically in a shared history).

Bit 3.14   Both sweb agency and e-commerce require secure authentication

Whether implemented authentication mechanisms become tightly centralized or distributed trust systems remains to be seen. Perhaps a middle ground can be found.

In this first discussion of trust, the focus is on the fundamental issue of authenticating identity assertions. Establishing trust rests on some process of **authentication**, and on some ultimate authority that can validate the assertion about a given identity. Opinion diverges about whether central authorities or distributed ones are most appropriate (or secure) for the Web.

- Using a central authentication proxy raises many concerns, from scaleability and technical reliability of systems not yet tested on such a large scale, to entrusting most matters of personal identity to a single arbitrator with unproven security and unknown agenda. Simply rolling out a single identity-authentication model devised for corporate use and expecting it to scale and function on the general Web for everyone in all situations seems foolhardy at best.

Instead most sweb perspectives assume distributed systems that can if needed interact.

Successful authentication results in some form of credentials (or certificates of trust, physical or electronic) that the user can present to querying parties to be granted authorization for intended actions. Implicit in this context is the requirement that the credentials are acceptable to a querying party (often custodians for protected data or access), and that low-overhead mechanisms exist to validate on demand the authenticity of the presented credentials.

Although, typically, we refer to 'identity' as if it always implied the concept of unique personal identity, a trusted identity can have other faces. The simplest real-world example is delegated authority, which means that the bearer of some issued credential becomes a proxy for the identity and is conferred with some measure of authority to act on the behalf of this person. Depending on circumstances, generally the degree of value at stake or 'risk' involved, the bearer may be accepted at face value (possession is sufficient), or in turn be required to authenticate and validate against for instance a class of permitted bearers.

As implied elsewhere in the context of sweb agents, delegated authority is expected to be the norm rather than the exception on the Web. Therefore, we had better get the whole identity-authentication model right when implemented in order to support securely this functionality. Table 3.2 exemplifies some alternatives to personal identity.

In no case is the trust system's knowledge of identity absolute. Instead, it is based on various situation-appropriate criteria for an 'acceptable' level of confidence that an asserted identity or type of access is correct, a level that can vary greatly between different contexts.

Table 3.2 Different identity concepts considered in authentication situations

Identity type	Question answered	Comments
Personal identity	Who am I to be trusted?	The basic identity query.
Knowledge identity	Do I know something that corroborates my asserted identity?	The basic PIN- or password-challenge situation.
Consistent identity	Am I the same person as in a previous authenticated context?	Enables authentication of anonymous yet trusted identity.
Purpose identity	Am I just here and now entrusted to perform a particular action?	In other words, the presented 'key' is trusted in context on its own merit, not any specific bearer.
Group identity	Do I belong to a trusted group with authority in this situation?	Simplifies the management of access rights and operation privileges.
Role identity	What am I, as part of a role group, allowed to do in a given context?	As above, and often combined with it.
Context identity	Am I seeking access in the appropriate context?	Can involve prior authentications, identification of associated software or hardware, trusted protocols, tracking mechanisms, or external methods.
Presence identity	Am I in the correct location to be trusted in this context?	May also include the issue of request occurring at appropriate time.

Bit 3.15   Signatures enable verifiable trust referrers and testimonials

Essentially, an unknown Web service can assert: 'Trust us because of the credentials of the following named individuals who vouch for us'. Digital signatures allow others to confirm the veracity of asserted identities and credentials.

In many authentication and certificate contexts today, the primary focus is on the personal identity of the user only in the context of a particular system, and hence on local mechanisms for the user to prove the asserted identity – typically, account name and password.

Any context that requires 'global' scope must satisfy higher demands:

- *Uniqueness of identity*. An e-mail address may therefore be better than a name or number, as it leverages an already globally unique Web URI that is personal and human readable.
- *Portability*. Unless specific security requirements dictate otherwise, it should not matter from where or over which media the access request comes. Ideally, the authentication dialog should occur encrypted, but there are perhaps situations where it is not possible.
- *Multi-factor components*. Realistic security can often be achieved by combining (hashing) several different and independent validation factors, such as asserted identity, PIN/ password, and certificate. One or more factors may be varied at random by the server.
- *Ease of authentication*. It is well-known that users will invariably find workarounds to defeat 'difficult' authentication measures. Yet ease of use should not compromise security.
- *Adequate safeguards*. The system must strike a reasonable balance to provide an acceptable level of confidence in the authenticated identity and the integrity of the subsequent session.

Table 3.3 summarizes some of the specific security issues raised, along with possible solutions.

Authentication of identity (and subsequent authorization) can be accomplished by any one of the following processes, in increasing degrees of formal security span:

- **Non-existent**. The system accepts your assertion at face value. For example, messaging or chat systems often require only a session-unique but arbitrary handle for each user and particular context. Underlying account identity is authenticated only by password sent in clear, which is generally stored in the context of the client system and sent automatically.
- **Implicit**. Authorization may be based on physical access to an identifiable instance of client hardware or software, or possibly to a stand-alone device that generates one-off certificates. Authentication is then assumed satisfied by external or informal processes that regulate such local access or device possession.
- **Derived**. Based on previous successful authentication (local or Web). The reasoning applied is that trusted users in one context are automatically trusted in another on the same network. Most intranets and many Web/sweb contexts support 'forwarded' authentication.
- **Explicit**. Authentication mediated by a remote server/service with well-regulated centralized authorization processes, exemplified by network login, typically with many access levels.
- **External**. Secure authentication mediated by trusted service (a central or distributed authentication service that issues encrypted certificates).

Table 3.3   Summary of authentication issues in different contexts

Identity type	Issues	Comments
Personal identity	Trust in certifying authority. Depth of authenticity check. Uniqueness.	Some systems and contexts might not care about more than local uniqueness of name handle.
Knowledge identity	User-security of known credential. Criteria for selection. System management.	Typically combined in the form identity-name plus PIN/password.
Consistent identity	Certification method.	Public key can prove consistency of secret key holder, directly or through key-generated hash of other data.
Purpose identity	Similar to a physical key.	Authority invested in the bearer.
Group identity	Shared access.	Often combined with personal identity.
Role identity	Usually shared.	Often synonymous with Group.
Context identity	Potentially shared.	Less often used, rarely alone. Must be adequately defined.
Presence identity	Not always valid for user. Personal integrity issues.	May be related to context.

While still rare, network or external 'public' authentication based on digital certificates is being deployed in newer solutions with increasing frequency. This solution might tip the balance towards the requirement always to use such strong authentication, even in casual Web client use.

The .NET Identity Model

The proposed .NET model of identity is a layered concept of personal identity, at minimum a three-tier structure that uniquely specifies the *individual*, the *application* that the individual is running, and the *location* where this software is running.

Instead of an application being the gateway to the desired data, the model sees the user as the gateway to the data. This shift in focus is good and well, as far as it goes; the particular application used should be incidental. However, for reasons evidently more to do with product licensing than user convenience, or Web functionality, the .NET identity model is unfortunately strongly tied to device and location – two aspects that should be transparent and largely irrelevant in the Semantic Web.

Bit 3.16   Identity on the Semantic Web should not be needlessly fragmented

A sweb-aware person is likely to use a multitude of devices from a multitude of locations. To deploy an authentication infrastructure that ties identity to particular instances of these two variables seems, to put it mildly, counter-productive. It is as if your legal name changed depending on whether you were home, at work, or in your car.

The .NET approach has other weaknesses – perhaps the greatest being the way the model tries to apply a 'one size fits all' authentication process for all situations, serious or trivial.

- Access to any Web service here requires a centrally issued certificate for each session. The system presumes a single authority for global trust, regardless of the session context. Perhaps more to the point, the proposed model is unproven at the scale required, and the trustworthiness of the system not universally accepted.

Different Trust Models

Alternative, distributed trust models have the benefit of corresponding trust structures already functional in the real world. For example, there are several different possible authentication models, modelled on corresponding ones in the physical world:

- *Single central authority*, which is/was the .NET model, with all user authentication certified by one instance and database. We have no single central authority for identity in the physical world to compare with, but the national identity registries in some countries come closest.
- *Independent regional authorities*, which we recognize from the physical world in identity credentials issued by each national government for its citizens. Such credentials are generally accepted on sight within the country without explicit validation, and usually also abroad (at least in 'internationalized' versions, such as a passport). The latter acceptance is due to international consensus to trust implicitly national authorities to define citizen identity.
- *Delegated authorities*, exemplified in the real world by various issuing authorities for public and special-purpose ID cards. Although the identity ultimately derives from national identity authentication, delegated certificates are used in their own right. Acceptance of such credentials is usually limited to entities within the country of issue.
- *Purpose-oriented authentication services*, which correspond in the real world to institution-issued or corporate certificates of identity or purpose. Acceptance for corporate-based credentials varies greatly, and these credentials are valid outside the nation or region domain in only a few cases – such as when due to organization alliances, such as VISA for bank-issued credit cards. A Web example is the system of Digital Certificate chains that emanate from some root Certification Authority (CA).
- *Distributed authentication services*, which are useful for generic identity validation in any context where it is considered adequate – perhaps only to prove 'the same identity' across sessions. This model is perhaps best exemplified in a working way by the distributed 'webs of trust' made possible by public key cryptography – PGP key server networks.
- *Community authentication*, which boils down to being vouched for by people who know you or trust your assertions. This level of personal authentication is informal but practical, and is historically the common basis for human trust systems in general. It is leveraged into the digital world of the Web by PGP certificates and digital signatures.

Not explicitly factored into these model descriptions is the issue of *trust history*, known as **reputability**. For example, you are personally more likely to trust a person or organization on the basis of successful past dealings (that *is, private reputability*) than an unknown.

An established community reputation (or *public reputability*) is a valuable but intangible asset for brick-and-mortar businesses. However, it proves hard to establish a comparable asset in the realm of online Web business, at least without broad brand recognition. Business reputation may be assessed by interpreting formal records such as registration, tax history, share performance, and published analysis, but such investigation and evaluation is non-trivial.

Bit 3.17   A trusted identity can be verifiable yet remain essentially unknown

Authentication of identity in the Semantic Web can be resolved adequately in many situations by just being able to prove the consistency of a verifiably 'the-same' identity across different contexts, not necessarily an absolute identity.

Table 3.4 summarizes the previous trust models and the issues relating to each.

The difference between models is where the line of trust is drawn and the how far that trust reaches. *Who do you want to trust today?* In all cases, there is an ultimate point in the recursive validation of authenticity beyond which the querying party must simply take the *identity assertion* at face value, on trust of the issuing authority.

Trust is not only about identity, but also about whether *published content* or *provided services* are trustworthy or can be validated. While many content-trust assessments are

Table 3.4   Summary of authentication models and related trust issues

Model	Problem	Possible solution
Central authority	Single source of vulnerability. Too much central control. Heavy demands on services.	Delegation. Alliance around common standards.
Independent regional authorities	Varying or unknown degree of trustworthiness. Validation difficulties.	Alliance of trust. Common minimum criteria for authenticity.
Delegated authorities	Varying or unknown degree of trustworthiness. Limited scope. Validation difficulties.	Common open standards. Trust chains. Vetting through central or regional clearing. Broad alliances of acceptance.
Purpose-oriented authentication	Might be 'trust the bearer' style. Varying or unknown degree of trustworthiness.	Used only within purpose domain. Combine with confirmation codes (such as PIN) and two-component login.
Distributed authentication services	Varying or unknown degree of trustworthiness. Validation difficulties. Too little control.	Webs of trust. Common open standards. Promote broad alliances of acceptance.
Community authentication	Circular chains of little span. Who to trust. Reputability.	Leveraging trust in community as such. Real-world analogy. Reputation influences trust.
Self-proclaimed	Trust on what basis?	Known history. Moderators. Peer voting (auctions, supplier comparisons). May evolve into community trust.

typically made with explicit or implicit reference to identity-trust validation, more generic models are required for semantic processing.

Trusting the Resources

Not only do we need to deal with trusted and untrusted identities, we also need to assess the trustworthiness of the information resources and services themselves. The issue is especially relevant to library-information application contexts that need to validate content 'truth' and dependencies. A growing research area deals directly with knowledge provenance and the development of technologies to identify, categorize, and process uncertain data automatically.

Bit 3.18   The Web will always offer uncertain and incomplete information

It is in the nature of the anarchy, so to speak. However, it should be possible to annotate Web content to at least create islands of local certainty.

Provenance level can range from strong (corresponding to high certainty and trust), to weak (that is, low certainty). The strongest category, *static*, concerns assertions (so-called 'facts') based on 'propositions' that do not change over time and can thus be validated once.

A weaker level of provenance, *dynamic*, concerns changing data, where validity might vary over time. Weaker still on the provenance scale are 'uncertain' data, where even the ability to validate the current data is uncertain.

The lowest level of provenance is context-relative, and typically called judgement-based since 'truth' then depends on human social processes and context-dependent human interpretations. Legal knowledge is one example of this category.

Although most real-world 'facts' are at least potentially variable and context-dependent, we can often treat them as 'static' for specific purposes and intervals, perhaps defined with formal constraints. This simplification makes the issue of provenance and validation easier, even though assessment must often rest on numerous implicit assumptions.

Bit 3.19   The Web will always offer undated and outdated information

The amount of undated information on the Web is surprising, even when researching academic writings. Sometimes, the only way to determine the topical relevance of a given source is through a careful study of the cited external references and that way make a reasonable assessment of 'not written before and not later than' dating interval.

For static provenance processing, we need to answer three questions:

- What is the truth value of any given proposition? (True, false, or uncertain.)
- Who asserted the proposition? (Person or organization.)
- Should we believe the person or organization that asserted the proposition?

In real-world cases (such as Web pages), typically we see complex assertions with implied dependencies on other Web resources: on commonly recognized knowledge (such as scientific laws); on sources not available on the Web (such as published papers or books); or on derived (or even anecdotal) knowledge. Relational links that define these dependencies then also require some form of provenance certification.

No complete solution to knowledge provenance has been proposed as yet. However, a number of technologies in development address automated processing of assertions on at least the level of static provenance, building on older truth maintenance systems (TMS) and working towards working ontologies.

A Commercial or Free Web

Although much of the public Web is free and open to anyone, and probably should remain that way for the common good, commercial interests must not be neglected. It is vital that as broad a spectrum of information as possible be made available, and some of that information is unavoidably connected with the business requirement to be a commercially viable offering; a return on investments.

The debate about commercial *contra* free is often a heated one, and the common cry that 'information wants to be free' is often a misrepresented stance – and ultimately not true, as it is instead various *people* who want it to be free. Recent legislation changes to further lengthen copyright and restrict usage rights, and litigation and technology to enforce the new restrictions, are causing both furore and consternation among developers and in broad public awareness.

Unfortunately, many of the restrictions and attempts to lock in perceived proprietary rights have already had or threaten to have a serious impact on the Web. For example, it can make it difficult or illegal (in some jurisdiction) to publish, re-use, or quote particular content. It can also make the development and deployment of sweb technology unexpectedly risk litigation due to claimed 'foreseeability by a hypothetical reasonable person' that users might infringe on content rights.

Bit 3.20   Information might become dead before it becomes free

It is not just about free/freedom. Much 'copyright management' today has the deplorable result of making vast tracts of previously published information inaccessible, indefinitely, in any format or medium, at any price – not just for years, nor even for decades, but for entire lifetimes, plus 75 years (or the next copyright extension).

It is not so much about such publicized issues as 'piracy' (a value-laden misnomer for intellectual property rights infringement), or 'file sharing' (and implication that p2p networking is inherently evil). These debated issues are admittedly legitimate concerns that suggest needed changes to how these rights are managed – or perhaps more properly, how issues of fair compensation to the creators and custodians of protected intellectual property be resolved. The real threats are more about outright grabs for **basic and established Internet technology**, such as the hyperlink mechanism, fundamental transport protocols, general architectural concepts, and established data formats or encoding types. The successful assertion of proprietary ownership of any of these could irreversibly cripple the general utility of the Internet.

Bit 3.21   Fundamentally free, for the greater public good

The W3C has taken the clear stance that insofar as implemented standards, protocols, and enabling technologies are concerned, the Internet as a whole must be free of patent claims – and remain free for the public good. This fundamental view apples equally to the emerging technologies of the Semantic Web. The landmark decision on W3C Patent Policy was formally taken in May 2003 (see summary at *www.w3.org/2003/05/12-director-patent-decision-public.html*).

Many participants in the original development of the Web could easily have sought patents on the work they contributed. They might subsequently have been tempted to secure exclusive access to these innovations or charge licensing fees for their use. However, they made the business decision that they and the entire world would benefit most from ubiquitous and royalty-free standards.

CERN's decision in 1993 (before the creation of the W3C) to provide unencumbered access to the basic Web protocols and software developed there, was critical to its rapid deployment and success. That the open platform of royalty-free standards enabled business to profit more from its use than from a Web based on proprietary ones is incontestable. Furthermore, the social benefits in the non-commercial realm are beyond simple economic valuation, in addition to laying the foundation for technical innovation, economic growth, and social advancement.

The new policy commitment to royalty-free (RF) standards means

- a stable, practical patent policy;
- a clear licensing framework;
- consistent disclosure obligations;
- an exception handling process when problems arise.

In practice, companies and individuals involved in the development of such components must therefore disclose and guarantee to waive any claims in order for these components to be accepted as part of the recommended specifications.

Bit 3.22   The underlying Web infrastructure must be open and free

The Web is such an important part of modern society that the policy of public 'greater good' takes precedence over proprietary interests. Voluntary adherence to this principle is better than enforced compliance, since enforcement always has undesired side effects.

Almost all companies, large or small, see the wisdom of this 'greater good' requirement and have signed on, which bodes well for the future of the Web. Otherwise, there is the established precedent that governments can go against business interests and by decree place critical patented technology in the public domain when proprietary interests threaten important public infrastructures.

However, beyond the basic infrastructure, the Web platform, considerable scope for commercial ventures is possible – necessary, even. It all hinges on payment models.

Payment Models

E-commerce has grown rapidly in the existing Web, despite being hampered by various problems ranging from unrealistic business models to inappropriate pricing. In most cases, the relative successes have been the traditional brick-and-mortar businesses offering their products and services online, perhaps largely because they already had a viable business plan to build on, and an established trust relationship with a large customer base willing to take their shopping online.

The advantage of real-world products and services in the Web context has been that transaction costs are drastically lowered, yet item prices are still comparable to offline offerings, albeit cheaper. It has, however, proven more difficult to construct a viable business model and pricing structure for digital content, in which category we include information 'products' such as music CDs, software and books, not just specifically Web-published content.

Bit 3.23   Nobody really objects to paying for content – if the price is right

And, it must be added, if the mechanism for ordering and paying is ubiquitous (that is, globally valid), easy to use, and transaction secure.

Arguably the greatest barrier to a successful business model for digital content has been the lack of a secure micropayment infrastructure. Existing pricing models for physical products are patently unrealistic in the face of the essentially zero cost to the user of copying and distributing e-content.

Credit-card payment over SSL connections comes close to providing a ubiquitous payment model, but suffers from transaction costs that make smaller sums impractical and micropayment transactions impossible.

Unfortunately, few of the large, global, content owners/distributors have realized this essential fact of online economics, and are thus still fighting a costly and messy losing battle against what they misleadingly call 'piracy' of digital content, without offering much in the way of realistic alternatives to sell the same content to customers in legal ways. Instead they threaten and sue ordinary people, who like most of their peers feel they have done no wrong. The debate is contentious, and the real losses have yet to be calculated, whether in terms of money or eroded public trust. It will take time to find a new and sustainable balance.

Even fundamental, long-established policies of 'fair use' and academic citation, to name two, are being negated with little thought to the consequences. It can be questioned whose rights are being served by in practice indefinite copyright that renders huge volumes of literature, music, and art inaccessible to the public as a matter of principle.

More disturbing is the insinuation by degree of 'digital rights management' (DRM) into consumer software and hardware, with serious consequences for legitimate use. DRM-technology can time-limit or make impossible the viewing of any material not centrally registered and approved, even the blocking ability of creators to manage their own files.

Perhaps the worst of this absurdity could have been avoided, if only one or more of the early-advocated technologies for digital online cash had materialized as a ubiquitous component integrated into the desktop-browser environment. For various and obscure reasons, this event never happened, though not for lack of trying. Instead, online payment is currently based on:

- real-world credit cards (for real-world prices);
- subscription models (to aggregate small charges into sums viable for credit-card transactions);
- other real-world payment means;
- alternate payment channels that leverage these means (such as *PayPal*).

The only relative success in the field has been the utilities-style rates model applied by ISPs for access and bandwidth charges, and some services with rates similar to phone usage.

Bit 3.24   Free content can complement and promote traditional sales

A brave few have gone against commercial tradition by freely distributing base products – software, music, or books – figuring, correctly, that enough people will pay for added value (features or support) or physical copies (CD and manuals) to allow actually turning a profit. *Baen Free Library* (*www.baen.com/library/*) successfully applies the principle to published fiction.

In the Semantic Web, an entirely new and potentially vast range of products and services arise, whose pricing (if set a price at all) must be 'micro' verging on the infinitesimal. Successful business models in this area must be even more innovative than what we have seen to date.

One of the core issues connected with payment mechanisms is that of trust. The Semantic Web can implement workable yet user-transparent trust mechanisms in varying ways. Digital signatures and trust certificates are but the most visible components. Less developed are different ways to manage reputability and trust portability, as are schemes to integrate mainstream banking with Web transactions.

Some technologies applicable to distributed services have been prototyped as proof-of-concept systems using virtual economies in the p2p arena (see for example *Mojo Nation*, described in Chapter 8 of *Peer to Peer*), and derivative technologies based on them are gradually emerging in new forms as e-commerce or management solutions for enterprise.

Bit 3.25   A Web economy might mainly be for resource management

It might be significant that the virtual economy prototypes turn out to be of greatest value as a way automatically to manage resources according to business rules.

The Semantic Web might even give rise to a global virtual economy, significant in scope but not directly connected to real-world money. The distribution and brokerage of a virtual currency among users and services would serve the main purpose of implementing a workable system to manage Web resources.

It is a known fact that entirely free resources tend to give rise to various forms of abuse – the so-called *tragedy of the commons*. Perhaps the most serious such abuse is the problem of spam in the free transport model of e-mail. It costs spam perpetrators little or nothing to send ten, or a thousand, or a million junk messages, so not surprisingly, it is millions that are sent by each. If each mail sent 'cost' the sender a small, yet non-zero sum, even only in a virtual currency tracking resource usage, most of these mails would not be sent.

Various measures of price and value, set relative to user priority and availability of virtual funds, can thus enable automatic allocation of finite resources and realize load balancing.

4

Semantic Web Collaboration and Agency

From information management in principle, we move on to the actors involved in realizing this management, mainly collaborating humans and automatic agents. The Semantic Web is slanted deliberately towards promoting and empowering both these actors by providing a base infrastructure with extensive support for relevant collaborative mechanisms.

For most people, collaboration processes might seem to be relevant mainly to specialized academic and business goals – writing research papers, coordinating project teams, and so on. But it is about more than this narrow focus. Strictly speaking, most human communication efforts contain collaborative aspects to some degree, so that embedded mechanisms to facilitate collaboration are likely to be used in many more contexts than might be expected. Such added functionality is bound to transform the way we use the Web to accomplish any task.

Planning and coordinating activities make up one major application area that involves almost everyone at some time or another. Current planning tools, such as paper agendas and electronic managers (**PIM**), started off as organizational aids specific to academic and business environments, yet they are more and more seen as essential personal accessories in daily life. They make daily life easier to manage by facilitating how we interact with others, and they remind us of our appointments and to-do items.

With the spread of home computing and networks, PIMs are now often synchronized with both home and work systems. It is not hard to extrapolate the trend, factoring in ubiquitous Web access and the new collaboration technologies, to get a very interesting picture of a possible future where networked collaboration processes dominate most of what we do.

Chapter 4 at a Glance

This chapter looks at a motivating application area of the Semantic Web, that of collaboration processes. *Back to Basics* reminds us that the Internet and the Web were conceived with collaborative purposes in mind.

The Semantic Web: Crafting Infrastructure for Agency   Bo Leuf
© 2006 John Wiley & Sons, Ltd

- *The Return of p2p* explores the revitalization of p2p technologies applied to the Web and how these affect the implementation of the Semantic Web.
- *WebDAV* looks at embedding extended functionality for collaboration in the underlying Web protocol HTTP.

Peer Aspects discusses issues of distributed storage, security and governance.

- *Peering the Data* explores the consequences and implications of a distributed model for data storage on the Web.
- *Peering the Services* looks at Web services implemented as collaborating distributed resources.
- *Edge Computing* examines how Web functionality moves into smaller and more portable devices, with a focus on pervasive connectivity and ubiquitous computing.

Automation and Agency discusses collaboration mechanisms that are to be endpoint-agnostic, in that it will not matter if the actor is human or machine.

- *Kinds of Agents* defines the concept and provides an overview of agent research areas.
- *Multi-Agent Systems* explores a field highly relevant to Semantic Web implementations, that of independent yet cooperating agents.

Back to Basics

For a long time (in the Web way of reckoning) information flow has been dominated by server-client 'broadcasting' from an actively-publishing few to the passively-reading many. However, peer collaboration on the Web is once again coming to the fore, this time in new forms.

As some would say: *It's about time!*

The original write-openness of the Web was largely obscured in the latter half of the 1990s, as millions surfed into the ever-expanding universe of read-only pages.

Bit 4.1   Not everyone sees themselves as Web content authors

The objection that most people are perfectly content to just read the work of others does, however, beg the question of how you define 'published content'. Much useful information other than written text can be 'published' and made accessible to others.

There were, of course, good reasons why the Web's inherent capability for anyone to 'put' new content on any server was disabled, but a number of interesting potential features of the Web were lost in the process. Reviving some of them, in controlled and new ways, is one of the aims of the Semantic Web initiative.

Content-collaboration became needlessly difficult when the Web was locked down. Publishing on the Web required not just access to a hosting server, but also *different* applications to create, upload and view a Web page. Free hosting made way for commercial hosting. The large content producers came to dominate, and portal sites tried to lock in ever more users in order to attract advertising revenue. In fact, for a brief period in the late 1990s, it almost appeared as if the Web was turning into a just another broadcast medium for the media corporations. Everyone was pushing 'push' technology, where users were expected to subscribe to various 'channels' of predigested information that would automatically end up on the user's desktop.

Collaboration became marginalized, relegated to the context of proprietary corporate software suites, and adapted to conferencing or 'net-meeting' connections that depended on special clients, special servers, and closed protocols.

Or so it seemed. Actually, free and open collaboration was alive and well, only out of the spotlight and awareness of the majority of users. It was alive between researchers and between students at the institutions and universities that had originally defined the Internet. It was also alive in a growing community of open-source development, with a main focus on Linux, even as the grip of commercial interests around the PC market and by extension the Web was hardening.

The concepts were also alive among the many architects of the Internet, and being refined for introduction at a new level. Developers were using the prototypes to test deployment viability.

Finally, there was the vision of a new Internet, a **next-generation Web** that would not just provide a cornucopia of content but one of functionality as well. The line between 'my local information' and 'out-there Web information' seems to blur in this perspective, because people tend both to want access to their own private data from anywhere, and to share selected portions of it for use by others (for example, calendar data to plan events). The server-centric role of old then loses dominance and relevance.

That is not to say that server publishing by sole content creators is dead – far from it. Ever increasing numbers of people are involved in publishing their own Web sites or writing Web logs (*blogs*), instead of merely surfing the sites of others. Such activity may never encompass the masses – the condescending comment is usually along the lines that most people simply have nothing to say, and therefore (implied 'we') do not need the tools.

But that is not the point.

Bit 4.2   The capability of potential use enables actual use when needed

Providing a good education for the masses does not mean that we expect everyone to become rocket scientists. However, we can never know in advance who will need that education, or when, in order to make critical advances for society as a whole.

We do not know in advance who will have something important to say. Therefore, the best we can do is to provide the means, so that when somebody does have something to say, they can do it effectively. Anyway, with increased interaction between user systems, distributed functionality and better tools, more users will in fact have more to say – possibly delegated to their agent software that will say it for them, as background negotiations with the remote services.

The Return of p2p

In the late 1990s, peer-to-peer technology suddenly caught everyone's fancy, mainly in the form of Internet music swapping. It was a fortuitous joining of several independent developments that complemented each other:

- The explosion in the number of casual users with Internet access – everyone could play.
- The development of lossy yet perceptionally acceptable compression technology for near-CD-quality music – leading to file sizes even modem users could manage (4 MB instead of 40).
- The new offering of small clients that could search for and retrieve files from other connected PCs – literally millions of music tracks at one's fingertips.

Legalities about digital content rights (or control) aside (an area that years later is still highly contentious and infected in the various camps), here was a 'killer application' for the Internet that rivalled e-mail – it was something everyone wanted to do online.

But the surge in p2p interest fostered more than file-swapping.

Bit 4.3   Peer-to-peer is fundamentally an architectural attitude

System architecture always profoundly reflects the mindset of the designers. In this case, p2p and client-server are opposing yet complementary attitudes that determine much more than just implementation detail.

New and varied virtual networks between peer clients were devised, developing and testing components that prove critical to many Semantic Web application areas, including collaboration. The reason they prove critical is that p2p networks faced many of the same design problems, albeit on a smaller scale. These problems are often related to issues of identity and resource management in distributed networks.

With client-server architectures, most of these issues are resolved through centralized registry and governance mechanisms depending on some single authority. Assuredly, there are situations when this rigid and controlled approach is appropriate, but the approach is not conducive to ease of use, nor to the establishing of informal and fluid collaborations over the Web.

Peer-managed Mechanisms

In implementation, we can envision two peer-like classes of entities evolving to meet the new paradigm of distributed functionality:

- *Controller-viewer devices*, sort of like a PDA with a screen and a real or projected keyboard, that are the user's usual proximate interface and local storage device. From it, control and views can be handed off to larger systems, such as a notebook, desktop, or room-integrated system with wall-projection screens. You might see it as a remote control on steroids.
- *Agent-actuator devices*, both physical and software, which negotiate data and interact with the user on demand, and function as the intermediaries between the user-local PDA views and the world at large.

Reviewer Austin David suggested, in the course of the first review of this book's text, the following vision of what this situation might be like for an average user:

My various views would interact with my personal agent, and I would carry the views on my person, or leave them attached to my car or bedroom wall, or in my toaster or radio. The agent would know my calendar, or how I like my toast, or what CD I want to hear in the morning, and would be 'out there' to interact with the rest of the Semantic Web.

Since its foundation, the Web has been eminently *decentralized* and based on *consensus cooperation* among its participants. Its standards are *recommendations*, its ruling bodies *advisory*, and its overall compliance driven by *individual self-interest* rather than absolute regulation.

Collaboration processes share these characteristics. Not surprising, since the ongoing development of the Internet as a whole rests on collaborative efforts. Why, therefore, should small-scale collaborative efforts be locked into tools and protocols that reflect a different, server-centric mindset?

New Forms of Collaboration

Many Web resources are repositories for information gathered from diverse, geographically separated sources. The traditional, centrally managed server model has the bottleneck that all updates must pass through one or more 'webmasters' to be published and become accessible to others.

Bit 4.4   In the new Web, sources are responsible for their resources

Although implicit in the original Web, this distributed responsibility became largely replaced by the webmaster-funnel model of information updating on servers.

When primary sources can author and update information directly to the server, without intermediaries, great benefits result. A small-scale, informal version of such co-authoring was explored in *The Wiki Way*, a technology that used existing Web and client-server technology but in an innovative way.

The new Web seeks to implement comparable features pervasively, as basic functionality built into the infrastructure's defining protocols.

WebDAV

Web Distributed Authoring and Versioning, **WebDAV** (see *www.webdav.org*), or often just DAV, is a proposed protocol extension to HTTP/1.1. It has for a time been gathering significant support in the Web community. DAV adds new methods and headers for greater base functionality, while still remaining backwards compatible with existing HTTP. As the full name implies, much of the new functionality has to do with support for distributed authoring and content management – in effect, promoting collaboration at the level of the base Web protocol.

In addition, DAV specifies how to use the new extensions, how to format request and response bodies, how existing HTTP behavior may change, and so on. Unlike many API and overlay protocols that have been developed for specific collaboration purposes, DAV intends to implement a generic infrastructure support for such functionality, thereby simplifying application development.

Bit 4.5   DAV is a basic transport protocol, not in itself semantic capable

The point of DAV in the context of the Semantic Web, however, is that it enables transparently much functionality that makes it easier to implement semantic features.

In the context of co-authoring, DAV is media-agnostic – the protocol supports the authoring (and versioning) of any content media type, not just HTML or text as some people assume. Thus, it can support any application data format.

The Basic Features

DAV provides a network protocol for creating interoperable, collaborative applications. Major features of the protocol include:

- **Locking**, which implements well-defined concurrency control to manage situations where two or more collaborators write to the same resource without first merging changes.
- **Properties**, based on XML, which provide storage for arbitrary metadata, such as a list of authors on Web resources.
- **Namespace** manipulation, which support copy and move operations for resources.

DAV implements DASL (DAV Searching and Locating protocol), to perform searches based on property values to locate Web resources.

DAV Extensions

Several extensions to the base DAV protocol are developed in the IETF:

- *Advanced Collections*, which adds support for ordered collections and referential resources (or pointers).
- *Versioning and Configuration Mangement*, which adds versioning support, similar to that provided by RCS or SCCS. The configuration management layer, based on Versioning, provides support for workspaces and configurations.
- *Access Control*, which is implemented as the ability to set and clear access control lists. In a generalized form, access control can be applied to the entire Web.

Access control will always be a factor, no matter how 'free' and 'open' the Web is styled. The difference in the open context is that access and governance methods are decentralized and firmly based on semantic models of identity/authority assertion. Strong public-key cryptography, with its associated digital signatures, is the enabling technology that makes such methods feasible. These issues are discussed in Chapter 3.

The future trust model is one of overlapping 'webs of trust' that complement existing central trust authorities; an interoperative system that can adequately establish secure identities on demand.

Peer Aspects

In the Semantic Web model, data and metadata are often moved away from central stores and into distributed storage all over the Web. In many cases, the primary user focus also shifts from the data as such to packaging services that deliver the data preprocessed or aggregated in various ways. Data governance then takes on a different cast, because the usual physical locality of data no longer applies, and the traditional security mechanisms become inadequate or inappropriate.

Bit 4.6   New consistent endpoint-to-endpoint security models are needed

A new Semantic Web infrastructure to handle security and integrity issues consistently is in fact being developed, leveraging existing endpoint encryption technologies.

The section on XML Security in Chapter 3 provides an overview of the new security infrastructure.

Many of the issues of distributed storage, security and governance may be explored from the perspective of existing p2p systems that are built on top of the current Web. Much of this material is still valid and applicable to many aspects of sweb architecture.

Despite the recent wave of hostility towards p2p systems, generally due to intense litigation and threats by representatives of the music industry against both hosts and developers, p2p as a technology (or rather a collection of technologies) is very much a part of the mainstream development of the Web towards the Semantic Web.

The following sections revisit and expand on some of these p2p themes.

Peering the Data

Distributed (and peer-managed) data have many advantages over centralized stores. However, the actuality is not an either/or situation, because server sourcing and distributed sourcing complement each other and can create a valuable synergy that leverages the best of both models.

Most content-creating users are for the foreseeable future still likely to publish to specified server locations. In only a few special situations is content published to (or 'inserted' into) the virtual network without any specified location.

Instead, it is the retrieval and subsequent processing of published data that will occur in a distributed and non-localized manner. In most cases, the user is not very interested in where the data reside, only that they can be found quickly and reliably, and presented in meaningful ways.

Bit 4.7   Usage patterns show that answers to 'what' are preferred to 'where'

The popularity of content-aggregating services (such as portal sites) suggests that users often prefer the convenience of subscribing to dereferenced compilations of desired content from a single source, instead of having to browse around to visit all the individual source sites.

Given a decent rules-following client, the same users will probably prefer to delegate 'find-and-fetch' tasks to the automations. In such a model, it is an advantage to have content addressed by identity rather than tied to locations. URI addressing in turn makes storage at consistent locations a non-issue, entirely optional.

Swarming

The synergy of server-peer is clearly evident in 'swarming' technologies built around sourcing data from a single publishing source. The collective distribution power of all the interested clients is used to reduce loading on the server by passing along data fragments already received to others requesting them.

- Practical demonstrations of the power of swarming technologies can be a dramatic experience for anyone with a 'fat pipe' connection (for example, multi-megabit/s ADSL). A client request is sent to some mediating server, which results in a download of the requested data starting. The transfer trundles along at a paltry few kilobyte/s for a while, but then the rate rapidly increases as more swarming clients are contacted in the background. Very soon, the download bandwidth is saturated, despite modest require- ments on the individual sourcing nodes.

Asymmetric connectivity (as in prevailing ADSL) provides significantly lower upload than download bandwidth, so swarming ensures maximum transfer rates to the downloading user. At the same time, as more fragments of the file are received, the receiving client begins to source them to other members of the swarm who lack just those fragments.

Swarming techniques may be more important in the context of adaptive load distribution in general, rather than just maximizing usage of individual connection bandwidth. A typical side-effect is that requested data are replicated and move closer to where the most demand is located. Subsequent requests from 'nearby' clients therefore need fewer node hops to access the same data. This caching is beneficial to the network as a whole.

Combining data fragmentation, encryption, and redundant replication of the chunks gains security and a measure of independence from individual servers. Clients instead rely on distributed services to locate and retrieve the data on demand from whatever stores are available at any given time.

Identity, not Location

A major design issue when distributing data and metadata over the network, especially in adaptive networks, is the desire to decouple access from a particular storage location and exposed access method on an identified server – the hitherto pervasive URL-based identity model.

Therefore, the Semantic Web instead presumes that the location-agnostic URI is the normal access identity for resources – and for that matter for the individual. Distributed retrieval services then handle the messy details of search and retrieval, dereferencing the URI to the specific Internet IP-location.

Such a URI can have many different forms, including the totally opaque digital-kcy, content-hash model used in *Freenet* (*freenet.sourceforge.net*). In fact, nowhere is the decoupling of data from locality more evident than in this model, where physical data location is smeared out though distributed caching and anonymization services to the extent that it is 'nowhere and everywhere' all at once.

A *Freenet*-like architecture, however, is an extreme solution for achieving very specific goals, including:

- **guaranteed anonymity** of both publisher and reader;
- **consistently verifiable identity** despite anonymity;
- **guaranteed immunity** to data change and censorship attempts.

As such, it is not likely to be the norm.

Storage Model

Nevertheless, we can expect a much more *virtual storage model* to proliferate. The user will still see 'discrete documents' in some kind of information space, and the metaphor of server storage in a hierarchical filesystem may remain a viable one, but the physical reality might look very different.

Unused storage space on various physical networks, such as corporate networks, might become a freely traded commodity. Securely encrypted fragments of public and private document data may then be redundantly spread across arbitrarily chosen rented disk capacity. Locator services would transparently handle storage and retrieval through distributed broker services.

The model is really not that strange – the local filesystem on a PC or a server already performs a similar dereferencing service, invisibly translating through several layers between the visual user model of a hierarchical system of directories and files displayed in a window, and the physical locator model of heads, cylinders, and sectors on the disk. The big difference is one of scale, of effectively turning the entire Web into an extended filesystem with an additional layer of descriptive metadata.

The future user of the Web will neither know nor especially care where the data are stored physically, except perhaps in special circumstances, as long as it is secure and reliably accessible. The immediate benefit is that the user can access data from anywhere on the Web; control of the data, instead of being is rather than being locked up in one or more specific machines, is mediated by some authority model based on personal identity.

Peering the Services

In addition to distributing data across the network, also distributing and delocalizing functionality reaps great benefits. This further step is an integral part of the Semantic Web model, because distributed data one just a special case of generic URI-addressed resources that also include functionality and services.

Implementing distributed services, however, does pose greater demands on the ability of the components to communicate and coordinate with each other when fulfilling requested tasks. The Semantic Web aims to provide the infrastructure and mechanisms to make such coordination easier.

Although many have written much about 'Web Services' in the past few years, not everyone means the same thing by the term. Nor do all implementations have very much to do with the Semantic Web model of distributed services (SWS, see discussion of WS in Chapter 3).

WS products are sometimes designed as traditional client-server architecture – centralized, monolithic, and controlled from a single location on the Web. While some services might be appropriate to implement in this way, most are probably not. The choice of implementation typically depends on several factors: the type of service considered, QoS required, and the risks at stake.

Bit 4.8   Web Services is a very broad term that covers many different kinds of implementations of greatly varying complexity

A Web service involved in medical applications (an early-adoption area for SWS) requires different constraints than a similar service for consumer product delivery.

Significant services are more likely to be implemented as highly distributed 'swarms' of adaptive components that can self-organize to meet the demands of the moment. It is more than likely that many of these will be realized in the form of connected devices near the users, while others might take the form of software manifesting virtual entities with ill-defined locality on the Web.

Edge Computing

As it happens, 'edge computing' devices are a central feature of the Semantic Web's focus on pervasive connectivity and ubiquitous computing. On the one side, the trend is towards smaller and more portable, yet ever more capable hand-held devices, starting to replace notebooks as they in turn replace desktops. Converging from the other side is the trend to add more computing power and wireless connectivity to other previously 'dumb' peripheral devices, until practically everything is programmable.

The sheer number and variety of such devices, and the many arbitrary combinations that might arise in a cooperative effort between them, precludes any way of effectively defining all situation parameters in advance. Therefore, the workable approach is to empower devices to discover each other independently and negotiate their own terms for cooperation.

Bit 4.9   Automatic connectivity, discovery, and negotiation will characterize devices

Embedding silicon and software is cheap. Typically, mass-produced devices appear with ever more 'free' functionality each year. Ubiquitous remote connectivity makes a transforming level of available features attractive to manufacturers.

The Semantic Web model addresses this functionality only indirectly, by defining the protocols and languages involved to implement it, and the overall framework for cooperative computing. Scc the following section on agents.

Smart and connected devices have not yet caught on in the market place. PDA/PIM-devices with wireless connectivity and the most recent 3G cellular phones are but precursors to the possibilities. As yet these devices suffer the problem that the more interesting potential services and nodes are just not there yet, or that connectivity is tied to proprietary solutions.

Early adopters are used to paying premium prices for first-on-the-block gadgets, yet some of the offered prototypes of Web-aware appliances have not caught even their fancy. Web-aware refrigerators, for example, have been around for years in one form or another, at ten times the price of normal ones, but clearly lack appeal in the absence of a complete home network to integrate more than just milk monitoring.

- I am reminded of early PC marketing touting large-footprint desktop machines for the home as just the thing to organize dinner recipes. Yet eventually a critical threshold *was* passed and several PCs in every home became a common situation.

Edge computing is an expression of 'little engineering', which concerns the common products that end up being produced in great volume. These products then cost comparatively little for each unit, and ultimately they redefine everyday life. Examples are legion: cars, household appliances, cameras, PCs, PIMs, and cellular phones – yes, especially the cellular, with units practically given away on promise of operator tie-in subscription.

Bit 4.10   'Little engineering' inherently drives significant social change
It does this not by solving any great problems (which is the purpose of 'big engineering'), but by driving down the cost of a known technology solution so that everyone (or at least the vast majority) can afford to use it.

Little engineering also critically depends on the existence of a standardized and available infrastructure, usually quite complex: cars need roads and filling stations and refineries, appliances need wall socket wiring and electric grid production utilities, cellular phones need operators and vast reaches of repeater masts every few miles, and so on. Without the infrastructure, the gadgets are useless – consider how much of what we use on a daily basis would not work on a remote desert island.

Therefore, the watershed event in edge-computing will not be any single, whiz-bang, Web-aware gadget (though assuredly over the years there will be a succession of hopeful candidates). No, in this context, it is simply the cumulative effect of a convergence of two developments:

- Ever more gadgets sold in new versions with always-on connectivity and agent software built in, ready to configure and use.
- An expansion of always-on or on-demand network connectivity to cover work and home environments, available as a normal infrastructure the same way as electricity.

Assuming the infrastructure exists for the devices and software, let us examine what may drive the device functionality.

Automation and Agency

Deploying generic mechanisms on the Web to facilitate collaboration between humans is important in its own right, but the Semantic Web advocates a much broader view.

Collaboration mechanisms are to be endpoint-agnostic – in other words, to facilitate the full span of possible collaboration processes:

- *human and human* (mediated by the tools and protocols used, often called PP in the sense of person-to-person, or even people-to-people, as a variant of the technical term p2p for peer to peer);
- *human and machine* (person to software agent, often called PA);
- *machine and machine* (agent to agent, often called AA).

Overall, collaboration is a more abstract and broader process than just interoperability – the latter concerns the fundamental technical ability to communicate, basically a matter of protocol and specifications, whereas collaboration processes are more concerned with the semantics of what is being communicated.

Bit 4.11   Collaborative processes imply meaningful conversations

Collaboration requires an exchange of information, of messages understood.

Idiomatically speaking, all three collaboration processes (PP, PA, and AA) should be seen as conversations imbued with meaning. What the Semantic Web brings to the situation are the predicated-based structures to express meaningful assertions, and the ontology back-end to parse meaning from these sentences. That is not to say agents will be able to 'think' like their human counterparts, but rather that they will be able to reason logically around relationships (explicit or implicit) and infer new valid ones.

Semantic reasoning capability is especially valuable in any conversations of type PA, because humans lapse easily into implicit referencing. Even such a 'simple' PA command as 'print this document on the printer to my left' would require considerable contextual inference by a software agent, regardless of whether it could interpret any supporting 'gestures' by the user.

Bit 4.12   Human semantic messages are complex and multi-modal

It is very difficult to get machines to understand lexically even artificially constrained human idiomatic usage. Conversely, humans are frequently frustrated by the constraints of artificial command languages and syntax. Semantic processing in machines can bridge the gap and hide the complexities.

Further supportive mechanisms include automatic identification and verification of authority (including delegation), automatic resource discovery (local devices, their capabilities, and availability), and appropriate levels of security (encryption). Granted that much of this process support is in some sense already available in the traditional Web, but it is then embedded in particular tools, special protocols, and special manual procedures. Invariably, it is an exception, an add-on, rarely convenient to use. The point of the Semantic Web is to make such support pervasive and transparent, a part of the fundamental protocols of the Web, always available from anywhere.

Kinds of Agents

'Agent' may need further expansion here. It is a term increasingly used (and at times misused), often without any special qualification, to describe a wide range of disparate computational entities. A selection of 'classic' definitions of agent include:

- a persistent software entity dedicated to a specific purpose;
- a program that emulates a human relationship by doing what another person could do for the user;
- an integrated reasoning process.

An e-business quip described an agent as 'a credit card with an attitude'. The decision to call any entity an agent ultimately rests on our expectations and point of view.

Sometimes it is just convenient for both developers and users to see a particular program as a 'rational agent' based on an interpretation of its behavior, rather than as an ordinary piece of software. Then, it becomes natural to talk in terms of *beliefs*, *desires*, *intentions*, and *goals* when designing and analyzing the behavior.

- In AI literature, this particular anthropomorphic attitude is sometimes referred to as '*Dennett's intentional stance*', as distinct from the earlier two predictive stances of physical object and functional design.

Bit 4.13   'Agent' is a descriptive term of convenience, not a well-defined one
Although there is some agreement on general characteristics of 'intelligent' agents, the decision to call a piece of software an agent is an arbitrary one.

Otherwise, a more formal description of agent capabilities tends to highlight many features that distinguish agents from ordinary application software:

- **autonomous behavior**, meaning they can act independently on behalf of the user to perform delegated, goal-oriented tasks;
- **reactive behavior**, meaning that they can perceive and respond to changes in their environment;
- **proactive behavior**, or goal-directed initiative, so they do more than just react to events or input;
- **social ability**, so they can communicate with the user, the system, and other agents, as required;
- **intelligence**, meaning reasoning at some level, from fixed rules to heuristic engines, that allows them to adapt to changes in their environment;

- **cooperation**, so they can work with other agents to perform more complex tasks than they themselves can handle;
- **mobility**, so that they may move between host systems to access remote resources or arrange local interactions with remote agents.

A given agent implementation need not exhibit all of these features; it depends as noted on the expectations and point of view. Simple cooperative agents might for example be entirely reactive.

A number of different graphical maps promoted in the literature attempt to place agents into more systematic groups, for example with schemes that overlap in nice geometric ways to indicate agent purpose areas.

Agent use cases are generally classified in two groups:

- intelligent resource managers that simplify distributed computing;
- personal assistants that adapt to user requirements.

Both of these areas are especially relevant to telecommunication companies, who not surprisingly also account for the largest portion of agent research. Evidently, they foresee enough of a return on the investment to continue.

A common approach is that agents provide intelligent interfaces that 'wrap' around legacy systems and provide new features to established functionality. Often the enhanced functionality comes from exposing previously local controls to remote access. Examples include distributed network management and self-help customer care systems. Such leveraging of existing technology seems a viable development path for much of the transition to the Semantic Web as well.

Bit 4.14   Agent technology implements better interfaces to existing resources

We already see powerful trends to expose Web-resource APIs to client-side programming (such as in a browser toolbar or applet). Such access can transform simple site functionality into more complex services that are called from user instances of customized interfaces.

Conceptually, agents have a long history, even in the context of the Web. As the name suggests, DARPA Agent Markup Language (DAML, see the DARPA section of Chapter 7) was considered as a special markup solution for agents. This approach changed as several strands of development joined to form more powerful descriptive structures incorporating meaning relations, and later the basis for OWL (described in Chapter 6). As yet, however, there are few indications of any dramatic 'killer application' to catapult agent technologies into the mainstream of user awareness – or for that matter, provider awareness.

Instead, we might be tempted to call them 'secret agents' because many are slowly diffusing out into undercover deployment. Garbed in traditional Web form interfaces, agents perform background AA tasks such as cooperative search without the immediate user realizing it.

As more small-device processing becomes the norm, such as users wanting to access WS from small hand-held devices, agents appear highly attractive. Mobile agents can take tasks into the network and to other hosts for processing, tasks that would overwhelm the capacity of the original hand-held host. Later they would return with results for display on the hand-held, or automatically forward more complex displays to more capable systems in proximity to the user.

Bit 4.15   Portable computing is likely to spawn many agent applications
This application area may be where we can expect the first killer applications to appear.

Agents Interacting

Agents are components in an agent infrastructure on the Web (or Web-like networks) – in the first analysis, interacting with services and each other. Such an agent infrastructure requires a kind of 'information food chain' where every part of the food chain provides information that enables the existence of the next part. The chain environment is described by the ontology (see Chapter 6) of the domain.

Each agent's own 'understanding' is not meant to describe all aspects of the domain, however, as it is sufficient to scope it to the agent's context and view. Views necessary for particular tasks may vary between different agents, perhaps designed around special ontologies, so the overall system design must handle and reconcile multiple viewpoints.

The problem-solving behavior of agents requires some kind of task decomposition approach, preferably as reusable components. The chosen approach, new to software agents, is essentially imported from task-method dichotomy from the field of Knowledge Engineering, where a task is decomposed into corresponding subtasks. This decomposition structure is iteratively refined down to a level at which primitive methods (so-called mechanisms) can solve the subtasks.

Interoperability between agents requires interoperability between ontologies, which is usually solved by mapping strategies between them. Various wrapper agents also need to be devised to transform data representations from locally stored structures at the information sources to the structures represented in the ontologies.

The *Ontobroker* project (see *ontobroker.semanticweb.org* for a W3C overview) was an early attempt to annotate and wrap Web documents and thus provide a generic answering service for individual agents. Figure 4.1 illustrates the general architecture of query agents in this service.

However, an infrastructure of interacting agents is increasingly seen as better implemented as a community of cooperating agents. This shift requires a different model of behaviour than simple query agents around information brokers. More fluid and adaptable responses are expected, and individual agents must be better equipped to appraise independently their situation in relation to specified goals.

Multi-Agent Systems

In the AA arena, we should look to current research in *Multi-Agent Systems* (MAS). The focus in MAS research is on systems in which many 'intelligent' agents (meaning autonomous entities with significant semantic logic capability) interact with each other to perform tasks. Such entities can be software programs or robots (devices).

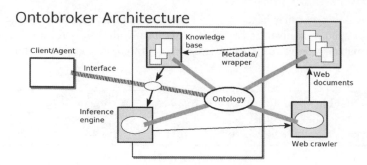

Figure 4.1 A schematic over Ontobroker architecture. The service functions as an automatic answering service for queries coming from external clients or agents, retrieving preprocessed, wrapped metadata

These systems are characterized by a number of basic tenants:

- Each agent on its own has incomplete information or capabilities for solving the assigned task – that is, a limited viewpoint.
- There is no system-wide, global control of the entities.
- Data sources are decentralized, and typically must be discovered.
- Computation is asynchronous, and communication is message-based.

Agent interactions may further be characterized as either *cooperative*, sharing a common goal, or *selfish*, pursuing agent-centric goals.

Bit 4.16   Multi-agent systems are comprised of autonomous agents
They may be seen as task-oriented swarms of activity, automatically forming and dissolving according to overall requirements and agent-goal fulfilment.

Figure 4.2 indicates the relationships from the MAS perspective.

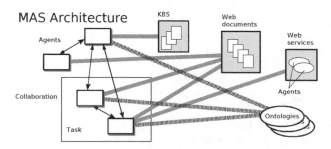

Figure 4.2 MAS architecture consists of independent agents interacting and collaborating with each other around tasks, and often implementing services for other agents

The research areas of multi-agent systems and distributed systems coincide, hence 'distributed agent computing' as a research area on its own. Multi-agent systems are often distributed systems, and distributed systems are platforms to support multi-agent systems.

MAS researchers develop communications languages, interaction protocols, and agent architectures that facilitate this development. They can leverage sweb work done on descriptive markup, message protocols, and ontologies.

Bit 4.17   Multi-agent design is often inspired by real-world interaction models

MAS research draws on ideas from many disciplines outside of AI, including biology, sociology, economics, organization and management science, complex systems analysis, and philosophy.

Interaction analogies are often made with models derived from corresponding interactions seen in nature or in human society – such as swarms, ant colonies, or the free economic market. A common experience from prototypes based on such designs is that surprisingly simple, agent-centric (that is, selfish) rules can generate complex behavior that benefits the collaborative goals.

For example, a MAS researcher may design rules so that a group of selfish agents will work together to accomplish a given task, basing the work on studies of how individual ants collectively gather food for the ant hill.

- A useful portal page to find important online publications and overviews on the various research areas of agents, MAS, and related fields is maintained at the AAAI (*www.aaai.org/AITopics/html/multi.html*). More news-oriented links are posted at *Multi-Agent Systems* (*www.multiagent.com*).

Information Agents

A special category of agents is identified with the purpose of managing the explosive growth of information. They are designed to manipulate or collate information from many distributed sources, and early implementations have been around on the Web without users even reflecting on the agent technology that powers the functionality.

Bit 4.18   Information agents often function as intermediary interfaces

In this capacity, they remove the requirement that the user finds and learns the specific interfaces for the actual information resources.

For example, meta-search engines parcel out user queries to many different search engines, process the many result lists, and return compiled and filtered results to the user. Web price-comparison services collect information from various online shops to present comparison tables for selected products. Other implementations search out and index published research papers, providing citation searching and other academic services.

- An overview of the seminal research projects that led to some of the services of this nature that are currently deployed on the Web may be found at the *Internet Softbot Research* site (*www.cs.Washington.edu/research/projects/WebWare/www/softbots/projects.html*).

Economic Agents

A particular model of cooperative agents centers on interactions governed by business logic in a virtual economy. Agents 'buy' and 'sell' resources to manage information in a kind of free market, with the overall goal of accomplishing set tasks.

Distributed swarms of agents can act in different roles depending on 'market pressure' and internal rules (for example, provide more instances of a scarce resource by more agents responding to a higher offered 'bid' for the service), or optimize critical performance (by bidding higher to contract premium services or bypass queues).

Bit 4.19   Market-based agent behavior can optimize resource usage/cost

Importantly, market-based solutions can virtualize and commoditize even the user's idle resources, and through background trading on the common agent market provide a transparent way to offset costs due to their own consumption of network resources.

The main advantages of such market-oriented programming are minimal communications for effective cooperation, automatic load balancing, and adaptive response to new requirements. Economic science has developed an extensive analytic framework for studying such systems.

The model is naturally well-adapted to management applications for physical network, but it has also been used extensively to build p2p virtual networks to provide adaptive storage and processing. The latter are in effect precursors to collaborative Web services.

Part II
Current Technology Overview

5

Languages and Protocols

Moving along from the general discussions about concepts and function, we need to look at how to express it all, and how to structure the data and metadata so that they are usable. Structure is an old concept, though most often it is something we are so used to that we take it for granted. Language and grammar, nuance and styling, layout and conventions – all make a 'texture' against which we paint our semantic meanings. Humans have an innate ability to analyze and structure sensory data, discovering and sometimes inventing patterns to order seeming chaos, and inventing conventions to impose corresponding patterns on recorded data.

The trouble is, most human patterns are not well-suited to the current requirements of machine-parsed patterns. Computer programming languages have therefore always come across as cryptic and non-intuitive – mainly because machines lack that ability we call intuition and must for this reason be directed in excruciating detail, step by step. Ongoing research has tried to address this deficiency by devising logic systems that exhibit something closer to the human mode of reasoning and that can reason and infer in limited ways from data and rulesets.

In the Semantic Web, it is no longer a question of isolated machines and their local processing or reasoning. Instead, a common approach is required so that the local software can not only reason about the data and metadata but also actively collaborate with other bits of software elsewhere. Languages and protocols must enable a free and implementation-independent exchange of structure based on some sort of common 'understanding' about what the data mean.

For this reason, part II is an exploration of the information technologies that define and enable structure and reasoning. We start with the basic building blocks and work up to more advanced structures – an overview but with some history.

Chapter 5 at a Glance

Chapter 5 is concerned mainly with the building blocks that structure the data, and enable the definition and expression of metadata assertions. Closely related to the structure are the protocols that define the process of managing the structures. *Markup* gives an overview of the fundamental methods for tagging structures and meta-information on the Web.

The Semantic Web: Crafting Infrastructure for Agency   Bo Leuf
© 2006 John Wiley & Sons, Ltd

- *The Bare Bones* (HTML) looks at the cornerstone of the published Web as we know it, the hypertext tag.
- *Disentangling the Logic* (XHTML and XML) explores the evolution of simple Web markup, from just tagging text to make it 'look good' into something that better supports codifying meaning.
- *Emerging Special-Purpose Markup* takes note of some of the newer, lightweight markup standards that bring XML to small devices.

Higher-Level Protocols takes up the message-based protocols that leverage XML to create more object-oriented and semantic-based exchanges.

- *A Messaging Framework* (SOAP versus REST) shows examples of how XML-based messages can define a common framework for packaging and exchanging XML-tagged content and functionality. It also critiques SOAP application by referring to the REST approach.
- *Expressing Assertions* (RDF) explains the logic and syntax behind the metadata structures (usually expressed in XML) that can define relationships between data elements and reference other resources to help define meaning for subsequent processing.
- *The Graphical Approach* introduces an alternative or complement to RDF, intended to deal with rich structures difficult to describe in a linear fashion.
- *Exchanging Metadata* (XMI) discusses an enterprise-level standard for exchanging metadata in distributed networks.

Applying Semantic Rules explores proposed standards for identifying and applying knowledge-base rulesets to knowledge-management databases ('expert systems').

- *RDF Query Languages* examines the process of extracting metadata from the RDF structures that define and express the relationships.
    Finally, *Multi-Agent Protocols* describes an emergent area of defining agent interaction using protocols capable of semantic expression.

Markup

And the Tag did impart structure upon the face of the Web.... Genesis, the Web version
Markup languages represent the technology for encoding metadata into the content itself. Markup tags are the specific codes or patterns that 'escape' the literal interpretation and rendering for the duration of the metadata blocks.

- Simple punctuation, as found in this text, is just another form of markup. More subtle forms are styling and layout in printed works. Electronic storage allowed much more intricate and well-defined markup solutions.

Bit 5.1   Markup is any arbitrary convention for embedded metadata

---

What 'embedding' means can vary greatly depending on the medium and the kind of 'processing' available to the end-user. Processing rules can be implicit (shared conventions), explicit (described literally as text), or self-defining (by reference).

Advanced markup languages were in fact devised long before the Web. The case can be made that HTML, a very simplified subset of SGML, is what really defined and empowered the Web. This markup gave a consistent structure to published content, and above all, it defined the powerful hyperlink mechanism that could express a simple form of relationships between documents.

So although many of the fundamental components of the Web were inherited from earlier application domains, including the hyperlink concept from *Hyperstack* cards and electronic texts, the synergy of combining them all with simple protocols for large-scale, open 'opt-in' networks proved unimaginably powerful.

The Bare Bones (HTML)

Although HTML was in principle a structural markup system, it quickly acquired a number of tags and usage conventions that targeted primarily visual presentation for human readers of Web content. This was an unfortunate, but perhaps inevitable, development in the circumstances. It was only partially rectified with the later introduction of a stylesheet layer (CSS) to separate visual layout attributes from the actual content markup.

A second problem with HTML is that the tag set was arbitrarily fixed according to an external standard. Over time this standard was extended in various ways for different reasons by different parties. Client development targeted mainly new visual presentation features using the new tags, with little regard for either usability or meaningful markup.

- In the HTML 3.x versions, for example, we saw these extensions become a part of the 'browser wars' as different vendors strove to dominate the Web. The deployment of browser specific features in the form of proprietary tags caused a fragmentation of the Web. Sites were tailored for a particular browser, and in the worst case ended up unreadable in another client.

Bit 5.2   HTML is structural markup used mainly for presentation purposes

Common usage subverts even the structural intent of some tags, raising non-visual rendering issues. Consider the pervasive misuse of complex nested 'table' markup used simply as a layout expedient, or 'blockquote' just to achieve visual indentation.

In HTML 4.0, and with the introduction of CSS, the purely visual elements were deprecated in the core markup. The intent was to achieve a new common standard for all browsers, with a clean separation of structural and visual markup (assuming consistent and proper application of the former). However, the basic premise and weakness remained of an externally defined and fixed set of tags, inflexible and ill-suited to serious metadata use.

Disentangling the Logic (XHTML and XML)

The *ad hoc* entanglement of logical and visual markup in HTML documents was a major reason for the development of stricter, yet more flexible, markup languages for the Web. Currently in deployment, and supported by the newer Web clients, the best-known of these is

XML. It provides a much richer and more logical approach to incorporating metadata about the content.

HTML, even considering just its structural-tag subset, can provide little meaningful metadata about the content apart from a few (and as noted, often inconsistently applied) logical markup elements. XML, on the other hand, allows for 'meaningful' and self-defining tags that can actually convey something specific about the nature of the tagged content – plus user-defined formats that are consistent. To define the tags, the document must contain a URI pointing to the resource where these definitions are found – an XML namespace, or possibly several.

XML is actually a *metalanguage*; a language that can represent other languages in a standardized way. It is a popular framework for various other metadata applications, a situation that often (and to advantage) blurs the border between local application spaces and a wider Web context. Web Services can thus exchange information using a common extensible format. It is even commonplace to use XML for storing application registry values. Therefore, XML is an important component in the Semantic Web context.

Bit 5.3   XML is a metalanguage markup framework, not just 'another set of tags'

As seen in later discussions, XML can describe relationships and meaning, besides providing structural markup in a more consistent and flexible way than HTML.

For the traditional Web at large, and importantly for the emerging Mobile Internet initiatives, the derived XHTML reformulation of HTML in the XML context is critical. Foremost, XHTML is the transitional vehicle for the predominantly HTML-tagged content on the Web today. It is the basis for the second incarnation of wireless application protocol (WAP2 or WML), which is essentially just XHTML Basic, a minimalist but strict version of XHTML (as noted later in the section on special-purpose markup).

Figure 5.1 is a simple graphic representation of the relationship between the three Web-markup languages mentioned here. As indicated, presentational markup (usually visual styling and layout) is separated and coded as CSS or XSL adjuncts.

Stylesheet Extension

While existing CSS can be used 'as is' with XML to specify presentational aspects of a document, the extensible albeit more complex XSL gives greater possibilities. Especially in

Markup Language Relationships

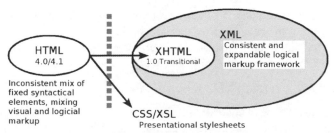

Figure 5.1 Relationship between HTML, XHTML and XML simplified and expressed as a graph depicting how XHTML is a subset of XML, recasting HTML in this more consistent framework

the context of Semantic Web applications and the manipulation of content and content metadata, such capability offsets the overhead of XSL.

In summary, using XSL invokes three parts:

- *XSL Transformations* (XSLT), which is a declarative language for transforming XML documents into other XML documents. Although XSLT can be used independently for arbitrary transformations, it is primarily intended for transformations relevant to XSL use.
- *XML Path Language* (XPath), which is an expression language used by XSLT to access or refer to parts of an XML document. This component, though not formally part of XSL, enables location-dependent functionality in applications. XPath is also used by the XML Linking specification, in XPointer, and in XML Query.
- *XSL Formatting Objects* (XFO), which is an XML vocabulary for specifying formatting semantics. This component expresses the main relevancy of XSL to the Semantic Web.

A clarifying note on the terminology used here might be in order. When originally conceived, the whole was simply XSL, comprising both stylesheet and formatting. The effort was later split more formally into two halves: the transforming language (XSLT) and the formatting vocabulary (XFO). The focus on transformation in the first part led developers to use the shorter term *XTL* (XML Transformation Language) in this context.

XTL is seen primarily as a conversion and transport protocol. Many developers tend to concentrate on XTL server-side transformations, from XML to display HTML or XHTML – in particular, to the more restrictive WML and WAP2 formats used in small mobile clients. The benefits of this approach are twofold: content can be published once and annotated in rich XML format, suitable for the more demanding applications, yet be customized on the fly to fit clients with limited capabilities.

XLink is a separate standard (a W3C recommendation, see *www.w3.org/TR/xlink/*) for extensible crafting and embedding in XML documents of described links between resources. In their simplest forms, XLinks resemble HTML links. In more complex forms, they allow XML documents to assert linking relationships at both endpoints, assert relationships among more than two resources, associate metadata with the links, express links that reside in a location separate from the linked resources, and point to only specific internal portions of the destination resource.

XML Pointer Language (*XPointer*) is an XLink implementation: the language to be used as a fragment identifier for any URI-reference that locates an XML-defined resource. One of its scheme parts is based on XPath.

The overall design goal is to capture and preserve a richer semantic content in XML on the server, irrespective of the served output of the moment, usually metadata constrained.

XML and the Semantic Web

Revisiting the W3C-based diagram (Figure 2.5) introduced at the end of Chapter 2, we can at this stage view a different continuation of the relationship chain from 'Agents' to cover the relationship that XML has with the path to the goal, and thus how it fits into the overall scheme of the Semantic Web.

A variation view is shown in Figure 5.2, extending the relationship chain into the lower levels of implementation can illustrate more detail of the markup space around the XML node and associated functionality.

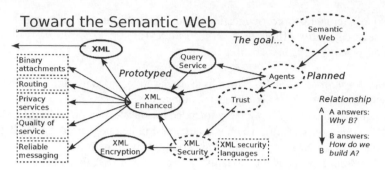

Figure 5.2   A variation of the W3C diagram provides a different continuation of the relationship chain from Agents, now with a focus on XML's markup space

We may note that most of the *How*? targets of 'XML Enhancements' (that is, components needed to implement it) are drawn as 'planned' because they have yet to be fully nailed down. Only XML has stabilized enough that it can be considered implemented. The arrow leading off to the left from XML implies a further continuation to be examined in a diagram in the next section.

Emerging Special-Purpose Markup

Specialized markup is continually being developed, fortunately often based on XML or subsets of it. This development is especially true and important concerning the new categories of edge devices that will form a significant component in the new Web.

Bit 5.4   Recent specialized markup is almost always based on XML

Communicating with the rest of the Web using a common protocol, or at least a subset or well-defined extension, ensures a solid basis for seamless interoperability. For this reason, Web markup that is not XML compliant is usually considered 'obsolete' today.

Other reasons to develop new protocol variants include explicit support for new media types supported by the device – for example, mobile-phone/PDA multimedia.

Compact Markup for Hand-held Devices

Due to constraints om processor, memory, power, and display, full-sized markup support is neither practical nor even possible in small, hand-held devices.

CHTML (Compact HTML, see *www.w3.org/TR/1998/NOTE-compactHTML-19980209/*) is a well-defined subset of HTML (2.0/3.2/4.0) recommendations, and is designed for small information appliances. Originally specified by the Japanese company *Access Company Ltd.*, it was adapted for use with small computing devices such as the I-mode PDAs, cellular phones, and smart phones. CHTML excludes support for the more demanding features of

HTML pages, such as tables, image maps, backgrounds, font varieties, frames, style sheets, and JPEG images.

The intent is to enable small devices to connect to the open Web, rather than merely a customized section of it. Although CHTML is gracefully compatible with HTML documents on the Web, it must be considered obsolete for the same reasons as HTML.

WML (Wireless Markup Language, *www.wapforum.org/what/technical.htm*) is then a better solution because it is based on XML, as a dialect specified by the WAP-forum to be used in poor network connections. It provides a complete set of network communication programs comparable to and supportive of Internet's set of protocols.

The main disadvantage of WML has been the lack of direct legacy-HTML support, thus denying the clients access to the larger portion of published Web content. This lack was addressed by reworking WML into v2 to support XHTML Basic, leveraging efforts to map HTML into XML.

Multimedia Markup

Legacy-HTML support may be one thing, but the current drive in both traditional Web publishing and hand-held devices (such as mobile phones) is to enable extensive multimedia support, including pervasive image capture (for example, phone-cams). This desire led to the development of special multimedia protocols based on XML.

Bit 5.5   Extensible markup allows easy upgrade paths for new devices

Leveraging the transform capabilities of XTL, upgraded devices can use the same XML-based markup to request higher-definition versions of the same content.

MMS (*Multimedia Messaging Service*) is based on the XML vocabulary SMIL (described next) and standardized by the *3rd Generation Partnership Project* (3GPP, *www.3gpp.org*) and the *WAP Forum*. Meant to be the successor to the popular *Short Message Service* (SMS) for mobile phones, for any device MMS can encapsulate and transmit multimedia presentations that contain images, sounds, text, and video clips. Multi-media in MMS documents are stored either as URI pointers to external resources, or as MIME-encoded blocks embedded in the document.

SMIL (*Synchronized Multimedia Integration Language*, see *www.w3.org/AudioVideo/*) was developed by a group coordinated by the W3C, which included representatives from CD-ROM, interactive television, Web, and audio/video streaming industries. SMIL is designed to control complex multimedia presentation, and it enables Web site creators easily to define and synchronize multimedia elements (video, sound, still images) for Web presentation and interaction. Each media type is accessed with a unique URI/URL, which means that presentations can be composed using reusable media objects arriving from distributed stores. Presentations can also be defined as multiple versions, each with different bandwidth requirements.

Numerous applications already implement extensive SMIL support, such as the popular media players and Web browsers, including the newest generations of mobile phones (which incorporate it in their MMS capability).

XUL/XML

Straddling the fields of basic markup and functionality programming, the *XML User Interface Language* (XUL, main resource at *www.xulplanet.com*) is a recent language that figures prominently in the development of Web clients based on the *Mozilla* project's *Gecko* rendering engine. Using XUL and other render components, developers create sophisticated applications and interfaces without special tools. Notable examples are the *Firefox* browser and *Thunderbird* e-mail client.

The specific attraction of XUL is that it can be used in any context where you might use a Web application. It leverages portable XML-based specifications (and other XML-derived standards) and core components to enable richer user interfaces that can be customized to particular contexts, for example to retrieve and process resources from the network.

In XUL you create an interface, use CSS style sheets to define appearance, and use *JavaScript* for client behavior. Programming interfaces support reading and writing to remote content over the network, calling Web services, and reading local files.

Higher-Level Protocols

Markup provides the building blocks, but first when these blocks are used to define higher-level protocols to accomplish some form of processing or message exchange, can we achieve a functionality greater than merely static representations of published content on the Web. Creating Web services without the proper messaging protocols and tools would mean much more hard work.

Messaging (so-called higher transport) protocols were previously designed somewhat *ad hoc*, without much in common except their reliance on the same underlying transport protocols for the network. However, they are now typically based on or derivative of XML, and are essential to Web Services because they formalize a common framework for how services and applications are to interpret and process exchanged data.

Figure 5.3, as a continuation of the previous diagram, illustrates how transport protocols fit into the overall scheme, and the way the relative dependencies and influences run.

Markup and transport are the workhorses, and it is at these levels that we have hitherto seen most development – extending, refining, and harmonizing the capabilities of the

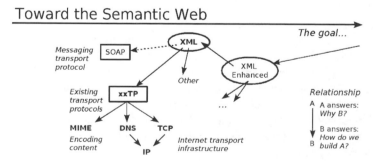

Figure 5.3 In this continuation of the previous diagram, we see how the focus shifts to transport protocols to convey information between endpoints. Here, we have largely backed into the existing pre-SW structures (the ultimate answers to How?), and away from the nebulous higher goal

infrastructure of the Internet. In this context, we additionally need to consider systems for codifying a declaration of the content type, such as MIME. At minimum, we need to allow for mapping between locally defined namespaces that define how blocks of data and associated metadata are to be interpreted. The later sections on ontology and schema explain the approaches taken in this regard.

A Messaging Framework (SOAP versus REST)

Brought to a greater public awareness because of its adoption in the. NET strategy, SOAP is an XML-based, 'lightweight' protocol for the exchange of information within decentralized, distributed environments. Its primary role is to expose existing application functionality as a WS, specifying API hooks for remote applications.

The original name was *Simple Object Access Protocol*, but this full expansion is now officially deprecated due to the change in focus away from just 'accessing objects' towards a generalized XML messaging framework. In fact, the acronym itself might well change in future.

The protocol was originally developed and implemented in 1999 by Dave Winer and was intended as an XML-RPC interface to the **CMS** package *Manila* (see *www.userland.com*). *Manilla* is one of the products that defined and popularized the now ubiquitous *weblog* (or 'blog') concept. However, that bit of protocol history now seems largely forgotten in the .NET context.

SOAP is discussed most often in the HTTP and HTTP Extension Framework – that is, in the Web context. The practical implementation of SOAP-like protocols turns out to have great potential reach, as is often the case with simple designs. In effect, it puts XML back into the message-passing business exemplified by the original Internet, but this time at a new level. The intent is that applications and Web services employ the extension as the common framework for packaging and exchanging XML-tagged content and functionality.

See it as a *Web Services transport layer* that can potentially be used in combination with a variety of other protocols. As it happens, SOAP is also an example of how XML can be used to define new protocols. We return somewhat later to a critique of how it has been implemented.

We may note that e-business figures prominently in early example scenarios, mainly because SOAP provides an application-agnostic way to exchange typical transaction information.

Characteristics and Specifications

Since June 2003, SOAP v1.2 has had W3C recommendation status, compliant with the W3C XMLP requirements. It is specified at the W3C site (see version links from *www.w3.org/TR/ SOAP/*), while general developers tend to visit *SoapWare* (*www.soapware.org*) for resources and the latest news.

The three defining characteristics that satisfy the basic XMLP requirements are:

- **Extensible**. While simplicity remains a primary design goal, features expected in more demanding environment, such as security, routing, and reliability, can be added to SOAP as layered extensions.

- **Protocol independent**. The SOAP framework is usable over a variety of underlying networking protocols, though it is commonly bound to HTTP.
- **Model independent**. SOAP is independent of programming models, as it just defines a way to process simple one-way messages.

In the Microsoft .NET context, we additionally see reference to GXA (Global XML Web Services Architecture) and WSE (Web Services Extensions), which are initiatives to implement extensions to the base protocol.

- Microsoft has moved its main SOAP Web resource location occasionally (at time of last revision, links can be found at *msdn.microsoft.com/webservices/*).

Variants of SOAP are implemented in several WS approaches, such as *Axis* (*ws.apache.org/axis/*) by the Apache Open Software Group. The aim here is to provide a reliable and stable base on which to implement Java *Web Services* (mentioned in Chapter 3).

Sun's J2EE platform, based on Java, implements a portfolio of protocols for WS. This work is portable to most platforms of sufficient capacity. As the API-bearing protocol names suggest, the developer focus is primarily on hacking the RPC-interfaces to services, not necessarily rethinking these services in URI contexts (see later discussion about REST).

Current SUN strategy appears to involve interoperability with .NET. Whether the focus on API will ultimately limit applicability to SOAP-wrapping existing RPC-resources remains unclear.

Functional Analysis

Analyzing SOAP, we can say that it consists of three parts:

- An envelope that defines a framework for describing what is in a message and how to process it.
- A set of encoding rules for expressing instances of application-defined datatypes.
- A convention for representing remote procedure calls and responses.

The envelope concept can be illustrated by depicting how a SOAP message is structured in its header and payload parts as part of an overall MIME-encoding. Figure 5.4 shows one such simple diagram of the encapsulation.

The framework is the specification of how to parse the result – for example, in order to identify and locate payloads. The protocol specifications also define a complete processing model that outlines how messages are processed as they travel through a path between endpoints.

In summary, there are two fundamental styles of SOAP messaging today:

- **Document style** indicates that the body contains an XML document. Sender and receiver must agree upon the format of this document.
- **RPC style** indicates that the body contains an XML representation of an encapsulated method call to the destination endpoint process.

SOAP Components

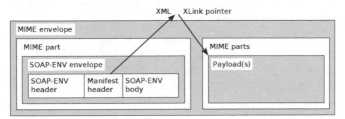

Figure 5.4   Schematic showing the internal structure of a MIME-encoded SOAP message, which in turn points to one or more payloads using the XML XLink pointer to provide necessary in-document locator information. The method is especially useful when sending non-XML payloads

Two techniques can be used to serialize the data into the body of the message:

- **Schema definitions**, which literally and unambiguously define the XML format for the body. This technique is most common for document style, and is called *document/literal*.
- **SOAP encoding rules**, which require the SOAP processor to walk through the various SOAP encoding rules at runtime to determine the proper serialization of the body. This technique, *rpc/encoded*, is more prone to errors and interoperability issues, but is commonly used in RPC messaging style.

Document/literal is the preferred mode in the Web Services context.

Rpc/encoded mode is commonly used when wrapping existing RPC-services instead of redesigning them to be schema oriented.

Applications and Implementation Techniques

The published implementations list (at *www.soapware.org/directory/4/implementations*) is long and slightly opaque. Closer inspection finds mainly projects to implement basic server support rather than any real WS applications. Revisiting the list in April 2005 finds no updates since January 2003, which makes it now seem rather dated.

Other (now also dated) implementation examples may be found in the 2002 *PerfectXML* article 'The XML Web' (at *www.perfectxml.com/artides/XML/TheXMLWeb.asp*).

Bit 5.6   The 'great potential reach' of SOAP remains somewhat elusive
Much of the reason for the seeming slow client-side adoption of SOAP must be ascribed to the general lack of deployed Web services on the public Web. SOAP is otherwise central to several architectural models of WS support, including .NET WS.

RESTian Critique

Developers have debated for years about whether SOAP, albeit simple in principle, is in fact an over-engineered, unnecessary addition to the WS infrastructure. The reasoning here is that the critics find very little 'Web' and much 'API' in the specifications.

At the center of this critique was Roy Fielding, chair of the *Apache Software Foundation* and co-writer of a number of the Web's core specifications. In his 2000 doctoral thesis, he identified an architectural style called REST (*Representational State Transfer*) as the basis of the Web. It deals with resources by using URIs to exchange resource representations (mainly as documents), and a handful of core methods (HTTP GET and POST).

Simplicity and generality are characteristic of this approach. Therefore, REST does not require you to install a separate tool kit to exchange WS data. The assumption is that everything you need to use WS is already available in the existing Web protocols if you know where to look.

The strength and flexibility of the early Web (based on HTML, HTTP, and URI) rests on the single fact that the primitive HTTP natively supported URI notation. All methods being defined as operations on URI-addressed resources led to the viral-like success of the Web.

Bit 5.7   REST is a demonstrated sound model for distributed computing

It is the model used by the world's largest distributed computing application, the Web.

The issue is largely about dereferencing the resources so that a user can concentrate on the 'what' rather than be distracted by 'where' and 'how'. It also explains why search engines, despite their current flaws, became the dominant way to find things on the Web. When the user just has to submit a form query and click on a result link, and such forms can easily be incorporated on any Web page using the existing infrastructure, it is hard to beat – it embodies 'the REST way'.

In the words of Paul Prescod, in a 2002 article 'REST and the Real World' (*webservices.xml.com/pub/a/ws/2002/02/20/rest.html*), applying REST principles to WS-design is essential to the growth, innovation, and success of Web services:

> *Unlike the current generation of Web services technologies, however, we make those three technologies central rather than peripheral, we rethink our Web service interface in terms of UPJs, HTTP, and XML. It is this rethinking that takes our Web services beyond the capabilities of the first generation technologies based on Remote Procedure Call APIs like SOAP-RPC.*

Effectively adopting the opposing stance, one major WS stakeholder, *Google*, forged ahead with exposing SOAP APIs, at once promoting more complex and more constrained solutions. It also discontinued experiments with simple REST-style services for metadata (such as the '*google.com/xml?q=*' query).

The RPC-oriented developer thus needs to 'hack the interfaces' to leverage the resources, hoping the site-specific APIs and URLs remain constant. This approach is *not* agent-friendly, and may require extensive documentation to figure out. It is also strongly technology dependent.

Regardless of the justified critique, however, enterprise *will* adopt (and *is adopting*) SOAP-RPC mainly because it presents a low-cost, non-disruptive way to migrate the existing application infrastructure into the WS environment. Rethinking business process is not a priority.

The approach of building complex API rules is sanctioned and promoted by the big players, so it will evidently always be with us to some extent. It is hoped it will not needlessly complicate, dominate, and restrict the Web resource technology.

The technology you use to move bits from place to place is not important. The business-specific document and process modelling is ... What we need are electronic business standards, not more RPC plumbing. Expect the relevant standards not to come out of mammoth software vendors, but out of industrial consortia staffed by people who understand your industry and your business problems.   Paul Prescod

---

Bit 5.8   Migrating existing infrastructure into a WS framework has profound consequences

'Virtually every software asset can now be offered, or soon will be able to be offered, as a Web service.' Jon Udell, in 2002 (*www.infoworld.com/artides/pl/xml/02/06/24/020624 plcomp.html*).

Slippery Security

The decisive characteristic of Web services is that they must be designed to work across organizational boundaries, across arbitrary and poorly controlled networks. This constraint has serious implications with respect to security, auditing, and performance.

Returning to REST, we might wonder what Web services look like when not constructed by exposing resource-server API structures and RPC-wrappers. Clearly, they too are based on a message protocol, and could well use SOAP, albeit in its more general document mode. However, the focus here is squarely on the network and identity, not the details of resource implementation.

Where RPC protocols try hard to make the network look as if it is not there, such selective fudging has serious security consequences. You cannot deal with resource access across a network in the same way as when accessing from the desktop in the corporate LAN. Automatically passing incoming network messages as RPC method parameters directly to local resource software is risky, to say the least. Many RPC exploits are known, and they all tend to be serious.

What the REST approach requires is to design a network interface in terms of URIs and XML resources. The usually object-oriented back-end implementation is therefore by necessity slightly distanced from the network side. Administrators can apply *Access Control Lists* (**ACL**) to secure services which use URIs as their organizational model, and to every URI-designated document that passes through the service. It is much harder to secure RPC-based systems where the addressing model is proprietary and expressed in arbitrary parameters.

Each step in a business process is then represented by a new document with an attached ACL. All actors have a shared view of the URI-space representing the process. The URIs point to shared documents on the respective Web servers. Standard HTTPS/HTTP authentication and authorization are sufficient to keep intruders from also being able to look at the documents.

Note also that session-generated unique URI-designated documents provide an easy way to track, link, and preserve all transaction contexts. This mechanism both ensures that involved parties stay in sync with each other, and that security audit requirements are met.

With regards to corporate firewalls, using plain HTTP is being very explicit about what is going on. By contrast, XML-RPC or SOAP on top of HTTP implements a tunnel, hiding the real actions in the XML body just to get through a firewall. Such obfuscation deliberately subverts security work and reduces the efficacy of audit and filter tools.

Bit 5.9   Anything possible with SOAP or any RPC can be implemented with URIs and the four base methods in HTTP (GET, PUT, POST, and DELETE)

As HTTP has only a few methods, developers can extend clients and servers independently without confusion and new security holes. Rather than invent new methods they find ways to represent the new concepts in XML data structures and headers.

URI-centric Web services are inherently easy to integrate. The output from one service can form the direct input to another, merely by supplying its URL. That WS specifications such as SOAP and WSDL lack any real support for this existing, simple yet powerful, data-sharing mechanism is a surprising discovery. Instead, they rely on complex translations to internal and implicit data models, ignoring the existing URI-resource of the Web and thus also many real-world requirements.

Essential WS

What HTTP needs to implement properly and fully secure WS, according to REST proponents, are concepts such as the following:

- transaction rollback mechanisms for distributed services;
- explicit and implicit store and forward relays;
- transformational intermediaries;
- validated return paths (call-back, notification).

These features may be implemented as native sweb infrastructure regardless. The extensions may in fact be borrowed from SOAP headers and SOAP routing, representing perhaps the only part of SOAP that is strongly compatible with the Web and with HTTP.

Ongoing discussions on REST can be found on *RESTwiki* (*rest.blueoxen.net/cgi-bin/wiki.pl*), among other places. Community support is evidenced by signs that REST-style WS is what people actually use – for example, when Amazon ran both SOAP and REST interfaces, the latter handled in the order of 85% of all requests (REST requests were also six times faster to process).

Implementing REST WS is also a less daunting task than to implement SOAP-RPC (the usual approach). Any site that presents a service process as a series of Web pages is easily modified to do the same thing with XML. Re-implementing the business logic is not required.

However, REST is not a plug-in solution to everything. One common issue is that it requires you to rethink the application in terms of manipulations of addressable resources. The usual design approach today calls for method calls to a specified component. Server-side implementation is arbitrary and behind the scenes, to be sure, but the API you communicate to your clients should be in terms of HTTP manipulations on XML documents addressed by URIs, not in terms of method calls with parameters.

Anyway, given any protocol for expressing messages, what shall the messages express?

Expressing Assertions (RDF)

To codify the meaning of Web content – in other words, do more than simply mark up syntactical content elements – the *Resource Description Framework* was devised as a way to mark up metadata. More to the point, **RDF** provided a means to exchange metadata, and the concept is fundamental to later constructs. The defining elements or rules are:

- **Resource**, which is anything that can be assigned a URI (for example as a Web-location URL).
- **Property**, which is a named *Resource* that can be used as *a property.*
- **Assertion**, which is a statement about some relationship, as a combination of a *Resource*, a *Property*, and a *value.*
- **Quotation**, which is making an *Assertion* about other *Assertions.*
- **Expressibility**, in that there is a straightforward method for expressing these abstract *Properties* in XML.

Assertions are like simple sentences in natural languages, the statement parts predictably denoted with the linguistic terms *subject, predicate,* and *object* – for example, '*Resource x (URI) is/has Property y (URI) value z (optional URI)*'.

The common programming notion of key-value pairs (**hash**) can be recast as a *triplet* by understanding that the predicate is in this case implicit in the context. **RDF triplets** can occur in an arbitrary order – something that seems natural in the context of metadata. By contrast, element order can be very significant in XML. Figure 5.5 provides a visual illustration of such three-part assertions.

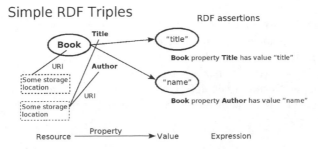

Figure 5.5 Example of simple RDF triplets applied to a typical case, asserting two properties for a book-type resource. A real-world example would have many properties, not just two. Libraries are big users of RDF

Simple RDF Extended Triples

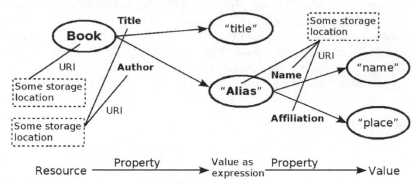

Figure 5.6 An example of a simple RDF triplet when a value is itself a resource about which an assertion can be made. Further indirection is also possible in that any value can be a resource

A concrete simple assertion using a resource variable might be '*MyLanIP is 192.168.1.3*' – or as URIs for each element of the triplet:

```
<http://example.com/currentMachine>
<http://www.commonverbs.org/terms/assignedIdentity>
<http://example.com/currentNodeIP>
```

Figure 5.6 builds on the previous example to show how indirection results when a value is itself defined as a resource about which assertions can be made.

Implied in the diagram is that the properties are defined in a common storage location given for the particular metadata application, but while common in practice, this assumption is not necessarily true. Each property (as resource) is associated with its own URI.

RDF is carefully designed to have the following characteristics:

- **Independence**, where as a *Resource*, any independent organization (or person) can invent a *Property* which can be properly referenced and interpreted by others.
- **Interchange**, which is expedited by conversion into XML, the new *lingua franca* of Web exchange.
- **Scalability**, where simple, three-part RDF records (of *Resource, Property*, and *value*) are easy to handle and use as references to other objects.
- **Properties are Resources**, which means that they can have their own properties and can be found and manipulated like any other *Resource*.
- **Values can be Resources**, which means they can be referenced in the same way as *Properties*, have their own *Properties*, and so on.
- **Statements can be Resources**, which allows them to have *Properties* as well – essential if we are to be able to do lookups based on other people's metadata and properly evaluate the *Assertions* made.

In passing, we might note that since properties and values can also be resources, they can be anything represented in a resource, not just words (although the latter is an understandable assumption made by most people).

Bit 5.10   RDF is a way of exchanging 'objects' rather than explicit elements

Business applications in particular find advantage in RDF because of the way business data often have complex structure best described as objects and relations.

The Containment Issue

It has been argued that RDF-element order is not totally arbitrary, but rather that there is some significance in an order-related 'containment' concept. Perhaps so, but it turns out to be elusive to capture, despite the fact that containment seems significant to human readers.

Why should 'RDF containment' be elusive? The answer is perhaps surprising, but RDF structures do not formally need the concept at all. A deeper analysis suggests that containment is significant only for the 'delete data' operation, which in any case many purists eschew on principle. The 'correct' way to amend data is to annotate, introduce new pointers, but never remove.

The containment concept turns out to seem relevant only because of the intuitive human understanding that deleting some object also deletes everything 'contained in that object'. Aha! We find that it is ultimately an implied, *human-semantic* interpretation.

RDF Application

By itself, RDF is simply an outline model for metadata interchange, defining a basic syntax in which to express the assertion relationships that make up the metadata. In fact, it is such a simple model that initially it takes a considerable amount of work to implement anything usable. But once defined, the broad capability for reuse makes the effort worthwhile.

Simplicity aside, an automated *RDF Validation Service* is provided by the W3C (found at *www.w3.org/RDF/Validator/*). Formal validation is always recommended whenever markup is involved – whether (X)HTML, XML, or RDF. Raw tagged text tends to be opaque even to experienced readers, and subsequent machine parsing is always highly intolerant of syntactical errors.

Application of RDF generally takes the form of creating XML templates for particular cases, where the resources and properties are structured in the assertions that will define the relationships to be mapped from the available data in some database.

The XML template is constructed as an abstract model of metadata, and it may also contain information about how to view and edit data in the model. Populating the template with actual values from a database record generates a viewable document to display as a response to an end-user request. Figure 5.7 illustrates a typical flow around a metadata database using populated templates.

Using RDF Vocabularies

Properties would normally not be defined in isolation, but would instead come in vocabularies – that is, ready-to-use 'packages' of several context-associated descriptors.

RDF Application

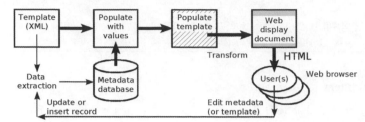

Figure 5.7   A possible application loop using RDF XML templates to access and manage metadata from a database. This model is fairly representative of many metadata management implementations

As an example, a set of basic bibliographic Properties for a simple book database might be *Author, Title*, and *Publication Date*, which in a more formal context can be extended with *ISBN, Publisher, Format*, and so on. Different vocabularies would be expected to arise for any number of Web contexts.

The nice thing about RDF vocabularies is that correspondences between similar vocabularies can easily (and globally) be established through new *Assertions*, so that one set's *Author* can be related to another's *Creator.* As anyone can create vocabularies, so anyone can create and publish correspondence definitions. It is unrealistic to think that everyone will (or can) use the same vocabulary; that is neither desirable nor necessary.

Bit 5.11   Custom RDF vocabularies can become valuable commodities

We can envision that opinions, pointers, indexes, and anything that helps people discover things on the ever-growing Web can become commodities of very high value.

With RDF (and appropriate tools) anyone can invent vocabularies, advertise them, and sell them – creating a free marketplace in vocabularies.

- Most niches of information would probably come to be dominated by a small number of specialized vocabularies, determined mainly by consensus and usage – or, if you will, by marketing and market-share. The really useful vocabularies would represent investments of expertise comparable to typeface designs, for example, or good technical dictionaries.

A ballpark estimate on the number of uniquely referenced objects on the Web today is in the order of half a million. It would clearly be a wasteful effort for everyone to reinvent vocabularies representing relationships between them all. Yet such wastefulness is precisely what happens when proprietary KB systems are implemented, and it risks happening if future RDF vocabularies are not made open, available for free or for modest licensing.

RDF is agnostic in terms of processing software. Different software implementations can process the same metadata with consistent results, and the same software could process (at least in part) many different metadata vocabularies.

Extended RDF Layers

As defined, RDF is both general and simple. The language at this point has in fact neither *negation* nor *implication*, and is therefore very limited. Consequently, further layers are required to implement a broader and more complex functionality than simply making statements about properties and statements about other statements.

Fortunately for early adoption, applications at this basic level of RDF are still very numerous. Such applications focus on the representation of data, which typically involves simple operations such as indexing, information storage, labeling, and associating style sheets with documents. This kind of processing does not require the expression of queries or inference rules.

While RDF documents at this level do not have great power, it is expected that these simple data will eventually be combined with data from other Web applications. Such extensions would require a common framework for combining information from all these applications – reason enough to go to the trouble of already now RDF-mapping the data in the simple applications.

Bit 5.12   Deploying RDF structure early is a way of evolving towards future reuse

Whether the local applications at present require RDF is immaterial, because once the data are mapped, new application areas for existing data become possible (and profitable) in ways that are hard to foresee.

The power of the greater whole will then greatly surpass what is possible in the local data representation language. Powerful query or access control logic implemented elsewhere can reference the local data through RDF exchanges.

Even though it's simple to define, RDF ... will be a complete language, capable of expressing paradox and tautology, and in which it will be possible to phrase questions whose answers would to a machine require a search of the entire Web and an unimaginable amount of time to resolve.
                                                                Tim Berners-Lee

Referencing Namespaces

XML makes the (re)use of predefined vocabularies, such as within custom RDF schemas (described later), fairly straightforward with its namespace concept. This capability has particular application when converting from one namespace to another.

Bit 5.13   RDF can correlate different properties that have similar meanings

The choice of a tag name depends on local usage and intended purpose, but the values assigned to one property can often fit for another property in another context. In this way, an application can import appropriate externally-maintained values.

- For example, suppose that a particular e-commerce implementation uses the tag *shipTo*, but could benefit from directly referencing another resource's definition of *Address*. The advantage would be that the former is then directly tied to any changes in the latter, which might reside in a central database for customer records.

Such referencing utilizes two steps:

- A **namespace declaration**, which introduces ('imports') the other schema to which reference is being made.
- A **type reference**, which embodies the actual reference to the other resource's stored value.

The namespace declaration in XML consists of an attribute '*xmlns:*' that defines a relation between a local prefix used in naming tags and an external schema referenced by a URI (often as a Web URL, and then often using a persistent PURL pointer):

```
xmlns: wherestuff ="http:// where.com / stuff .xml"
```

The URI is the exact reference, which is tied to the arbitrary prefix that can be used locally to consistently reference the remote schema in the rest of the XML document. The external schema types can thereafter be used almost as if they were defined in the local schema:

```
<element name= shipTo type= wherestuff :Address />
```

While 'Address' alone would uniquely reference a name in the local schema, each prefixed version is a different unique reference to an item in a another URI-specified schema. A practical schema typically leverages at least several external references, starting with the basic W3C schema (or equivalent) for defining the terms used to define the tags of interest.

Element referencing is simple (and human-readable despite the many tags), as shown by a really simple example RDF file:

```
<?xml version="1.0" encoding="utf-8"?>
<rdf:RDF
  xmlns:rdf="http://www.w3.org/1999/02/22-rdf-syntax-ns#"
  xmlns: dc="http://purl.org/dc/elements/1.1/">
  <rdf:Description rdf:about="http://leuf.com/TheSemanticWeb">
  <dc:creator>Bo Leuf</dc:creator>
  <dc:subject>
  Information page for book The Semantic Web
  with support page links.
  </dc:subject>
  </rdf:Description>
</rdf:RDF>
```

This snippet of metadata imbues a resource (the specified Web page) with two 'about' properties: *creator* and *subject*. Published at some public location on the Web, it thus makes assertions about these properties for the resource.

- The meaning of the 'about' term is defined in an external resource maintained by W3C, and referenced after the *xmnls* declaration by the prefix in the *'rdf:about'* container.
- The *creator* and *subject* properties are standard elements defined by a *Dublin Core v1.1* (*'dc:'*) reference that was created for publication metadata. This DC base list contains the 15 most relevant properties for this purpose: *Title, Creator, Subject, Description, Publisher, Contributor, Date, Type, Format, Identifier, Source, Language, Relation, Coverage*, and *Rights*.

Of course, other application areas can leverage these properties for other purposes, simply by referencing the DC definitions.

RDF Schema

The next layer up from just mapping the data into new formats is a schema layer in which to make *meta-Assertions* – statements about the language itself. For example, at this next level we can do a number of useful things, such as:

- declare the existence of a new Property;
- constrain the way a Property is used, or the types of object to which it can apply;
- perform rudimentary checks on Resources and associated Property values.

This layer is formally adequate for dealing with conversions between different schemas, even though it too lacks logic tools, because most conversion is merely a matter of identifying and mapping similar Properties.

Bit 5.14   A schema is both a defining and a mapping description

A schema describes types of objects and types of properties (attributes or relations), both of them being organized in two hierarchies of types.

The full power of a schema layer, however, will ultimately depend on yet another layer: the *logical layer*, which can fully express complex logical relations between objects. Because of the self-defining nature of the XML/RDF edifice, the logic needs to be written into the very documents to define rules of deduction and inference between different types of documents, rules for query resolution, rules for conversion from unknown to known properties, rules for self-consistency, and so on. The logical layer thus requires the addition of predicate logic and quantification, in that order, to expand the simple statements possible in the previous layer.

RDF Schema Sample

Despite the wordy length, a selection of simple RDF schema samples might illustrate these conventions in their proper context. At the core is the W3C syntax specification, often referenced using the suggestive local prefix *'rdf:'* when applying its terms:

```
<?xml version="1.0"?>
<RDF
  xmlns="http://www.w3.org/1999/02/22-rdf-syntax-ns#"
  xmlns:rdf="http://www.w3.org/1999/02/22-rdf-syntax-ns#"
  xmlns:s="http://www.w3.org/2000/01/rdf-schema#">
   <! -
  This is the RDF Schema for the RDF data model as described in the
  Resource Description Framework Model and Syntax Specification
  http://www.w3.org/TR/REC-rdf-syntax
   - >
<s:Class rdf:ID="Statement"
  s:comment="A triple consisting of a predicate, a subject, and an
    object." />
<s:Class rdf:ID="Property"
  s:comment="A name of a property, defining specific meaning for
    the property" />
<s:Class rdf:ID="Bag"
  s:comment="An unordered collection" />
<s:Class rdf:ID="Seq"
  s:comment="An ordered collection" />
<s:Class rdf:ID="Alt"
  s:comment="A collection of alternatives" />
<Property ID="predicate"
  s:comment="Identifies the property used in a statement
  when representing the statement in reified form">
  <s:domain rdf:resource="#Statement" />
  <s:range rdf:resource="#Property" />
</Property>
<Property ID="subject"
  s:comment="Identifies the resource that a statement is
    describing when representing the statement in reified form">
  <s:domain rdf:resource="#Statement" />
    </Property>
    <Property ID="object"
  s:comment="Identifies the object of a statement when
    representing the statement in reified form" />
  <Property ID="type"
  s:comment="Identifies the Class of a resource" />
  <Property ID="value"
  s:comment="Identifies the principal value
    (usually a string) of a property when the
    property value is a structured resource" />
</RDF>
```

This basic set of defined relational properties comprises just ten: *Statement, Property, Bag, Seq, Alt, predicate, subject, object, type,* and *value*. The container comment provides an informal definition for human readers (that is, programmers) who are setting up correspondences in other schemas using these standard items.

- A schema for defining basic book-related properties is found in a collection of Dublin Core draft base schemas (it is named *dc.xsd* at *www.ukoln.ac.uk/metadata/dcmi/dcxml/ examples.html*). It renders several pages in this book's format, so it is listed for reference in Appendix A. A so-called 'Simple DC' application (using only the 15 elements in the '*purl.org/dc/elements/1.1/namespace*') might import only this basic schema, yet be adequate for many metadata purposes.

The namespace definitions provide the information required to interpret the elements. An excerpt from the '*purl.org/dc/elements/1.1/*namespace', including the first property-container (for 'title', which follows the initial namespace description container) should be sufficient to show the principle:

```
<?xml version="1.0" encoding="UTF-8"?>
<!DOCTYPE rdf:RDF [
<!ENTITY rdfns 'http://www.w3.org/1999/02/22-rdf-syntax-ns#'>
<!ENTITY rdfsns 'http://www.w3.org/2000/01/rdf-schema#'>
<!ENTITY dens 'http://purl.org/dc/elements/1.1/'>
<!ENTITY dctermsns 'http://purl.org/dc/terms/'>
<!ENTITY dctypens 'http://purl.org/dc/dcmitype/'>
]>
<rdf:RDF xmlns:dcterms="http://purl.org/dc/terms/"
  xmlns:dc="http://purl.org/dc/elements/1.1/"
  xmlns:rdfs="http://www.w3.org/2000/01/rdf-schema#"
  xmlns:rdf="http://www.w3.org/1999/02/22-rdf-syntax-ns#">
<rdf:Description rdf:about="http://purl.org/dc/elements/1.1/">
<dc:title xml:lang="en-US">
  The Dublin Core Element Set v1.1 namespace providing access
  to its content by means of an RDF Schema
  </dc:title>
<dc:publisher xml:lang="en-US">
  The Dublin Core Metadata Initiative
  </dc:publisher>
<dc:description xml:lang="en-US">
  The Dublin Core Element Set v1.1 namespace provides URIs
  for the Dublin Core Elements v1.1. Entries are declared using
  RDF Schema language to support RDF applications.
  </dc:description>
<dc:language xml:lang="en-US">English</dc:language>
<dcterms:issued>1999-07-02</dcterms:issued>
<dcterms:modified>2003-03-2 4</dcterms:modified>
<dc:source rdf:resource=
  "http://dublincore.org/documents/dces/"/>
```

```
<dc:source rdf:resource=
  "http://dublincore.org/usage/decisions/"/>
<dcterms:isReferencedBy rdf:resource=
  "http://www.dublincore.org/documents/2001/10/26/dcmi-
    namespace/"/>
<dcterms:isRequiredBy
  rdf:resource="http://purl.org/dc/terms/"/>
<dcterms:isReferencedBy
  rdf:resource="http://purl.org/dc/demitype/"/>
</rdf:Description>
<rdf:Property
  rdf:about="http://purl.org/dc/elements/1.1/title">
<rdfs:label xml:lang="en-US">Title</rdfs:label>
<rdfs:comment
  xml:lang="en-US">A name given to the resource.</rdfs:comment>
<dc:description xml:lang="en-US">
  Typically, a Title will be a name by which the resource is
  formally known.</dc:description>
<rdfs:isDefinedBy
  rdf: resource="http: //purl. org/dc/elements/1.1/"/>
<dcterms:issued>1999-07-02</dcterms:issued>
<dcterms:modified>2002-10-0 4</dcterms:modified>
<dc:type rdf:resource=
  "http://dublincore.org/usage/documents/principles/
    #element"/>
<dcterms:hasVersion rdf:resource=
  "http://dublincore.org/usage/terms/history/#title-004"/>
</rdf:Property>
  ...
```

This entry formally defines the DC Property 'title', with description and pointers to further resources that provide usage guidelines. Each of the other properties has a corresponding entry in the complete definition file.

The previous examples show some variation in style and might seem off-putting in their length in relation to what they provide. The important point, however, is that *they already exist as a Web resource.* You do not have to write them all over again – just reference the resources and properties as needed in application RDF constructions. In any case, most of the tag verbosity in RDF definitions is generated automatically by the tools used to construct the RDF.

How-to

Adding RDF metadata to existing Web pages is actually rather simple. An example is given in Appendix A, right after the *dc.xsd book* schema mentioned earlier.

The points to note are the addition of the *about.xrdf* file to establish provenance, and metadata blocks to affected existing pages. Optional steps leverage public-key technology to

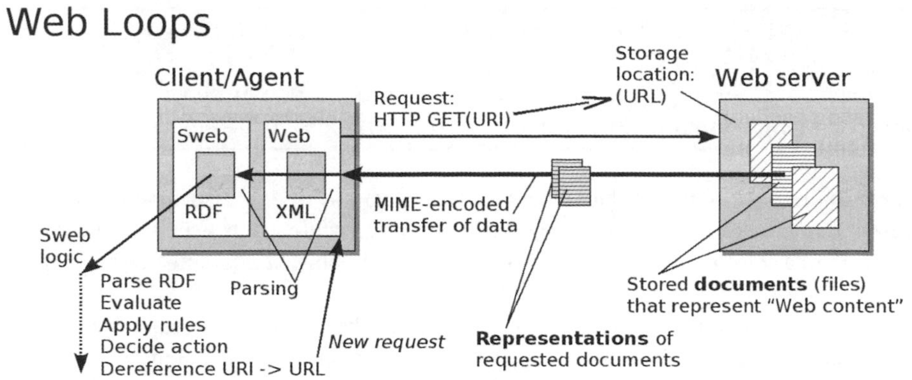

Web Loops

Figure 5.8 Conceptual overview of how client software might parse a served document into XML and then extract RDF information. The logic part parses and processes the RDF assertions perhaps to arrive at a new URI, which dereferenced becomes another Web request to a server

secure and validate content: a reference to a public key stored on the site, and digital signing of each page using the corresponding private key, storing the result as referenced signature files.

Web Loops

Putting XML and RDF into the loop, as it were, we can provide a sketch of how the Web and Semantic parts can intermesh and work together, shown in Figure 5.8. This sketch is illustrative only at the conceptual level.

The Web part proceeds from an initial URI and sends a request (using the HTTP GET method), which is turned into a transfer representation of a document by the responding server – a returned stream of bits with some MIME type. This representation is parsed into XML and then into RDF.

The Semantic part can parse the RDF result (graph or logical formula) and evaluate the meaning of statements contained in the document. It might then apply weighting based on signed keys, trust metrics, and various other rules, before selecting a course of action. The result might lead to the dereferencing of one or more other URIs, returning to the Web part for new requests.

Application can be varied. For example, consider how someone might be granted access to a Web resource in the Semantic Web. A document can be issued which explains to the Web server why that particular person should have access. The underlying authority and authentication would then rest on a chain of RDF assertions and reasoning rules, involving numerous different resources and supported by pointers to all the required information. *Distributed logic and agency at work*!

The availability of strong public key encryption and digital signatures is one of the factors that can make this kind of decentralized access control viable. Trust processing and signed digital certificates implement the Web versions of access keys and letters of authority.

The Graphical Approach

As XML evolved to become the standard format of exchanging data, some developers found the RDF approach to metadata somewhat cumbersome to implement.

Bit 5.15   RDF has a useful data model but a problematic XML syntax

This view is said to be a widely held one within the XML developer community, one
claimed to raise the threshold to achieving practical implementations.

Their main objection was that heterogeneous data (from say distributed systems) may
represent complex relationships among various entities, with a rich structure that is difficult
to describe in a common, serialized, XML format. An important goal must be to preserve the
exact structure in the serialization to XML.

Reasoning that complex relationships can usefully be modeled as graphs with multiple,
directed paths (and their inverses) between entities (graphically depicted as lines between
nodes or edges), these developers suggest that the graph model is better than RDF (the
'grammar' map) as the primary representational vehicle.

Technology refinements to make RDF easier to use are sometimes based on the *Directed
Labeled Graph* (DLG) model, taking the approach of serializing graphs into XML
descriptions. For example, consider a Web page linking to other pages, as in the DLG in
Figure 5.9.

This trivial example can be serialized in XML as follows:

```
<child>
  <name>Page1</name>
  <linkto>
    <name>Page2</name>
  </linkto>
</child>
<child>
  <name>Page1</name>
  <linkto>
    <name>Page3</name>
  </linkto>
</child>
```

An application reading this file can reconstruct and display a valid graph (Page1 linking to
Page2 and Page3), or it can process the information in other ways (for example, to determine
the relationship between the siblings Page2 and Page3).

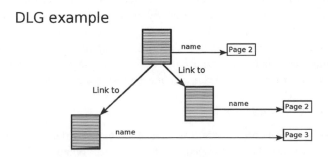

Figure 5.9   A DLG representation of a Web page that links to other pages

Note that all the information is here given as relationships. From a certain point of view, RDF is just a special instance of DLG – you can map all the relationships as a graph. Many of the simple figures that illustrate the concepts in this book are in effect DLG representations of resources and properties connected by relationship arrows.

Bit 5.16   Graphs do not replace grammars but do provide another view

The claim is sometimes asserted that graphs capture the semantics of a communication while grammars do not. However, graphs are just another grammar.

DLG systems define rulesets to serialize graphs of data into a **canonical form.** Other representations of the same data can be mapped into and out of the canonical form as required. Formulating such rulesets provides a way to generate mechanically a particular grammar from a schema describing a database or graph.

Actually a DLG system is capable, not just of serialization, but of many other 'services' as well, most of which are essentially the same as for any 'grammatical' RDF system, for example:

- viewers and editors for the mapped structures;
- compositing, or the ability to merge views from multiple graphs;
- persistent storage for the modeled data/metadata;
- query mechanisms;
- inferential services such as type checking and inheritance.

It might be interesting to note that a browser client such as Mozilla (*www.mozilla.org*) is designed on top of an RDF-DLG model of the API.

One might reasonably ask why bother with a canonical syntax at all, instead of just providing direct mappings to the graph's schema. The reason is that the seeming indirection by way of a canonical form, while not actually providing any added functionality, has the significant benefit of not requiring new vocabularies for each mapping. The canonical form has the further advantages of decreasing the amount of mapping required and of leveraging future XSL developments.

Bit 5.17   If something can be modeled as a DLG, it can be queried as a DLG

For every type of data, there is an associated data source – a data source that responds to queries from an RDF engine. Hence the generality of the model.

Designing a DLG

Typical DLG design criteria can be summarized as a number of guidelines:

- **Readable**. Typical syntax must be readable by humans.
- **Learnable**. The rules must be simple enough to be easily learned, and also to be implemented easily in agents.
- **Compatible**. The method must use only facilities available in current XML – which usually has meant the XML 1.0 specification plus the XML Namespaces specification.

- **Expressible**. Instances must be capable of expressing fairly arbitrary graphs of objects and directed relations.
- **Usable**. The chosen syntax must support a clear query model.
- **Transparent**. It should be possible to embed the syntax within Web pages, ideally without affecting content rendering.
- **Mappable**. A well-defined mechanism must exist for mapping other syntax families to and from the canonical form.

In practical models, it is deemed acceptable if full decoding might sometimes require access to the corresponding schema. In other words, it is not a requirement that the ruleset must cover all possible graphs.

Canonical syntax can be defined to obey five rules:

- Entities (depicted as nodes) are expressed as elements.
- Properties (edges) are expressed as attributes.
- Relations (edges) to other objects are expressed as single attributes with a particular datatype, which can codify if they are of the same type.
- The top-level element is the name of a package or message type, and all other elements are child elements – order does not matter.
- If an element cannot exist independently and can only be related to one other element, it may be expressed as a child of either that element or the top-level element.

A fully-explicit, canonical syntax makes it easy to convert from syntax to a graph of objects. Procedures to convert to or from the canonical syntax can usually be summarized in two or three iterative steps.

An alternative or abbreviated syntax may also be used for serialization. Such a choice might be due to historical or political factors. One might also wish to take advantage of compressions that are available if one has domain knowledge. The abbreviated syntax is converted to the canonical one by using either some declarative information in the schema to restore the missing elements, or a transform language such as XSL.

Exchanging Metadata (XMI)

XML Metadata Interchange (XMI) was developed as a response to the Object Management Group (OMG, *www.omg.org*) request for proposals for a stream-based model interchange format (the SMIF RFP). Outlined in 1998, XMI was from its onset identified as the cornerstone of open information model interchange, at least at the enterprise level and when developing Web Services.

XMI specifies an open information interchange model and is currently available as formal specifications in v1.2 and v2.0 (see *www.omg.org/technology/documents/formal/xmi.htm*). It is intended to give developers working with object technology the ability to exchange programming data over the Web in a standardized (XML-managed) way.

The immediate goals of XMI are consistency and compatibility for creating secure, distributed applications in collaborative environments, and to leverage the Web to exchange data between tools, applications, and repositories. Storing, sharing, and processing object programming information is intended to be independent of vendor and platform.

The hundreds of member companies of the OMG produce and maintain a suite of specifications that support distributed, heterogeneous software development projects. As an industry-wide standard, directly supported by OMG members, XMI integrates XML with the other OMG standards:

- **UML** (*Unified Modeling Language*) is pervasive in industry to describe object oriented models. It defines a rich, object oriented modeling language that is supported by a range of graphical design tools.
- **MOF** (*Meta Object Facility*) defines an extensible framework for defining models for metadata. It provides tools with programmatic interfaces to store and access metadata in a repository. MOF is also an OMG meta-modeling and metadata repository.

The XMI specification has two major components:

- The *XML DTD Production Rules*, used to produce XML Document Type Definitions for XMI-encoded metadata. The DTDs serve as syntax specifications for XMI documents, and also allow generic XML tools to be used to compose and validate XMI documents.
- The *XML Document Production Rules*, used to encode metadata into an XML-compatible format. The same rules can be applied in reverse to decode XMI documents and reconstruct the metadata.

Overall, OMG deals with fairly heavy-weight industry standards, mainly intended for enterprise realization of distributed computing systems, usually over CORBA (*Common Object Request Broker Architecture*), often according to the newer specifications for *Model Driven Architecture*. In particular, XMI and the underling common MOF meta-model enable UML models to be passed freely from tool to tool, across a wide range of platforms.

Although it is easier to integrate a collection of computers if they are all 'the same' according to some arbitrary characterization, such is not the reality of the enterprise. What the industry needs, in the OMG view, is a computing infrastructure that allows companies to choose and use the best computer for each business purpose, but still have all of their machines and applications work together over the network in a natural way, including those of their suppliers and customers.

Data Warehouse and Analysis

A related standard is the *Common Warehouse Metamodel* (CWM), which standardizes a basis for data modeling commonality within an enterprise – across its many databases and data stores. CWM adds to a foundation meta-model (such as one based on MOF) further meta-models for:

- relational, record, and multidimensional data support;
- data transformations, OLAP (*On-Line Analytical Processing*), and data mining functionality;
- warehouse functions, including process and operation;
- CWM maps to existing schemas.

Also it supports automated schema generation and database loading.

OLAP technology has definite application in the Semantic Web, and builds on some of the same ideas. A simple distinction in this context might be useful:

- Data Warehouse (DW) systems store tactical information (usually in a relational database) and answer *Who?* and *What?* questions about past events. Data processing is on the level of summing specific data fields.
- OLAP systems store multidimensional views of aggregate data to provide quick access to strategic information for further analysis. Although OLAP can answer the same *Who?* and *What?* questions as DW, it has the additional capability to answer *What if?* and *Why?* questions by transforming the data in various ways and performing more complex calculations. OLAP enables decision making about future actions.

The strategies are complementary, and it is common to have a DW as the back-end for an OLAP system. The goal is to gain insight into data through interactive access to a variety of possible views.

Applying Semantic Rules

Once the data are described and the metadata published, focus turns to utilizing the information. The basic mechanism is the query – we ask for results that fulfil particular requirements.

A 'query' is fundamentally a collection of one or more rules, explicit or implied, that logically define the parameters of (that is, usually the constraints on) the information we seek.

Rules in general can be modeled in different ways, with particular sets able to be reduced to others in processing in order to trigger specific events. Query rules may in some models be seen as a subset *of derivational* rules, for example, forming part of a transformational branch in a rules relationship tree.

Bit 5.18   Query languages provide standardized ways to formulate rules

It stands to reason that XML and RDF provide all the features required to construct consistent and useful sets of rules to define the query and the process.

The special case of a rule with an empty body is nothing other than a 'fact' (or in *techno-speak*: 'a positive ground relational predicate' assertion). Query systems start with some ground facts (that is, given constants or axiomatic relations) when setting up rules, from which they can derive other and variable assertions. We may further distinguish *extensional* predicates, which assert relations stored in the database, and *intensional* predicates, which are computed when needed by applying one or more rules.

A rule is 'safe' if every variable occurs at least once in a positive relational predicate in the body. Translated, this qualification means that each referenced variable is actually assigned a defined value somewhere. Without this requirement, a strict logic process might hang in the absence of a required value. Alternatively, it might proceed with a null value which leads to incorrect or incomplete results.

Bit 5.19   'Garbage-in, garbage-out' applies even to query rules
Even with the best rules, it is easy to construct flawed queries that do not express the underlying intentions. The results can be anything from misleading to indeterminate.

Query technologies have a long history, especially in the context of database access, but have most often been concerned with relatively simple lexical pattern and value matching under given logical constraints. They mainly applied to a well-defined database with a uniform fixed structure and a known size. Therefore, the search process can be guaranteed terminated with a handful of closure conditions.

Query on the Web about online resources is inherently a different situation. Current generations of search engines using traditional technology cope by essentially recasting the query to apply to a locally maintained database created from the Web by sampling, conversion, and indexing methods.

In the Semantic Web, on the other hand, queries should primarily trigger resource discovery and seek information directly from the Web resources themselves – possibly by way of distributed Web resources that can act as intermediaries. Such a search process is an open-ended and less precise situation than a simple database query. In addition, the expectation in Semantic Web contexts is that the query mechanisms can reason and infer from metadata, perhaps returning valid results that need not be present explicitly in the original documents.

Bit 5.20   Semantic query technologies must be *everything*-agnostic
Ideally, any query need only describe the unique 'pattern' of the information we want to match – not from where, in what format, or how it is accessed. Nor should it depend on any predefined limits on redirection or multi-source compilation.

Queries are satisfied by applying rules to process relevant information. Rules processing forms an action layer distinct from browsing (by a human) and processing (by an agent). Rules might be embedded or referenced in the markup (as an extension to the basic XML), or might be defined in an external database or knowledge base (typically RDF-based structures). Figure 5.10 summarizes the relationships.

Much information published on the Web does contain considerable knowledge, but expressed in 'common sense' ways that only a human reader can properly read. Typically, contextual references are implied between parts, relationships that require explicit rules for an automated agent to process.

The RDF model confers many advantages in this context, including mapping, conversions, and indirection. In addition, and unlike the traditional relational database model where you have to know the structure of tables before you can make the query, and where changes to the data may affect both database and query structure, RDF data are only semi-structured. They do not rely on the presence of a schema for storage or query. Adding new information does not change the structure of the RDF database in any relational sense. This invariance makes pure RDF databases very useful for prototyping and for other fast-moving data environments.

Markup Layers

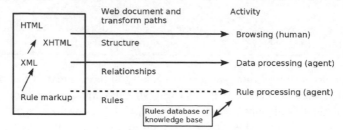

Figure 5.10 Relationship between HTML, XML and possible rule markup in terms of subsequent processing of Web-published content. Rules make explicit the implicitly embedded data relationships in a document

RDF Query Languages

While RDF is the core specification in the Semantic Web, most of the development activity is currently in the area of RDF *query languages* (QL). The former defines the structure, while a QL defines the process of extracting information from it. Two general approaches to query RDF metadata can be distinguished:

- SQL/XQL style approach, which views RDF metadata as a relational or XML database, and devises API methods to query the object classes.
- KB/KR style approach, which views the linked structure described by RDF metadata as a Web knowledge base, and applies knowledge representation and reasoning technologies on it.

The chosen approach has a critical effect on the scope of the solutions.

In the SQL/XQL-database approach, the QL is implemented as predefined API methods constructed for specific schemas – a transformation, as it were, of classic relational database query operations (SQL to XQL). Researchers with experience from database representations appear to favor this view.

An underlying assumption here is that the resource properties and the corresponding relationships between them are all known in advance. The RDF structure is thus viewed more as an XML instance of metadata rather than a model, which poses a number of problems in the broader RDF context, even though the XQL may work well in the specific instance.

- For example, the API view *disregards RDF schema relationships*, on which RDF instance documents are based, and therefore loses a great deal of the semantics in the descriptions. As shown earlier, referencing chains of schema resources is a common practice that enables inference and discovery of unstated but implied relationships (for instance, the document property 'creator' mapped to another schema's property 'person' to gain in the first context the initially unstated property 'home address'). This kind of implied and *ad hoc* relationship is inaccessible to the XQL approach.

Bit 5.21   API-coding of query rules is an inherently static affair

Like SQL, XQL becomes based on an application-level coding of the basic methods and
the rules restricted to specific instances. By contrast, RDF-coded solutions can leverage a
multitude of self-defining resources to construct and adapt rules.

The second approach, viewing RDF as a 'Web' structure, is supported mostly by the W3C
RDF working group and other related researchers, as well as the founders of RDF itself.

- Part of the reason for their choice may be that they come mainly from different
  communities that deal with KB/KR rather than database representations. Since the initial
  motivation that led to RDF was the need to represent human-readable but also machine-
  understandable semantics, the driving impulse is to use this representation to do clever
  things.

The Requirements

In this context, the following requirements for an RDF query language were identified in
early RDF position papers (see QL98, *purl.org/net/rdf/papers/*):

- *Support the expressive power of RDF.* The underlying repository of the RDF descriptions
  should support the expressive power of both the RDF model (that is, the assertions), as
  well as of the RDF Schemata (the class and property hierarchies).
- *Enable abstraction.* The QL should abstract from the RDF syntax specifications (that is,
  from the XML encoding of RDF metadata).
- *Provide multiple query facilities.* Typical QL-supported options should include simple
  property-value queries, path traversal based on the RDF graph model, and complex Datalog-
  like queries (Datalog is a database QL based on the logic programming paradigm).
- *Enable inference of subsumption, classification, and inverse fulfilling.* The QL should be
  able to infer implied but unstated relationships. Examples include at least that subordinate
  groupings between classes or properties in RDF Schemata (using *rdfs:subClassOf and
  rdfs:subPropertyOf*) can be identified, that a shared class-related property implies the
  resources belong to the same class, and that a relationship between resource and property
  implies a converse relationship expressible in some other property.
- *Automatic query expansion.* The QL should be able to explore generalization and
  specialization in relations between property values (such as broader, narrower, synonym,
  near, and opposite), much as a thesaurus allows exploration of related dictionary words.

A complete consideration of QL should also include the requirements of the different
kinds of query clients. In the RDF context, these can be grouped as other RDF services,
custom agents, and markup generators (implemented using languages such as PHP or Ruby).

Semantic Web QL Implementations

As noted earlier, we can distinguish between different implementation approaches (that is,
query API, protocol and language). Another important practical distinction is between

specification and implementation, either of which might not even exist for any given instance of the other.

Bit 5.22   Current RDF QLs are often defined by just a single implementation

Call it the early-development syndrome. Note that QL definitions outside of a specific implementation may have different characteristics than the implementation.

A number of published papers have undertaken in-depth comparisons of existing partial implementations, such as the recent ISWC'04 paper 'A Comparison of RDF Query Languages' (*www.aifb.uni-karlsruhe.de/WBS/pha/rdf-query/*). The eight implementations compared here were *RDQL*, *Triple*, *SeRQL*, *Versa*, *N3*, *RQL*, *RDFQL*, and *RxPath*.

Two handy dimensions enable evaluating QL implementations:

- **Soundness** measures the reliability of the query service to produce only facts that are 'true' (that is, supported by the the ground facts, the rules, and the logic). Failures here suggest that the rules (or logic) are incomplete or incorrect in design.
- **Completeness** measures the thoroughness with which the query service finds all 'true' facts. Failures here would suggest that the rules (or logic) are incomplete or inefficient in implementation.

As soon as one starts writing real-world RDF applications, one discovers the need for 'partial matches' because RDF data in the wild tend to be irregular and incomplete. The data are typically expressed using different granularity, and relevant data may be deeply nested.

More complex ways of specifying query scope and completion are required than in the typical database environment. On the bright side, RDF QLs are usually amenable to test and refine iterations of increasingly sophisticated rules.

To assist developer validation, the W3C has a policy of publishing representative use cases to test implementations. Use cases represent a snapshot of an ongoing work, both in terms of chosen samples and of any QL implementation chosen to illustrate them. Originally published as part of the *W3C XML Query Requirements* document as generic examples, query use cases have been republished with solutions for XML Query (at *www.w3.org/TR/xquery-use-cases/*).

XQuery (XML Query, *www.w3.org/XML/Query*) is designed as a powerful QL in which queries are concise and easily understood (and also human readable). It is also flexible enough to query a broad spectrum of XML information sources, including both databases and documents. It meets the formal requirements and use cases mentioned earlier. Many of its powerful and structured facilities have been recognized as so fundamental that they are incorporated into the new version of *XPath* (v2.0).

So, what is 'the standard' for QL applied to the Semantic Web? For several years, up to 2004, the verdict was 'none yet'. Revisiting the scene in early 2005, a W3C working draft suggested that for RDF query, SPARQL is now a proposed candidate *(www.w3.org/TR/rdf-sparql-query/)*.

An earlier W3C survey of the QL field (updated to April 2004, see presentation slides at *www.w3.org/2003/Talks/0408-RdfQuery-DamlPI/*) enumerated and compared many evolving proposals for semantic (that is, RDF) query languages. The following list was originally derived from it:

- *SquishQL* (*ilrt.org/discovery/2001/02/squish/*, the name stands for 'SQL-ish') is aimed at being a simple graph-navigation query language for RDF, which can be used to test some of the functionality required from RDF query languages (also see *ilrt.org/discovery/2002/05/squish-iscw/index.html*).
- *N3* (Notation3, *www.w3.org/DesignIssues/Notation3.html*) is a stripped-down declarative RDF format in plain text. Developed by Tim Berners-Lee, it is a human-readable and 'scribblable' language designed to optimize expression of data and logic in the same language. Results are mapped into the RDF data model.
- *Algae* is another early query language for RDF written in *Perl*, recently updated to *Algae2*. It is table-oriented, does graph matching to expressions, and can be used with an SQL database, returning a set of triples in support of each result. (Algae was used to power the W3C's *Annotea* annotations system and other software at the W3C (see *www.w3.org/1999/02/26-modules/User/Algae-HOWTO.html*).
- *RDQL* is an SQL-like RDF query language derived from *Squish*. Implementations exist for several programming languages (Java, PHP, Perl). A similar 'family' member is *Inkling*.
- *RuleML* (Rule Markup Language Initiative, *www.dfki.uni-kl.de/mleml/*) intends to 'package' the rules aspect of each application domain as consistent XML-based namespaces suitable for sharing. It is extended RDF/XML for deduction, rewriting, and further inferential-transformational tasks.
- *DQL* (DAML Query Language, *daml.semanticweb.org/dql/*) is a formal language and protocol specifying query-answer conversations between agents using knowledge represented in DAML+OIL.
- *OWL-QL* (*ksl.stanford.edu/projects/owl-ql/*) is implemented by *Stanford KSL* to fit with OWL, hence it is a successor candidate to DQL.
- *XDD* (XML Declarative Description) is XML with RDF extensions; a representation language.
- *SeRQL* (Sesame RDF Query Language, pronounced 'circle', *sesame.aidministrator.nl/publications/users/ch05.html*) is a recent RDF QL that combines the best features of other QLs (such as *RQL, RDQL, N-Triples, N3*) with some of its own. (Sesame is an *Open Source RDF Schema-based Repository and Querying* facility.)

There are quite a few more (see, for example, *www.w3.org/2001/11/13-RDF-Query-Rules/*). Tracking the development field and assessing the status of each project is difficult for anyone not actively involved in developing RDF QL. Many of these evolving QLs may remain prominent even in the face of a future W3C recommendation, such as for SPARQL.

A Rules Markup Approach, XRML

A recent extension to the HTML-XML layering approach to making more sense of existing Web content is XRML (*eXtensible Rule Markup Language*). One goal of this effort, formulated by Jae Kyu Lee and Mye M. Sohn (see *xrml.kaist.ac.kr*), is to create a framework that can be used to integrate knowledge-based systems (KBS) and knowledge-management systems (KMS). To understand why the integration of KBS and KMS is so interesting, we need to examine the role and history of each discipline.

- Traditionally, the main goal of KBS technology is the automatic inference of coded knowledge, but practical knowledge processing has been constrained by the inability to handle anything but clearly structured representations of well-defined knowledge domains.
- On the other side, KMS technology started as support for search engines, targeting human users of interactive search. The primary issues here have been effective sharing and reuse of the collected knowledge, leveraging the human understanding of the knowledge.
- Convergence of KBS and KMS requires maintaining consistency between the processing rules (in KBS) and the knowledge structures (in KMS).

What XMRL adds to the mix is a way to make explicit, and therefore possibly to automatically process, the implicit semantic rules that are embedded in traditional Web documents. Much published Web document data cannot be processed automatically even when recast into XML, because XML does not deal with these implicit rules – it deals only with the defined relationships.

XRML is a lightweight solution that requires three components:

- **RIML** (Rule Identification Markup Language) identifies the implicit rules in documents and associates them with explicit rules formulated elsewhere.
- **RSML** (Rule Structure Markup Language) defines an intermediate representation of KBS-stored rules that can be transformed into the structured rules required when processing the document.
- **RTML** (Rule Triggering Markup Language) defines the conditions that should trigger particular rules, embedded in both KBS and software agent.

A concept architecture might look similar to Figure 5.11.

Semantic Rule Markup, SWRL

In the context of rule-markup, the DAML is developing a specific Web version to integrate *RuleML* (mentioned earlier) with the now 'standard' OWL as the **Semantic Web Rule**

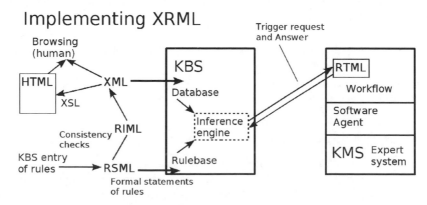

Figure 5.11 How XRML can be implemented and maintained in, for example, a workflow management system

Language (**SWRL**, see *www.daml.org/rules/*). This effort is likely to have important consequences for ontology work.

SWRL extends OWL with first-order (FOL) logic rules based on *RuleML*, and uses XML syntax and RDF concrete syntax based on the OWL RDF/XML exchange syntax.

The DAML proposal for a FOL language (see submission at *www.w3.org/Submission/2005/01/*) was accepted by the W3C in April 2005, thus making it a good candidate for a future W3C sweb working group, and perhaps ultimately a sweb recommendation.

Multi-Agent Protocols

Multi-agent systems are based on simple autonomous agents that interact to show more complex behaviors in the resulting 'society' than implemented in any particular individual. One may speak of a distributed intelligence defined by the interactions. Agents communicate with each other according to specified guidelines, more or less flexible depending on the interaction model.

Systems that comprise more than a small number of agents commonly rely on so-called *Middle Agents* (MA) to implement a support infrastructure for service discovery and agent interoperation – they can go by various implementation names, such as matchmakers, brokers, or facilitators. The MA layer is intermediary and exists only to make agent-to-agent communication more efficient. The general move to higher-level representations makes it easier to avoid implementation assumptions.

Lately, the preference is to model agent behavior in terms of high-level concepts: goals, schedules, actors, intent, and so on. The models often contain implicit relationships derived from the organizational setting in which the agents belong. This situation encompasses relationships to peers, teams, projects, and authorities.

Multi-agent protocols formally regulate the interactions between collaborating independent agents, ensuring meaningful conversations between them. In addition, these protocols should also respect agent autonomy, enabling the flexible capability of agents to exploit opportunities and to handle exceptions.

Early protocols were syntactic descriptions and thus lacked the capability for semantic expressions to deal with typical real-world situations, for example:

- Decisions need to be made at run-time about the nature and scope of the interactions.
- Unexpected interactions must be expected – all interactions cannot be defined in the design.
- Coordination between agents occur dynamically, as needed.
- Competition may be a factor in interaction.
- The inability to achieve all goals means the agent must commit resources to achieving a specific and useful subset of the goals.

Multi-agent protocols distinguish between non-committed behavior (user may confirm or reject proposed plans) and committed behavior (full delegation and contractual binding of tasks).

Contract Net Protocol (CNP) implements agent coordination and distributed task allocation on the *bid and contract* model. Bid values can be defined in whatever common 'currency' that seems appropriate. Manager agents ask for bids on tasks (negotiations or auctions) from selected contractee agents and award tasks to the 'best' bid. The winning contractee performs the task under manager monitoring.

CNP is fully distributed for multiple heterogeneous agents and easy to implement, and therefore forms the basis for many subsequent MAS protocols. Its known problems include manipulation risks, possible loops, and the lack of performance checks.

Ideally, a contractee agent should commit to task performance and the manager, and vice versa, but simple self-interest rules might terminate a contract in special situations. For example, during execution a contractee may receive an offer for a more profitable task, or a manager a better bid. More complex rules add various forms of enforcement by specifying termination penalties or price hedging against probable risks. An additional complication is that although the assumption is that agents are truthful when bidding, lying can be 'beneficial' in the context of self-interest rules.

Auctions may reduce the risk of agents lying but introduce other problems. Several different auction models exist with differing bidding procedures, in analogy to real-world auctions. No single solution seems optimal, but a full analysis is beyond the scope of this book. An overview is available as a presentation (at *www.cs.biu.ac.il/~shechory/lec7-00-6up.pdf*).

Other models of multi-agent interactions may define the concept of 'social commitments' that are similar to those found in human interaction. The metrics differ from bid and contract, and may instead focus on some form of reputation history. Bids and commitments made by one agent to another agent to carry out a certain course of action therefore become qualified by a calculated trustworthiness factor. Business logic uses the reputability factor when selecting a contractee – or for that matter when selecting manager agents for bid submission.

Reputability tracking has become a common strategy when managing p2p networks and protecting it against rogue nodes. Several p2p solutions implement resource management through virtual bid-and-contract protocols. Some prototype examples were examined in detail for both cases in the book *Peer to Peer.*

6

Ontologies and the Semantic Web

In the broader context of the Semantic Web, applications need to understand not just the human-readable presentation of content but the actual content – what we tend to call the meaning. Such capability is not covered by the markup or protocols discussed so far. Another layer of functionality is thus needed and defined here, that of *Ontology*. It addresses ways of representing knowledge so that machines can reason about it to make valid deductions and inferences.

We really cannot yet adequately explain or construct true artificial intelligence, despite decades of research and models. Different models of the human brain (and by implication, intelligence) arise to be in vogue for a span of years before being consigned to historic archives. Oddly enough, these models all tend to bear striking resemblances to the conceptual technological models of the time. Models in the last half century have clearly built on contemporary computer designs. One might be forgiven for presuming the normal state of affairs is that the latest (computer) technology inspires models for understanding human intelligence, rather than the other way around.

Nevertheless, it is meaningful to speak of 'intelligent' machines or software, in the sense that some form of machine reasoning using rules of logic is possible. Such reasoning, based on real-world input, delivers results that are not preprogrammed. The machine can at some level be said to 'understand' that part of the world. When the input is human-language text, the goal must be to semantically 'understand' it – perhaps not the intrinsic and full meaning of the words and assertions, as we imagine that we as humans understand them, but at least the critical relationships between the assertions and the real world as we model them.

Chapter 6 at a Glance

This chapter looks at how knowledge can be organized in the Semantic Web using special structures to represent property definitions and meaningful relationships. *Ontology Defined* starts with an overview of what an ontology in the modern, computer-related sense is, and why it is an important concept.

The Semantic Web: Crafting Infrastructure for Agency   Bo Leuf
© 2006 John Wiley & Sons, Ltd

- *Ontologies for the Web* describes the aspects specific to the Web.
- *Ontology Types* is an overview of different types of ontologies and their purpose, noting that they will vary depending on application and view.
- *Building Ontologies* outlines a methodology for how ontologies are built.
- *Web Ontology Language* describes the core W3C efforts to design OWL, a common generic ontology language for the Web.
- *Other Web Ontology Efforts* samples a few industry-related ontology initiatives.

Knowledge Representation examines a number of languages used to describe formally concepts and relationships.

- *Conceptual Graphs* is a kind of counterpoint to DLG representation of RDF but is different in a few critical ways.
- *Promoting Topic Maps* describes another way to organize mapping of semantic meaning.
- *Description Logics* is another formal logic system that has successfully been used to implement a variety of KR and query systems.

Ontology Defined

The dictionary meaning of *ontology* concerns *the metaphysics of the nature of being*. Interesting to be sure, but abstract, and seemingly out of place in technology discussions about the Web. However, in the Semantic Web context, a narrower and more precise meaning of casual usage is intended, namely:

> *Computer ontologies are structures or models of known knowledge.*

The term 'ontology' has in fact been used in this way for a number of years by the artificial intelligence (AI) and knowledge representation (KR) communities. It has lately become part of the standard terminology of a much wide community that includes object modeling, XML, and the Semantic Web. In its own way, this computer-oriented definition is as valid as any 'classical' use of the term, and has unlike the latter very immediate and practical applications.

The original definition of ontology in this new sense, as 'a specification of a conceptualization' (proposed by Tom Gruber in 1992, *www-ksl.stanford.edu/kst/what-is-an-ontology. html*) was expanded and qualified in later analysis (such as by Nicola Guarino and Pierdaniele Giaretta, *www.loa-cnr.it/Papers/KBKS95.pdf*) with a weaker intentional focus in order to be more useful in the field:

> *An ontology is a partial (and explicit) account of a conceptualization ... The degree of specification of the conceptualization which underlies the language used by a particular knowledge base varies in dependence of our purposes ... An ontological commitment (is thus) a partial semantic account of the intended conceptualization.*

Ontology-based semantic structures effectively replace the jumbles of *ad hoc* rule-based techniques common to earlier knowledge representation systems. KR languages thus become an easier-to-manage combination of ontology and logic.

These structured depictions or models of known (and accepted) facts make applications more capable of handling complex and disparate information. Semantic structures of this nature go one step further than what can be achieved with relational and XML databases, valuable as these are, by providing at least some underpinning for meaningful reasoning around semantic relationships. Such ontological approaches appear most effective when the semantic distinctions that humans take for granted are crucial to the application's purpose.

A different, and perhaps too broad, purpose definition is provided by *OntologyOrg* (*www.ontology.org/main/papers/faq.html*), an organization founded in 1998 to highlight the need for ontologies (as defined here) in Internet commerce:

> *The main purpose of an ontology is to enable communication between computer systems in a way that is independent of the individual system technologies, information architectures and application domain.*

This formulation is ultimately satisfied by the result of implementing an ontology in the previous 'commitment' sense, but it does not define what an ontology is or what problems it tackles. Therefore, see it perhaps more as an intentional guideline.

Analysis of semantic structures in human language forms the bulk of the field of computer ontologies, because most semantic meaning is expressed in this way (for example, on the Web). Structured semantic analysis, however, has an interesting and surprisingly old history.

A precursor to modern computer ontologies can be found as early as the 1600s in *An Essay Toward a Real Character and a Philosophical Language*, by *Bishop John Wilkins* (1614–1672). This remarkable work contains a large ontology, a written and spoken language derived from the ontology, and a dictionary that maps terms in the ontology to English.

Numerous insights into issues of ontology and knowledge representation make it more than just of academic historic interest to the modern reader. The scope and scholarship of the work are impressive, albeit the learned language of the 1600s can be difficult to a reader unused to the wordy style and implicit references. The work is available from various sources, including on the Web (for example, see *reliant.teknowledge.com/Wilkins/* for parallel pages of HTML and facsimile images).

When most successful, the new ontologies model those semantic relationships and distinctions that humans take for granted, but which are crucial to the purpose of the intended application. Typically, so-called 'expert systems' cannot reduce all the human-expert knowledge being modeled to simple rules. (For that matter, some knowledge or skills cannot even be modeled at all.) Some of the expertise being modeled depends on 'common sense' lurking in the natural language that expresses the knowledge. Human reasoning can cope with such implicit relationships and distinctions; machine reasoning on its own cannot.

Bit 6.1   Expertise is not just knowledge but also requires reasoning

Semantic-based systems using formal relationship descriptions, rulesets, and logic can however go a long way to emulating human expert reasoning.

Externalizing such reasoning around implied and inferred relationships enables significant advances in creating applications that can leverage many traditional knowledge stores – and, in fact, help automate the creation of better ones.

Since the mid-1990s the domain of expert systems has broadened significantly and segued into 'knowledge-based systems' (KBS), acknowledging the fact that the goal is not always to produce a fully autonomous and artificial expert, but rather to model and represent knowledge and inferences that can be used in 'lesser AI' systems to assist humans in some tasks.

Examples of applied ontologies span many areas of research, some of which are described in popularized accounts rather too often just as 'examples of AI':

- Semantic Web research in general, mainly to study and develop better representational models and inference engines;
- e-commerce systems, to enable the automated exchange of electronic information among commercial trading partners;
- engineering design, especially to mediate collaborative design efforts;
- health care systems, to generate best-practice medical guidelines for managing patient health;
- biological systems, to automate the mapping of plant and animal genomes, and more critically, to describe, manage, and share large amounts of biological data;
- information systems, to assist searching for specific public information resources;
- security systems, to perform automated in-depth security analysis.

In all these cases, the application of ontologies provides an attempt to specify objectively information (the 'expert' knowledge) within a particular domain. In particular, the life sciences have proven eager to adopt the emerging sweb ontology technologies in order to in some way automate the management of the enormous amounts of research data being produced.

Any ontology model represents only a *consensual agreement* on the concepts and relations that characterize the way knowledge in that domain is expressed, albeit that the notion of 'domain' is relative to the adopted application point of view and is often fuzzy. Different models are possible, perhaps applicable to different contexts and points of view. A higher-level (or top) ontology may just model 'common knowledge' or semantic basics, not specific data.

Ideally, the domain representation is expressed in an *open language* that enables broad access and operability – for example, as Web-published RDF. This specification can then be the first step in building semantically-aware information systems on the public Web to support services and activities at all levels: personal, commerce, enterprise, and government.

This broad context motivates a deeper examination of ontologies in general, and implemented Web-based ontologies in particular, because these are the mechanisms that promise to deliver all manner of stored knowledge in a consistent way.

Bit 6.2   Open knowledge presupposes well-defined dissemination methods

Stored expert knowledge must be both easy and affordable to access, yet the traditional KB and KR stores present formidable barriers to use outside their respective domains. Semantic Web technology aims to change this situation.

Expert knowledge in humans is both expensive and difficult to develop and access, and it is subject to serious constraints of time and location. Therefore, it makes good business sense to invest heavily in developing useful, sufficiently generic ontologies, in order to later reap the benefits of online access, automation, and synergy.

Ontologies for the Web

The key ingredients in the 'new' Internet-related ontology are:

- a *vocabulary* of basic terms;
- a precise *specification* of what those terms mean.

Domain-specific terms are thus combined with domain-independent representations of relationship. Ideally, consensus-standard vocabularies can be handled by defining reusable vocabularies, and assembling, customizing, and extending them.

The idea is that by sharing a common ontology expressed in this way, unambiguous and meaningful 'conversation' can occur between autonomous agents or remote applications to achieve a specified purpose, even in the absence of detailed step-by-step instructions and interpretations at the endpoints. *Application logic* and *rules of inference* should then be able to fill in the blanks.

A practical ontology actually goes further, in that it requires more than just an agreed vocabulary, and more than just a taxonomy or classification of terms:

- On the one hand, it provides a set of well-founded constructs that can be leveraged to build meaningful, higher-level knowledge from basic definitions. The most basic (or abstract) concepts and distinctions are identified, defined, and specified. The resulting structure forms a complete set with formally defined relationships, one element to another, which provides the semantic basis for the terminology.
- On the other hand, ontologies include richer relationships between terms than the simply taxonomic contribution to the semantics. These relationships enable the expression of domain-specific knowledge without requiring domain-specific terms, and hence also support easier translation between domains.

Ontological knowledge is characterized by *true assertions* about groups, sets, and classes of things that exist, irrespective of the current state of affairs in the specific occurrences of these things.

In a Semantic Web environment, a major goal is to enable agent-based systems to interoperate with a high degree of autonomy, flexibility, and agility – without misunderstanding. Ontologies in the current sense seek to achieve this goal.

Relying on, for example, XML language frameworks alone provides only syntactic representations of knowledge. Such frameworks do not address ontology and therefore cannot convey any semantic meaning. Interoperation is in such cases dependent on each implemented agent complying to a standardized usage of particular tag sets, externally defined, and using these consistently. While it is true that one can get quite far in this way, results are fragile with external dependencies.

Many different communities have an interest in ontologies, although they differ with respect to the nature of the knowledge dealt with and the purpose to which it will be put. The natural language community, for instance, uses ontologies to map word meaning and sense. The AI community is concerned with knowledge-based systems, where ontologies capture domain knowledge, and problem-solving methods capture the task knowledge. The database community, when concerned with semantic interoperability of heterogeneous databases, uses ontologies as conceptual schemata.

In semantic-based information retrieval, in particular, *ontologies directly specify the meaning of the concepts to be searched for*. By contrast, XML-based systems in this context are severely limited in utility unless the independent site-content authors happen to agree on the semantics of the terms they embed in the source metadata. Ontologies reduce such semantic ambiguity by providing a single interpretation resource, and therefore fill an important role in collaborations (human and machine), interoperation contexts, education, and modeling.

Ontologies can also enable software to *map and transform information* stored using variant terminologies, thus enabling Information Broker Architecture to satisfy human or agent queries by searching a far wider selection of online sources.

Ontology Types

Current ontologies can vary greatly in a number of different ways, from content to structure and implementation:

- **Content Type**. Content-defining terms can be typed and defined at different levels of precision, from abstract (such as concepts) to specific (or itemized) object detail, within the same conceptual domain.
- **Description Level**. This classification reflects the formalization span of models from simple descriptive lexicons and controlled vocabularies, to the more complex models where properties can define new concepts and relationships between terms.
- **Conceptual Scope**. Some models are restricted to more general, 'upper-level' relationships, essentially free of domain-specific relationships. Upper ontologies are used to bootstrap or graft more specific ones that target very specific domains in great detail. These differently-scoped models can work together, complementing each other for particular applications.
- **Instantiation**. Each model can draw its own distinction between the classical terminology concept and some point where it deals with instances of objects that manifest the terminology (that is, the model's assertional component). In practical implementation, the chosen instantiation line is where a knowledge database of specific items can be extracted and maintained as a separate entity.
- **Specification Language**. Although ontologies can be built using ordinary programming languages, it is more common to use languages specially designed to support ontology construction or description logic. Some examples are given later in this chapter.

As this list suggests, all ontologies are not created equal. Rather, each is constructed according to particular design goals and developer demands.

It is fortunate therefore that the underlying intent of Web ontologies is for *interoperability*. If managed in the appropriate languages (XML-based), applications can translate between them, or use several at once to complement each other.

Bit 6.3   Existing ontology work moved into the Web enables sharing and synergy

Much Semantic Web work targets the design and implement of the required tools.

Building Ontologies

For the sake of completeness in this overview, the following list outlines the general steps involved in constructing an ontology:

1. *Acquire the domain knowledge.* This stage consists of identifying and assembling appropriate expertise and information resources, to define in a common language all descriptive terms with consistency and consensus.
2. *Design the conceptual structure.* Identify the main concepts within the domain, along with their associated properties. Thereafter identify the relationships among the concepts, perhaps creating new abstract concepts as organizing features, and distinguish which concepts have instances. Reference or include any supporting ontologies as needed, and apply other guidelines appropriate to the chosen methodology.
3. *Develop the appropriate detail.* Add concepts, relationships, and instances to achieve the level of detail to satisfy the stated purpose of the ontology.
4. *Verify.* Check the structure for consistency, and reconcile any syntactic, logical, or semantic inconsistencies among the elements. New concepts may have to be defined at this stage.
5. *Commit.* After a final verification by domain experts, the ontology is committed by publishing it within its intended deployment environment.

The construction process is in practice not as linear or simple as this list suggests. It is more often highly iterative and may proceed from many different points at once, both top-down and bottom-up. The process is similar to yet fundamentally different from software development. A major difference is that while software is generally concerned with functionality and flow control, ontologies are declarative and span more levels of semantic abstraction.

Once published, the ontology can be referenced and used within the environment, as is the case described for the languages (such as XML) and schemas (RDF).

Bit 6.4   Published shareable ontologies are valuable online resources

Anyone can link to – and through the power of XML/RDF leverage – the information.

- A step-by-step narrative of an ontology-building process, accessible even to non-expert readers, is found in 'Natural Language Processing with *ThoughtTreasure*' by Erik T. Mueller (published at *www.signiform.com/tt/book/Rep.html*). The example used in the paper is based on Usenet-published cinema reviews and resulted in a downloadable ontology consisting of 21,521 concepts and 37,162 assertions. The example is part of the *ThoughtTreasure* artificial intelligence program and toolset project, which contains much else besides the ontology.

- A useful resource for avoiding simple mistakes in ontology construction is 'Ontology Development Pitfalls' (at *www.ontologyportal.org/Pitfalls.html*). The maintainers remark that most other (that is, more complex) pitfalls are not a problem in practice because too many people are still making the simple ones. Required reading to gain a feeling for the subject.

Ontology-building Tools

The first and most reasonable action when starting out on an ontology project is to find a suitable ontology software editor. Tools are available to accomplish most aspects of ontology development, actual ontology editors being but one type.

Other tools can help acquire, organize and visualize domain knowledge. It is common to find different tools for the construction of the terminology component and of the assertional component. Browsers can inspect the ontology during and after construction.

What about tool availability? A survey published in November 2002 ('Ontology Building: A Survey of Editing Tools' by Michael Denny, see *www.xml.com/lpt/a/2002/11/06/ontologies.html*) identified about 50 editing tools. Even though the survey included everything from unfinished prototypes to commercial offerings, the result was still termed 'a surprising number' by its author.

In the years since, the number and utility at tools has steadily increased. A follow-up survey published in July 2004 ('Ontology Tools Survey, Revisited', see *www.xml.com/pub/a/2004/07/14/onto.html*) noted first that OWL is rapidly replacing its predecessor DAML+OIL, as expected. Available tools, of which 94 were examined, also more often directly support the emerging Web ontology standards.

Bit 6.5   Tools for building, editing, and integrating ontologies are in abundant supply

Although the field still remains fragmented, 'semantic integration' tools are increasingly being offered to help migrate or exchange data with legacy KB repositories and unstructured streams.

Several measures of increasing technology maturity may be noted. Vendors of mainstream enterprise-application-integration (EAI) now commonly refer to published taxonomies and ontologies, and often provide tools to help integrate existing data with the new structures. Requests for new tool features are now commonly concerned with making the construction of full-blown ontologies easier and more foolproof for the domain experts.

As projects tend to involve solutions based on several ontologies from various sources, management tools and integrated environments (IDE) become just as important as the tools for constructing them, if not more so. Of particular interest are tools that can map and link between different models, or transform to and from XML database and UML schemas.

- Major web-search services now use ontology-based approaches to find and organize content on the Web. For example, *Google* acquired *Applied Semantics, Inc.* (one of the leading vendors of semantic extraction tools) which suggests ontologies will play a major part in future search-technology solutions, probably to integrate results for text and other media.

Mapping, linking, and transformation are important capabilities, especially when considering the fact that many editing tools still tend to be fairly specific to their development environments – typically, data integration or specification.

Otherwise, the key to editor usability lies in its ability to organize and manage an emerging ontology in all its complexity of interlinking concepts and relations. The standard approach is to use multiple tree views with expanding and contracting levels. Less common, though often more effective for change editing, is a graph presentation.

Some tools were noted in the first survey to lack useful import or export functions for broader use, navigational and overview aids, visualization capabilities, and automation features. In the ensuing year, toolmakers have addressed such requests, but are often finding the task unexpectedly demanding given the scale and variability of the ontologies studied.

Ontologies are commonly treated as standalone specifications during development, even though they are ultimately intended to help answer queries to a knowledge base. Some editors incorporate evaluation capability within the development environment by allowing the addition of new axioms and rules of inference to the ontology. Standard rule languages with this capability are not yet available, so these extensions are usually proprietary, but it is expected that emerging rule-markup standards (such as SWRL) will address this matter in new versions of the tools.

The first survey concluded that ontology building is still a fragmented development space, arising from diverse environments, many logic languages, and different information models – something that tends to cause conflict with the fundamental concept that ontologies are for sharing. The wider the range of applications and other ontologies that can use an ontology, the greater its utility. Interrelating ontologies provide mutual utility.

The interoperability aspect, however, is seldom addressed beyond the level of import and export in different language serializations. The lack of formal compatibility on syntactic as well as semantic levels greatly hinders the ability to accommodate domain interoperability, or to access specialized XML languages or standards being adopted in industry.

Bit 6.6   Interoperability is promoted by shared environments

Development communities centered around common ontology languages, online toolsets, and shared repositories of ready-to-use ontologies are therefore valuable Web resources that promote moving specialized solutions into the Semantic Web.

The follow-up survey noted increased support for ontology languages built on RDF (OWL) and the use of URIs as identifiers for referring to unique entities. At the same time, controversy exists whether RDF is in fact the best base language for implementing ontologies on the Web, or elsewhere. Some professionals doubt whether it affords the same scalability, representational power, or expressiveness as known proprietary solutions, necessary to implement very large ontologies and webs of ontologies, and necessary for demanding applications.

Nevertheless, RDF does offer compelling advantages in the universal use of URI and XML namespace protocols on the Web. This unifying aspect should make it easier to establish, through collaboration and consensus, the ontologies needed (as useful vocabularies) to implement applications for extensive integration and cooperation on the Web.

Current tools offer a wide and varied range of capabilities. In the absence of an established IDE for ontology construction, still the case, the practical approach is to rely on different tools to fashion different aspects of the desired ontologies and to manage the development process.

The focus in this book is the general and the greater sweb interoperability, so despite their often wide usage base, some more specialized tool sets may receive little mention here. Anyone with specific interest in evaluating and comparing tools in general should consult the earlier mentioned surveys, at least as a starting point. The second survey gives much good advice on choosing tools.

Chapter 9 explores some practical applications and related sweb-ontology tools.

Barriers to Re-use

A mention should be made of the barriers typically encountered when trying to develop re-usable ontologies. For example, implementation-dependent semantics are still common, which needlessly limit application to a single domain. Re-use requires formal definitions on a more general level, applicable to many domains.

If heterogeneous representation languages are designed, specialized for specific methods and tasks, application in other domains then requires translation between languages. In a similar way, heterogeneous ontologies specialized for specific methods and tasks also require translation.

Specialized ontologies may still have many concepts in common, however, so instead the solution is to base the specialized versions on shared foundation ontologies, suitably extended to cover the specialized situations. Such harmonization depends largely on having the appropriate tools to manage and develop ontologies based on shared libraries.

Resources

An *Interlingua* project, **KIF** (*Knowledge Interchange Format*), can be used to represent ontologies. By itself, KIF includes only a fairly basic vocabulary, but it can be extended by defining relations, functions and objects, and adding axioms. The result of combining extended KIF and Ontology is *Ontolingua*, a common framework and Web service for developing shared ontology libraries. The project and its hosted ontology projects is described in Chapter 9.

Also related to KIF is the *Standard Upper Ontology* (**SUO**), a top-level semantic structure expressed in a simplified version of the format, *suo-kif*. The SUO home is at *suo.ieee.org*, with further pointers to **SUMO** (Suggested Upper Merged Ontology), a model that merges upper and lower ontologies. SUMO is also featured in Chapter 9.

An early resource for exploring KR systems and ontology development was *The Ontology Page* (*TOP, www.kr.org/top/*). The following is from the introductory page:

> *TOP identifies world-wide activity aimed at developing formalized ontologies as the basis for shared and modularly-structured knowledge. TOP covers every aspect of work on ontologies, including the construction, specification, formalization, representation, analysis, and documentation of ontologies, as well as their use at all levels in communication, computation, and reasoning.*

The projects listing can still provide insight even though it is now perhaps mainly of interest for delving into the history of such development, as most of the material derives from cira 1995.

Reasonably up-to-date listings of ontologies are found at *the DAML Ontology Library* (*www.daml.org/ontologies/*), including the *Ontologies by Open Directory Category* list that leverages the *DMOZ Open Directory* project (see *www.dmoz.org*). Several of these tables render very wide, so a casual visit might miss the right column associating each entry with a corresponding resource URI.

Bit 6.7   Computer ontologies benefit from a study of philosophical issues
Practical KR and model building often requires an understanding of deeper theory.

A more general resource is *The Stanford Encyclopedia of Philosophy* (at *plato.stanford.edu*), which can provide insight and pointers relating to specific *a priori* issues of knowledge representation and ontology building. For example, modelling the concept of 'holes' turned out to be far more difficult than might have been expected, and a background analysis is provided in the article by Roberto Casati and Achille Varzi (see *plato.stanford.edu/archives/ fall1997/entries/holes/*).

Web Ontology Language, OWL

Early on, the W3C saw the need for a language to facilitate specifically greater machine readability of Web content than that supported by just XML, RDF, and RDF Schema. The *OWL Web Ontology Language* (OWL, current home: *www.w3.org/2004/OWL/*) was designed for this purpose by the W3C Web Ontology Working Group *(WebOnt*, now closed). The motivation was clear enough:

> *Where earlier languages have been used to develop tools and ontologies for specific user communities (particularly in the sciences and in company-specific e-commerce applications), they were not defined to be compatible with the architecture of the World Wide Web in general, and the Semantic Web in particular.*

OWL is a *semantic markup language* for publishing and sharing ontologies on the Web, using URI notation and RDF technology, suitable for applications that need to process the content of information instead of just presenting information to humans.

Standardized formal semantics and additional vocabulary enable OWL to represent explicitly term descriptions and the relationships between entities. It implements a rich language to describe the classes, and relations between them, that are inherent in Web documents and applications.

Bit 6.8   The intent of OWL is to enable developers to publish and share sets of terms (specific ontologies) on the Web in a consistent way
Published ontologies with a consistent model structure can then be leveraged to provide advanced Web search, software agents, and knowledge management.

As noted earlier, it is this kind of mapping that defines the common use of 'ontology' in computer science. Such ontologies are not new, and OWL builds on earlier (and now superseded) W3C efforts to create a semantic markup language for Web resources, primarily what came to be known as 'DAML+OIL'. All these efforts attempted to build on the RDF and RDF Schema languages to provide richer modeling primitives.

The OWL specification set moved from draft status to W3C recommended v1.0 candidate status in August 2003, and to W3C recommendation in February 2004. In practical terms, such a move signals that OWL is ready for wide-spread implementation.

As the W3C 'standard', OWL currently enjoys much more attention than any other ontology language, even off the Web. It seems set to dominate at least Web-based ontology construction, regardless that detractors still tend to single out perceived 'shortcomings', such as its limits of expression, inelegant syntax, occasional cumbersome constructs, and reliance on the RDF model of representation using triples. The major design features in OWL are:

- ability to be distributed across many systems;
- scalability to Web needs;
- compatibility with Web standards for accessibility and internationalization;
- openness and extensibility.

It is further specified with three increasingly expressive sub-languages: *OWL Lite, OWL DL*, and *OWL Full*. These optional adoption levels go from allowing quick migration under easy constraints, to enabling maximum functionality – though perhaps with no guarantee of practical computability in all cases for the full version.

Each level is a valid subset of the higher ones, both in terms of expressible ontologies and computed conclusions. Expressed another way, *OWL Full* is an extension to RDF in its entirety, while *OWL Lite* and *OWL DL* should be seen as extensions to a restricted view of RDF.

In the 'stack' model, the W3C approaches OWL in the following progression from underlying sweb components:

- XML, to provide a surface syntax for structured documents;
- XML Schema (XMLS), to restrict the structure of XML documents in well-defined ways, and to provide the data types;
- RDF, to model object data ('resources') and relations between them, and provide a simple semantics for this data model;
- RDF Schema (RDFS), to describe properties and classes of RDF resources in a defined vocabulary, with a semantics for generalized hierarchies;
- OWL, to add more vocabulary, as described earlier.

This progression has been implied earlier, but it is spelled out here for clarity.

A series of design goals for OWL are summarized in Table 6.1. A core requirement is that ontologies are resources, specifiable with URI notation.

Development Path

A brief explanation of the older cryptic names, and something of their history and relationships might be in order, to see where OWL came from. Most development can be

Table 6.1 Design goals for OWL

Design Goal	Motivation	Comments
Shared ontologies	Interoperability and explicit agreement about common terms needed.	Sharing ontologies commits to using the same basic identifiers and meanings
Evolution of ontologies	A changing Web means changing ontologies.	Ontology language must be able to accommodate ontology revision.
Interoperability	Different ontologies may model same concepts in different ways.	Primitives must map concepts between different models.
Inconsistency detection	Different ontologies or datasets may be contradictory.	Inconsistencies must be detected automatically (RDF and RDFS do not allow for contractions).
Balance between expressivity and scaleability	Large ontologies and rich expression represent opposing forces.	The Web is huge, and still growing, so practical solutions must balance requirements.
Ease of use	Low learning barrier and clarity in concept required; must be human readable.	Basic philosophy of the language must be natural and easy to learn.
Compatibility	Other industry standards are in common use.	Support exchange of data and ontologies.
Internationalization	Must support multilingual and multicultural contexts.	The Web is an international tool; allow all to use same ontologies.

traced back to a series of ontology language releases based on DAML (see DARPA in Chapters 3 and 7).

The original DAML project officially began in August 2000, with the goal to develop a language and tools to facilitate the concept of the Semantic Web. As the name component 'Agent Markup' suggests, a major goal was to empower autonomous agent software. The rich set of constructs enabled machine-readable markup, along with the later creation of ontologies – a basic inference infrastructure. Efforts to incorporate the ontology descriptions led to the expanded DAML-ONT releases of the language.

The immediate precursor to OWL was DAML+OIL, a semantic markup language for Web resources that built on earlier W3C standards, such as RDF and RDF Schema, and which used *XML Namespaces* (XMLNS) and URIs. It provided extensions with richer modeling primitives from frame-based languages, and with values from *XML Schema* datatypes. The point was to better describe Web-object relationships.

DAML+OIL was itself an extension to the previous DAML-ONT release, in an effort to combine many of the language components of OIL (Ontology Inference Layer, *www. ontoknowledge.org/oil/*), and be more consistent with OIL's many application areas. The European OIL project was in turn a proposal for a Web-based representation and inference layer for ontologies, compatible with RDF Schema, and it included a precise semantics for describing term meanings, and thus also for describing implied information.

DAML itself is still a widely referenced and used language, despite the fact that it is technically obsolete now that OWL is the W3C recommendation. A number of experimental application areas are still referenced (at *daml.semanticweb.org/applications/*) that show various ways to interlink RDF data, DAML, and small bits of code to generate interesting results in various contexts.

- For example, *DAML Agenda* comprises an ontology and supporting tools for large meeting agendas. It can accommodate shuffling and resizing of talks during meeting planning, and can generate an archival HTML record of the meeting (by providing links to speakers and briefings). The Agenda is available as a service and, using the URL, can be referenced to a DAML representation of the talk parameters.

The Use Cases

The W3C motivated OWL (or Web ontologies in general) by referring to six use cases as a cross-section of representative use in the Semantic Web.

- **Web portals**, in the context of defining community ontologies that enable users to declare and find related-interest content more efficiently. A cited example portal is *OntoWeb* (*www.ontoweb.org*).
- **Multimedia collections**, in the context of semantic annotations for collections of images, audio, or other non-textual objects.
- **Corporate Web site management**, in the context of broader and semantic indexing of stored documents to provide more intelligent retrieval.
- **Design documentation**, in the context of managing large, multi-typed collections of engineering documentation that are part of different hierarchical structures.
- **Agents and services**, in the context of providing delegated task fulfilment for the user. An example given is a social activities planner that can compose proposed programs for the user based on user preferences and Web-published calendars and events.
- **Ubiquitous computing**, by which is meant a shift away from dedicated computing machinery to pervasive computing capabilities embedded in our everyday environments.

We explore many of these topics in Part III when considering future implementations.

Demonstration Implementation

The OWL Working Group request for recommendation (see *www.w3.org/2001/sw/WebOnt/ rqim.html*) listed a number of then working implementations of OWL that could be used to validate the design, in addition to noting that OWL tools had been used to generate the supporting comment lists.

Until late 2003 the WG maintained a page detailing testing by implementation category (at *www.w3.org/2001/sw/WebOnt/impls*). The results of this testing led the WG to believe the implementation experience was sufficient to merit widespread deployment. Some examples follow:

- The *Horns* project (described in *www.semanticweb.org/SWWS/program/position/soi- kettler.pdf*), developed for the intelligence community by the DARPA DAML project.
- The *AKT* Portal (*www.aktors.org/akt/*) at the University of Southampton, largely ontology- based and using OWL.

- Several OWL-based demo applications developed at the University of Maryland Baltimore County, such as *Travel Agent Game* (TAGA, see *taga.umbc.edu*, a multi-agent automatic trading platform for *Agentcities*) and *CoBrA* (Context Broker Architecture, see *users.ebiquity.org/~hchen4/cobra/*, currently modeling intelligent meeting spaces).
- *BioPax* (Biopathway Exchange Language, at *www.biopax.org*), a data exchange format for biological pathways data, implemented using OWL.
- Automating W3C tech reports (*www.w3.org/2002/01/tr-automation/*), related to the 'multimedia collections' OWL use case.
- The *Mindswap Project* (*owl.mindswap.org*), uses OWL to generate Web pages and 'custom home pages' for members of the research group – self-styled as 'the first site on the Semantic Web' (it was originally based on the seminal SHOE ontology language, and was the first OWL-compliant site).

The *Mindswap* site has pointers to a number of useful RDF, ontology, and OWL tools (at *owl.mindswap.org/downloads/*). *Agentcities* is mentioned in Chapter 9 as an early prototype of a distributed platform for a variety of agents and agent-driven services.

The OWL Approach

OWL is used in a three step approach to improve term descriptions in a schema vocabulary:

- Formalize a domain by defining classes and properties of those classes.
- Define individuals and assert properties about them.
- Reason about these classes and individuals to the degree permitted by the formal semantics of the OWL language.

In this last context, recall that OWL defined three sub-languages, or levels of implementation, targeting different needs. The choice determines the scope of reasoning possible.

The Overview document for OWL is found at the W3C site (as *www.w3.org/TR/owl-features/*), along with a wealth of technical reference detail (as *www.w3.org/TR/owl-ref/*). Both sources are classed 'W3C Recommendation' from 10 February 2004.

As can be seen in Chapters 8 and 9, and in the DAML lists mentioned earlier, numerous ontologies either already exist or are under development. How do these alternative ontologies relate to or conflict with OWL?

Other Web Ontology Efforts

While OWL is a generic Web ontology framework, independent of domain, other practical and deployed ontologies commonly have a more specific focus on a particular application area. The result may therefore be more or less interoperable with other areas.

Nevertheless, from basic principles and a generic framework such as OWL, specific ontologies can be mapped onto others. Such mapping is important for the general Semantic Web goal that software agents be capable of processing arbitrary repositories of Web information. This task is made easier when the specialized ontologies and related data structures are constructed using XML or XML-like languages.

Several examples of industry-anchored ontology initiatives are given below.

Ontologies for Enterprise Modeling

Established in the early 1990s (by Fox 1992, *Fox and Grüninger* 1994), the *Toronto Virtual Enterprise Project* (TOVE, see *www.eil.utoronto.ca/enterprise-modeling/*) set out to develop a rich set of integrated ontologies to model KR in both commercial and public enterprises. The KR models span both generic knowledge (such as activities, resources, and time), and more enterprise-oriented knowledge (such as cost, quality, products, and organization structure).

TOVE takes a 'second-generation knowledge-engineering' approach to constructing a Common Sense Enterprise Model – 'engineering the ontologies' rather than extracting rules from experts. Such a common-sense approach is defined by the ability to deduce answers to queries using only a relatively shallow knowledge of the modeled domain. It is but one aspect of the work at the Enterprise Integration Laboratory (*www.eil.utoronto.ca*) at the University of Toronto.

Bit 6.9   TOVE provides important tools for enterprise modeling

Its approach of engineering from principles rather than just collecting expert knowledge has broad and adaptive application.

The TOVE *Foundational Ontologies* category currently comprises two primary representation forms, or top-level ontologies:

- **Activity**, as set out by a theory of reasoning about complex actions that occur within the enterprise (that is, organization behavior).
- **Resource**, as set out by a theory of reasoning about the nature of resources and their availability to the activities in the enterprise.

The *Business Ontologies* category has a primary focus on ontologies to support reasoning in industrial environments, especially with regard to supply-chain management. For instance, it extends manufacturing requirements planning (MRP) to include logistics, distribution, and 'concurrent engineering' (that is, issues of coordinating engineering design). The category includes four subgroups of ontologies:

- **Organization**, here considered as a set of constraints (rules and procedures) on the activities performed by agents.
- **Product and Requirements**, which in addition to the generic process of requirements management also considers communication, traceability, completeness, consistency, document creation, and managing change.
- **ISO9000 Quality**, which provides the KR structures to construct models for quality management, and ultimately for ISO9000 compliance.
- **Activity-Based Costing**, which spans several general representation areas to manage costs in either manufacturing or service enterprises.

The latest area of research appears to be the issue of how to determine the validity and origin of information (or knowledge) on the Web, with applicability to the corporate intranet, as well.

Four levels of Knowledge Provenance are identified: *Static, Dynamic, Uncertain*, and *Judgemental*. So far, only the first has been addressed (see *www.eil.utoronto.ca/km/papers/ fox-kp1.pdf*), with a proposed ontology, semantics, and implementation using RDFS.

Ontologies for the Telecommunication Industry

The Distributed Management Task Force, Inc. (DMTF, see *www.dmtf.org*) assumes the task of leading the development of management standards for distributed desktop, network, enterprise, and Internet environments. It coordinates research and proposes industry standards for a large group of major corporations in telecommunications and IT.

Of the three focus DMTF areas, the third is the ontology:

- **DEN** (*Directory-Enabled Networking*) provides an expanded use of directories for data about users, applications, management services, network resources, and their relationships. The goal is integration of management information enabling policy-based management.
- **WBEM** (*Web-Based Enterprise Management*) is an open forum for the ongoing development of technologies and standards to provide customers with the ability to manage all systems using a common standard, regardless of instrumentation type.
- **CIM** (*Common Information Model*) is a conceptual information model for describing management that is not bound to a particular implementation. It enables interchange of management information between management systems and applications.

CIM represents a common and consistent method of describing all management information that has been agreed and followed through with implementation. It supports either 'agent to manager' or 'manager to manager' communications – for Distributed System Management.

The CIM Specification describes the language, naming, Meta Schema (the formal definition of the model), and mapping techniques to other management models. The CIM Schema provides the actual layered model descriptions:

- **Core Schema** information model that captures notions applicable to all areas of management.
- **Common Schemas** that capture notions common to particular management areas, but independent of technology or implementation.
- **Extension Schemas** represent organizational or vendor-specific extensions.

A defining characteristic of the DMTF initiative is that its focus is implementing distributed and intelligent (agent-driven) Web Services for an industry.

Business Process Ontology

The *Business Process Management Initiative* (BPMI, see *www.bpmi.org*) stresses the importance of integrating entire business processes, rather than simply integrating data or applications. Corporate motivation to integrate the people, applications, and procedures necessary to do business in a BPM model, is streamlined operations and a noticeable return on IT investments. Standardized BPM facilitates interoperability and simplifies electronic business-to-business transactions.

BPMI defines open specifications, such as the *Business Process Modeling Language* (BPML, published as a first draft in 2001), and the *Business Process Query Language* (BPQL, published in 2003). These specifications enable standards-based management of business processes with future *Business Process Management Systems* (BPMS, under development). The analogy suggested is with how the generic SQL (Structured Query Language) specifications enabled standards-based management of business data with off-the-shelf Database Management Systems (DBMS).

BPML provides the underpinning to define a business process ontology. It is a meta-language designed to model business processes, in a similar way to how XML can model the business data. The result is an abstracted execution model for collaborative and transactional business processes based on the concept of a transactional finite-state machine.

Knowledge Representation

Related to ontologies and the issue of modeling, and central to communicating, is the study of *Knowledge Representation*. The main effort of KR research is providing theories and systems for expressing structured knowledge, and for accessing and reasoning with it in a principled way.

Both RDF and DLG (described in Chapter 5) are KR languages, and RDF is at the very core of the Semantic Web. Other languages for KR are also used and promoted in different modeling contexts. All have their fervent advocates. They are not always that different, instead just emphasizing different aspects of expressing the logic. The KR structures they represent can be transformed to and from RDF, or serialized into XML – typically with a minimum of trouble.

The following overview is intended mainly to familiarize the reader with other formal KR languages, and to make a few observations about their applicability in the SW context.

Conceptual Graphs

Conceptual graphs (CGs, see *www.cs.uah.edu/~delugach/CG/*) define a system of logic based on semantic networks of artificial intelligence. They were originally conceived as existential graphs by *Charles Sanders Peirce* (1839–1914). CGs can serve as an intermediate language for translating computer-oriented formalisms to and from natural languages.

The graphic representation serves as a readable yet formal design and specification language. CG notation may be 'web-ized' for sweb use (mainly a syntax modification to allow URI referencing). A detailed sweb comparison is on the W3C site (at *www.w3.org/DesignIssues/CG.html*). One notable restriction compared with RDF is that a relation in CG cannot express something about another relation. Concepts and types in CG are disjoint from relationships and relationship types.

CGs are used in two logically equivalent forms:

- **Display Form** (DF) is essentially a DLG diagram. Concepts are rectangles, relations are ovals linked to the concepts.
- **Linear Form** (LF) is a compact serialized notation that line by line expresses the relations between concept and relation.

CG LF is suited for both human and machine parsing.

Conceptual Graph

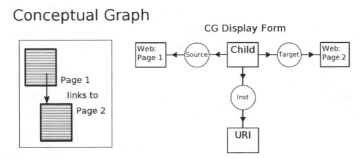

Figure 6.1. A CG DF representation of a Web page that links to another page. A slightly expanded model that specifies the concept of URI is used to provide more syntactic detail in the LF version in the text

Re-using the simple DLG example from Chapter 5, a hyperlink relationship between just two pages in CG DF is shown in Figure 6.1.

In CG LF, the equivalent relationship can be expressed in text in a variety of equivalent styles. Declaring it as a 'child' relationship, we identify three concepts:

```
[Child]-
(Source) ->[Web: Page1]
(Target) ->[Web: Page2]
(Inst)-> [URI].
```

CG supports nesting of more complex relations – in DF by concept boxes enclosing subordinate complete relationships, in LF by deferring the closing concept bracket until all subordinate relationships have been specified.

The more compact (and 'official') Conceptual Graph Interchange Form (CGIF) has a simpler syntax when multiple concepts are involved:

```
[Child *x] (Source ?x [Web 'Page1']) (Target ?x [Web 'Page2'])
(Inst ?x [Link])
```

CGIF is mainly used for contexts when machines communicate with machines; it is hard to read for humans. Transformation between CGIF and the widely supported Knowledge Interchange Format (KIF) enables a CG-based machine to communicate with machines that use other internal representations than CG.

CGs have been implemented in a variety of projects for information retrieval, database design, expert systems, and natural language processing. A selection of tools to develop and maintain CG resources are referenced from the CG Web site mentioned earlier.

Promoting Topic Maps

Topicmaps (*www.topicmaps.org*) is an independent consortium developing the applicability of the 'Topic Maps Paradigm' to the Web, leveraging the XML family of specifications as required. This work includes the development of an XML grammar (XML Topic Maps, or XTM) for interchanging Web-based Topic Maps. The 'Topic Maps' (TM) concept was fully described in the ISO 13250 standard published in 2000 (see *www.y12.doe.gov/sgml/sc34/ document/0129.pdf*).

So what is a topic map?

The short answer is that it is a kind of index, or information overlay, which can be constructed separate from a set of resources, to identify instances of topics and relationships within the set.

The longer and wordier answer is that it is document (or several) that uses a standardized notation for interchangeably representing information about the structure of information resources used to define particular topics. The structural information conveyed by topic maps thus includes:

- **Occurrences**, or groupings of addressable information objects around topics (typically URI-addressed resources);
- **Associations**, or relationships between topics, qualified by roles.

Information objects can also *have properties*, as well as assigned values for those properties. Such properties are called *facet types*, and can represent different views.

Technically, any topic map defines a multidimensional topic space, where locations are topics and the distances between topics are visualized as the number of intervening topics along the path that represents the relationship. Therefore, you can visually graph at least subsets or particular views (projections) of the map, which is similar to the DLG approach to RDF structures described in Chapter 5.

The Case for Interoperability

The defining goals for TM were inherently different from that of the RDF effort, instead being mainly to support high-level indexing to make information findable across resources. A considered view is that topic maps and RDF representations are in fact complementary approaches:

- RDF needs TM in order to make scalable management of knowledge from disparate sources simple, practical, and predictable.
- TM needs RDF in order to have a popular, widely-accepted basis upon which to describe exactly what a topic map means, in a fashion that can be immediately processed by a significant number of existing and well-funded tools.

These motivations are from a longer illustrative discussion by Steven R. Newcomb in 2001 (see *xml.coverpages.org/RDF-TopicMaps-LateLazyVersusEarlyPreemptiveReification.html*).

The Essential Differences

One major difference easy to spot is that RDF is predicative; it can *ad hoc* define 'verbs' in the role of direct relationships. In topic maps, however, connections can be made only between 'events' in this context – an extra state node needs to be defined to take the place of an RDF verb relationship. In some contexts, this extra step is an advantage, as it simplifies the process of amalgamating facts and opinions when the things anyone will want to express a new fact or opinion about are not known in advance.

Although topic maps are in the formal ISO standard defined in terms of SGML notation, it is noted that XML, being an SGML subset, may also be employed. This variant is exactly what XTM grammar for TM-RDF interchange is based on.

Lars Marius Garshol of *Ontopia* (*www.ontopia.net*) has published papers outlining the essential differences. 'Living with topic maps and RDF' (*www.ontopia.net/topicmaps/materials/tmrdf.html*) looks at ways to make it easier for users to live in a world were both technologies are used. He looks at how to convert information back and forth between the two technologies, how to convert schema information, and how to do queries across both information representations.

The existence of two independently formalized standards, TM and RDF, is largely an unfortunate accident of history. The two developer communities did not really communicate until it was too late to back out of commitment to disparate standards. Therefore, given two established and entrenched, yet complementary tools, the sensible thing is to devise ways to make them work together to leverage each other's strengths.

Both standards can be viewed in the context of three layers: syntaxes, data models (the standards), and constraints (proposed TMCL for TM, RDF Schema and OWL for RDF). Interoperability makes sense for at least the latter two, allowing for the fact that at present only a lower-level schema language is used to formulate constraints: OSL (*Ontopia Schema Language*).

The three most fundamental TM-RDF differences are identified:

- The ways in which URIs are used to identify things (two ways in TM, one way in RDF).
- The distinction between three kinds of assertions in TM, but only one in RDF.
- The different approaches taken to *reification* and *qualification* of assertions.

These three problems make it technically very difficult to merge topic maps and RDF into a single technology. TM also turn out to give higher-level representations than RDF, in the sense that a topic map contains more information about itself than does an RDF model.

Mapping Strategies

After considerable comparisons and analysis, the previously cited *Ontopia* paper offers practical mechanisms for moving data between RDF and topic maps, providing at least some level of interoperability exists between the two in terms of their data models (and also underlying syntax).

It is also proven possible to map between any RDF Schema document and a paired TM and OSL representation. The assumption is that once TMCL is defined, this mapping becomes easier.

Finally, concerning top-level conversions, the paper concludes that converting the constraints in an OWL ontology to the future TMCL should not be too difficult, provided an RDF-TM mapping is provided. The best outcome might then be if TMCL is created only to support constraints, and OWL is reused for the ontology aspects which might otherwise have become part of TMCL.

Query languages are also part of the standards families. ISO has defined work goals for TMQL (ISO 18048), and interoperability at the QL level provides an alternative approach to mapping between the families. The difficulties here are not so much technical as political, as two rather different standards-defining agencies are involved, ISO and W3C.

Description Logics

Description Logics (DL main resource site at *dl.kr.org*) are considered an important Knowledge Representation formalism to unify and give a logical basis to the traditions of *Frame-based systems, Semantic Networks, Object-Oriented representations, Semantic data models*, and *Type systems*.

Typically given declarative semantics, these varied traditions may all be seen as sub-languages of predicate logic. DL seeks to unify KR approaches and give a solid underpinning for the design and implementation of KR systems.

- The basic building blocks are *concepts, roles*, and *individuals*. Concepts describe the common properties of a collection of individuals. Roles are interpreted as binary relations between objects.
- Each description logic also defines a number of language constructs (such as *intersection, union, role quantification*, and others) that can be used to define new concepts and roles.
- The main reasoning tasks are *classification* and *satisfiability, subsumption* (or the 'is a' relation) and *instance checking*.

A whole family of knowledge representation systems have been built using these languages. For most of these systems, the complexity results for the main reasoning tasks are known.

The Description Logic Handbook, published in January 2003, provided a unified resource for learning about the subject and its implementation. In its applications part, descriptions include examples from the fields of conceptual modeling, software engineering, medical informatics, digital libraries, natural language processing, and database description logic.

Lists of systems, applications, and research groups are provided with Web links on the site. Another resource is the *'Patrick Lambrix DL page' (www.ida.liu.se/labs/iislab/people/patla/DL/)*.

If a DL-based KR system were nothing more than an inference engine for a particular description logic, there might be little point in including DL in this sweb overview. However, a KR system must also provide other services, including adaptive presentation and justification for performed inference. It must also allow easy access to application programs, external agent software, and more. DL solutions to these services run parallel to sweb development.

The main area of commonality probably lies in ontology development, where DL brings much of the classic philosophy to the field, such as the epistemic aspects of individual knowledge.

7

Organizations and Projects

Some indication of where to look for ongoing development in core areas of the Semantic Web seems prudent, given that present-day development is both rapid and somewhat unpredictable. During the course of writing this book, several tracked projects either terminated abruptly or were assimilated into other projects with perhaps variant aims.

The sudden lapse of a particular project, especially at universities, is sometimes due to obscure funding decisions, or simply that the involved researchers moved on. The same or similar projects may therefore reappear in other contexts or with other names. Sometimes the overall direction of an effort changes, making previous research goals inappropriate. A shift in specifications or adoption of a new standard can also marginalize a particular project.

European projects in particular have episodic lifespans. Often funded under one or another EU initiative, typically they will run for one to three years, leaving behind a published site but at first glance seemingly little else. Therefore, it is difficult to track conceptual continuity and the degree of immediate relevance to sweb technology.

Projects and initiatives may come and go, but the major players do remain the same, or largely so. The best Web sites associated with the corresponding players or community groupings are updated with relevant links to the latest news within their interest area. Although some sites undergo drastic revisions in the face of project changes, it is usually possible to drill down through the new structure, or use qualified searches to pick up lost threads. The main thing is recognizing where the hotbeds of activity are, and thus being able to localize the search to something considerably less than the eight billion plus indexed Web pages on search engines.

Chapter 7 at a Glance

This chapter is an overview of the organizations that are active in defining specifications and protocols that are central to the Semantic Web. *Major Players* introduces the organizations at the hub of Semantic Web activity.

- *W3C* and its related project areas are often the initiative takers for any development direction, and the W3C retains the overall control of which proposals become recommended practices.

The Semantic Web: Crafting Infrastructure for Agency   Bo Leuf
© 2006 John Wiley & Sons, Ltd

- *Semantic Web Communities* describes the interest area groupings that often coordinate corporate and institutional efforts.
- *Dublin Core Metadata Initiative* defines an important area in organizing metadata that is greatly influencing the large data producers and repositories, such as governments and libraries.
- *DARPA* manages and directs selected basic and applied research and development projects for DoD, often within sweb-related areas.
- *EU-Chartered Initiatives* explains something of how European research is conducted, and uses the example of the OntoWeb Network and Knowledge Web to describe how European-centric consortia may explore and promote sweb technologies.
- *AIC at SRI* provides an overview of the *Artificial Intelligence Center* at SRI (formerly *Stanford Research Institute*) and its research into core sweb technologies.
- *Bioscience Communities* gives examples of ontology-adopting communities that need sweb technologies to cope with the vast and complex amounts of data that they produce and want to share.

Major Players

Different aspects of technologies relating to the Semantic Web are concentrated to a number of large coordinating bodies. A number of other organizations are involved in researching and developing various areas applicable to The Semantic Web – sometimes overtly, sometimes in the context of specific applications.

It is useful to summarize identified major players, along with project overviews.

W3C

In many eyes the actual 'home' of the Web, the *World Wide Web Consortium* (W3C, see *www.w3.org*) is perhaps the first place to go when researching Web development, in particular the design of languages, protocols, and descriptive models. It coordinates, recommends, and takes the initiative for many of the relevant efforts.

The W3C was the first context in which the Semantic Web concept was promoted. Historically, the Semantic Web is the successor to the *W3C Metadata* initiative to enhance the Web. Under W3C auspices, we find numerous framework projects that in one way or another relate to the Semantic Web and associated technologies. Even when the cutting edge of development moves on to other projects and locations, the defining discussions and documents remain. Needless to say, the W3C site has extensive collections of links to relevant resources elsewhere.

The general prototyping policy of W3C has always been: 'Our approach is Live Early Adoption and Demonstration **(LEAD)** – using these tools in our own work'.

Bit 7.1   Live early adoption leads to rapid development of useful tools

The principle is to develop and test the new tools by from the start using them in the kinds of real-world environments in which the final product will be used.

Like in other environments, from Open Source to large service providers, the 'eating your own dogfood' principle (as it is often known) can be a vital force in the rapid development of viable functionality. Nothing motivates a developer more to get things right than the requirement to use the ambitious new functionality on a daily basis, and to experience personally the failings of a design not quite ready for prime time.

Other descriptions of the development policy at W3C include: '80/20: 80% of functionality 20% of complexity' and 'KISS: Keep It Simple and Stupid'.

The W3C runs much of its own organizational flows using established and emerging sweb technologies. One presentation gives an interesting view of why it is better to use Semantic Web approaches to manage information and workflow than a single MIS SQL database (see *www.w3.org/Talks/2002/02/27-tpsw/*). One of the slide items makes a significant point:

> *[To have the] Goal of making it cost-effective for people to record their knowledge for machine consumption.*

A common problem facing MIS tasked with finding and entering the data is answering the question: *Where is the truth*? Is it in the database, in the Web page, or in the e-mail archives?

Sometimes a single source provides multiple views, and sometimes consistency checks must be applied to multiple sources. Ideally, all can be tracked by agent software if only it is Web-accessible in machine-readable formats. Sweb natively provides this kind of multi-view access.

Bit 7.2   Sweb models must account for a plurality of 'truth' assertions

The truth is not only 'out there' but also a value defined by the context where the information is required – it is often relative, not absolute. Therefore, assertions commonly require qualifications, sometimes only implied, about validity contexts.

SWAD

Semantic Web Activity: Advanced Development (*www.w3.org/2000/01/sw/*) was an early W3C core effort – a catch-all for early prototyping of sweb functionality. It devoted resources to the creation and distribution of core components intended to form the basis for the Semantic Web, just as earlier W3C-developed components formed the Web. The aim was to facilitate the deployment and future standards work associated with the Semantic Web.

Often weaving together several different areas of research and implementation, SWAD initiatives typically involved collaboration with a large number of researchers and industrial partners, and the stimulation of complementary areas of development.

In particular, the *SWAD-Europe* project (*www.w3.org/2001/sw/Europe/*), which ran from May 2002 to October 2004, supported the W3C Semantic Web initiative in Europe. With EU funding, it provided targeted research, demonstrations, and outreach to ensure Semantic Web technologies move into the mainstream of networked computing. An overview report is at the W3C site (*www. w3. org/2001/sw/Europe/200409/ercim-news/*).

The motivation for SWAD can be summed up in two quotes:

Now, miraculously, we have the Web. For the documents in our lives, everything is simple and smooth. But for data, we are still pre-Web.
Tim Berners-Lee, in 'Business Model for the Semantic Web'

The bane of my existence is doing things that I know the computer could do for me.
Dan Connolly, in 'The XML Revolution'

Some project examples can illustrate the span of involvement. Several of these projects are described in Chapter 9.

- The *Annotea* project was an early and seminal initiative concerning Web Annotation. As W3C test-bed browser, the *Amaya* client gained native annotation support. Related efforts in collaboration meshed in the goal of creating a 'web of trust' for annotations.
- *SKOS* is an open collaboration developing specifications and standards to support the use of knowledge organization systems (KOS) on the Semantic Web. It defines RDF vocabulary for describing thesauri, glossaries, taxonomies, and terminologies. The project was a product of the outreach and community-building efforts of SWAD-Europe.
- The *Zakim* (facilitating) and *RRSAgent* (record keeping) teleconference IRC agents are tools that provide presence for real-time teleconference audio in the meeting IRC co-channel and record the progress of the meeting. They jointly capture meeting data and make them available in RDF/XML for other tools, such as for workflow analysis.
- Work with the MIT/LCS *Project Oxygen* collaboration technologies enabled the formation of spontaneous collaborative regions that provide support for recording, archiving, and linking fragments of meeting records to issues, summaries, keywords, and annotations.
- *Semantic Blogging* and *Semantic Portals* also arose in SWAD-Europe activities. '*Semblogging*' entails the use of sweb technologies to augment the *weblog* paradigm, and also is applied to the domain of bibliography management

Monitoring SWAD and post-SWAD activity can provide some indication of potential application areas before broader support brings them to public awareness. Such areas include:

- Descriptive metadata for multimedia content, integrating photos, music, documents, e-mail, and more.
- Access control rules, proofs, and logic to express authority and validity states coupled to specific presented documents.
- Tools for authoring, extraction, visualization, and inference, to extract data from legacy sources, to adapt existing authoring tools to produce RDF, and to visualize the results.

P3P

The *Platform for Privacy Preferences Project* (P3P, see *www.w3.org/P3P/*) is an emerging industry standard to support simple, automated solutions for users to gain more control over the use of personal information on Web sites they visit. The current basis of P3P is RDF Schema and protocol specification that provides a mechanism that allows users to be informed of the major aspects of a Web site's privacy policies and practices.

Site administrators use online or offline tools to answer a standardized set of multiple-choice questions to generate a P3P-compliant declaration of the site's policy and publish the resulting document on the site. From this machine-readable document users, or software operating on their behalf, can then make informed decisions about whether to proceed with requested or intended transactions on the site, or warn when site policies conflict with user preferences.

The P3P 1.0 specification has been a W3C recommendation since April 2002, but compliance is voluntary on the part of any Web site. Although still relatively rare in site deployment, modern browsers do provide varying degrees of support. For example, viewing cookies in verbose mode before accepting them from a site might display an extract of any found P3P policy on the site.

Future enhancements to the protocol could allow software to negotiate automatically for a customized privacy policy, typically based on user-defined rules, and come to an agreement with the site which will be the basis for any subsequent release of information. Future P3P specifications will support digital-certificate and digital-signature capabilities, which will enable a P3P statement to be digitally signed to prove that it is authentic, and perhaps by extension form the basis for a legally binding contract with the user to govern any transactions of user information.

Bit 7.3   Privacy issues remain a thorny problem on the Web

It seems likely that the user/client-negotiated model of rules-based acceptance may be the only reasonable way to allow the use of individual information by sites. Such 'filtering' is currently performed *ad hoc* by user actions or configured cookie filters.

Work on Privacy is being managed as part of W3C's Technology and Society domain. While it may seem peripheral in current implementations, some of the P3P concepts have fundamental consequences for potential Web Services and agent capabilities, especially ones managing delegated user authority. P3P also interacts with issues of trust and reputation tracking.

WAI

The *Web Accessibility Initiative* (WAI, see *www.w3.org/WAI/*) may also at first seem peripheral to the Semantic Web, as it apparently addresses only human readability. It is, however, important in light of one of the core Web aims. Guidelines are mature and expected to reach v2 in 2005.

The power of the Web is in its universality. Access by everyone regardless of disability is an essential aspect.   Tim Berners-Lee

One might suggest that software agents actually constitute a growing group of 'semantically challenged' (that is, disabled) users, and therefore some WAI recommendations for human-readable content are just as motivated to make Web content easier to parse by machines.

The growth of the Semantic Web does raise the issue of a *semiotic* level (that is, dealing with signs and symbols to represent concepts) to present and manipulate the underlying semantic structures. Sweb and WAI communities may soon have shared concerns.

In coordination with organizations around the world, WAI pursues accessibility of the Web through five primary areas of work: *technology, guidelines, tools, education and outreach*, and *research and development*.

Semantic Web Communities

A number of different and partially overlapping communities have grown up around Semantic Web activities. In some cases, such as for digital archives, the activity in question might be a side-issue to the main concern; in others, it is the primary mission.

Community efforts do more than simply organize interested parties for discussions and conferences. Quite often, they themselves form an early user base to test the proto-type systems that are being developed. They may coordinate disparate corporate efforts. They may also hammer out by consensus the specifications and standards proposals that will be considered by the more formal bodies – for example, the working groups and the W3C.

The Semantic Web Community Portal

To better coordinate efforts and facilitate member and interested party contact at all levels of development, a portal site was deployed (see *www.semanticweb.org*). The site collects and lists different technology approaches, explains them, and functions as a forum for people interested in the Semantic Web, under the motto 'together towards a Web of knowledge'.

It is the stated view of the *'SemWeb'* community that we already have the technology available for realizing the Semantic Web; we know how to build terminologies and how to use metadata. The vision thus depends on agreeing on common global standards to be used and extended on the Web.

Several organizations and initiatives conveniently use sub-domains under *semanticweb.org*, sometimes in addition to their own proper domains. Some examples that illustrate the usage follow:

- *Business.semanticweb.org* discusses sweb business models.
- *Challenges.semanticweb.org* collects sweb-related challenges for Open-Source Software.
- *P2P.semanticweb.org* has a focus on developing a peer-to-peer sweb infrastructure.
- *Ontobroker.semanticweb.org* promotes the creation and exploitation of rich semantic structures for machine-supported access to and automated processing of distributed, explicit, and implicit knowledge.
- *Triple.semanticweb.org* describes an RDF query, inference, and transformation language for the Semantic Web.

Other sub-domains include *OpenCyc* (ontology), *DAML* (agent markup), *Protege* (toolset), *COHSE* (navigation), *HealthCyberMap* (mapping health resources), *NetworkInference* (application interaction), *SWTA* (enterprise deployment), *OntoWeb* (portal), *OntoKnowledge*

(KMS), and so on – either complementing the 'official' sites for the corresponding technology, or forwarding to it.

In late 2004, the portal site was to 'relaunch' itself as a SWAD-inspired *Semantic Portal*, based on RDF and community maintained. Presumably, previous sub-domain site content will by then have migrated back to the respective organization-maintained sites.

A few of the 'community collections' are described in more detail in the following sections.

Semantic Web Science Foundation

The *Semantic Web Science Foundation* (SWSF, see *swsf.semanticweb.org*) has the stated mission to help to bring the Web to its full potential. Its specific focus is to promote and organize research and education on Semantic Web issues. It is organized as a non-profit foundation and runs a number of sweb initiatives:

- *Organization of the International Semantic Web Conference* (ISWC, *iswc.semanticweb. org*), a major international forum at which research on all aspects of the Semantic Web is presented. The first (ISWC'02) was held in 2002, preceded in 2001 by a working symposium (SWWS'01).
- Collecting educational material in co-operation with *Ontoweb*.
- Maintaining research journals of high quality in co-operation with the *Electronic Transactions on Artificial Intelligence* (ETAI, *www.etaij.org*).
- Support the organization of workshops and tutorials around the Semantic Web.

The main area of activity appears to be ISWC, providing a meeting ground for experts in the field.

Dublin Core Metadata Initiative

The Dublin Core Metadata Initiative (DCMI, *www.dublincore.org*) is an open forum dedicated to promoting the widespread adoption of interoperable metadata standards that support a broad range of purposes and business models. It is a 'community' effort that attracts professionals from around the world to focus on the technologies, standards, and other issues related to metadata and semantics on the Web.

Although an international organization, the name counter-intuitively has nothing to do with Ireland, but instead derives from the fact that the original 1995 metadata workshop for the initiative was held in Dublin, Ohio, U.S.A. This event stemmed from ideas at the *2nd International World Wide Web Conference*, October 1994, in Chicago, and concerns the difficulty in finding resources when working on semantics and the Web. In the later workshop it was determined that a core set of semantics for Web-based resources would be extremely useful for categorizing the Web for easier search and retrieval.

DCMI develops specialized metadata vocabularies for describing resources that enable more intelligent information discovery systems. Its more prominent activities include consensus-driven working groups, global workshops, conferences, standards liaison, and educational efforts. It actively promotes a widespread acceptance of metadata standards and practices.

Bit 7.4   URI-declared Dublin Core vocabularies exist for many useful contexts

The advantage of referencing a DC schema by its URI is that the application can leverage well-defined properties (using common terms) in custom schemas.

The active Working Group names are descriptive of the current span of the work:

- **Accessibility**, to propose adoption of *AccessForAll* information model for user preferences.
- **Agents**, to develop functional requirements for describing agents, and a recommended agent data set for linking records and data elements.
- **Architecture**, to develop a model, strategy, and roadmap for the practical deployment of DCM using mainstream Web technologies (XML/RDF/XHTML).
- **Citation**, to agree on mechanisms for generating a DCM-compliant set of properties to record citation information for bibliographic resources.
- **Collection Description**, to develop a DC-based application profile for collection description and explore possible mappings.
- **Date**, to foster the adoption of standards and practices that will enhance the interoperability of date and time information conveyed in DCM.
- **DC Kernel/ERC**, to explore the ultra-simple ERC 'kernel' approach to metadata and develop an XML representation of kernel elements.
- **Education**, to develop proposals for the use of DCM in the description of educational and training resources, and for advancing the goal of metadata interoperability.
- **Environment**, to implement DCM in the environmental domain.
- **Global Corporate**, to promote the use of DCM by enterprise, and to develop best practices, case studies, and examples.
- **Government**, to identify commonalities in current government metadata implementations, and make recommendations for future DCM application within and between government agencies and IGOs.
- **Libraries**, to foster increased operability between DC and library metadata by identifying issues and solutions, and explore the need for cross-domain namespaces for non-DC elements and qualifiers used by libraries.
- **Localization and Internationalization**, to share information and knowledge gained from experiences in local or domain-specific applications of DC with global community, and to exchange ideas on adapting DC to the local or domain-specific applications.
- **Persistent Identifier**, to recommend application of digital identifiers that are globally unique and that persist over time.
- **Preservation**, to inventory existing preservation metadata schema, and investigate the need for domain-specific ones.
- **Registry**, to evolve a distributed metadata registry of authoritative information regarding DCMI vocabulary and the relationships between DCM terms.
- **Standards**, to monitor standardization efforts and the progress of DC as it makes its way through the different standards processes.
- **Tools**, to support planning and discussion for workshops devoted to developing tools and applications based on DCM.
- **User Documentation**, to discuss user guidance issues and maintain the Web-published 'Using Dublin Core' guide.

Additional *Special Interest Groups* address specific problems or areas, such as accessibility, global corporate, environment, localization, and standards. DCMI also maintains a historical record of deactivated WGs and SIGs (see list at *dublincore.org/groups/*).

The governments of Australia, Canada, Denmark, Finland, Ireland, and the UK have adopted Dublin Core, in whole or in parts, Tools and metadata templates are available from many countries and in many languages.

Whereas many other efforts involve implementing sweb structures from the technology on the Web, the DCMI efforts are complementary. They deal with transforming and bringing existing data and metadata from large national, academic, or library repositories into machine-accessible formats on the Web.

A projects listing (at *www.dublincore.org/projects/*) indicates the implementations and prototype systems to date. Dublin Core has attracted broad international and interdisciplinary support, as the range of projects show.

Core Resources

Central to DC are the collected *DCMI Recommendations* that signify stable specifications supported for adoption by the Dublin Core community. The core Semantic Recommendations comprise a series of periodically updated documents for 'one-stop up-to-date' information:

- 'DCMI Metadata Terms', the main reference document, periodically regenerated;
- 'DCMI Type Vocabulary', a subset of the previous as a separate document;
- 'DCMI Metadata Terms: a complete historical record', for historical interest or for interpreting legacy metadata.

In addition, a series of other documents discuss implementation guidelines, RDF/XML expression, namespace policy, and other specialized developer topics.

Of lesser significance are the *Proposed Recommendations*, which signify that the document is gathering growing support for adoption by the Dublin Core community and that the specifications are close to stable. *Working Drafts* collects documents that might lead to later proposed recommendations. *Process Documents* and *Recommended Resources* are also useful to the DCMI-involved developer.

A *Notes* category may collect individual documents that have no official endorsement by the DCMI, but form a basis for further discussion on various issues.

Tools and Software

Another major function of the DCMI is to provide a sorted listing of tools and software that are recommended or useful to members who work in the field. The general categories are:

- Utilities of a general nature for metadata applications;
- Creating metadata (predefined templates), the naturally largest group, which also gives a fair indication of the primary application areas and users (libraries, universities, medical metadata for health care, and government databases);

- Tools for the creation/change of templates, which simplify the later generation or collection of metadata;
- Automatic extraction/gathering of metadata, which is interesting from the perspective of using existing published material on the Web;
- Automatic production of metadata, usually based on populating templates provided for a particular purpose;
- Conversion between metadata formats, which may be required when leveraging different resources;
- Integrated (tool) environments, to design, create, and manage metadata;
- Commercially available software, which has been found useful.

DARPA

The Defense Advanced Research Projects Agency (DARPA, *www.darpa.mil*) is the central research and development organization for the *U.S. Department of Defense* (DoD). It manages and directs selected basic and applied research and development projects for DoD.

DARPA's policy is to pursue research and technology where risk and pay-off are both very high, and where success may provide dramatic advances for traditional military roles and missions. In that capacity, DARPA is often involved in cutting-edge sweb-related technology, allocating and funding research projects, just as it earlier figured prominently in the then-emerging technologies of the early Internet and Web.

DARPA formally organizes projects under eight Program Offices:

- *Advanced Technology Office* (ATO), which includes active networks and tactical mobile robotics;
- *Defense Sciences Office* (DSO), which includes an assortment of defence technology projects;
- *Information Awareness Office* (IAO), which deals with various information gathering and assessment technologies;
- *Information Processing Technology Office* (IPTO), which manages research in mobile robotic agents, next-generation Internet, and a variety of embedded agent systems;
- *Information Exploitation Office* (IXO), which targets areas such as battle management, sensor and control, and identification surveillance systems;
- *Microsystems Technology Office* (MTO), which researches core electronics, photonics, and micro-electromechanical systems;
- *Special Projects Office* (SPO), which is involved in large-scale military systems, such as missile defense;
- *Tactical Technology Office* (TTO), which explores future (unmanned) combat systems architectures for space operations.

Several of these offices (notably IPTO and ATO) are thus likely to be involved with emerging sweb technology, directly or indirectly, typically by awarding and funding university research projects.

Forward links from DARPA to funded projects are hard to find. Instead, respective projects may mention a DARPA connection somewhere in their overviews. A 'legacy' sub-web (*www.darpa.mil/body/legacy/index.html*) does however showcase new products made available as a result of DARPA investment, where early DARPA demonstrations and prototype programs shifted to a military customer for additional development and eventual purchase.

Bit 7.5   To a great extent, DARPA projects have formed the Internet as we know it

Except for specific projects classified under military security, most projects (pursued by universities, for example) resulted in open specifications and products, made freely available for the greater good of the global information community.

Some project technologies, such as DAML and the work leading up to OWL (see Web Ontology Languages in Chapter 6), show their DARPA origins more clearly than others.

A European counterbalance does exist, however, to the many decades of advanced IT research sponsorship and direction by DARPA and the U.S. DoD. The EU has political directives that mandate concerted efforts to bring together researchers and industrials, promote interdisciplinary work, and strengthen the European influence on technology development, especially in IT.

EU-Chartered Initiatives

The reasons for pushing strong European initiatives is not just one of 'Euro-pride'. It has long been recognized that there are considerable and important differences in infrastructure, legislation, politics/policy, economy, and attitude. A Web, and by extension a Sweb implementation based only or even mostly on the perceptions of stakeholders and developers in North America, would simply not be acceptable or viable in Europe.

As mentioned at the start of this chapter, many European IT and Web/Sweb projects have episodic lifetimes covering only a few years, being defined as chartered projects under the umbrella of one EU-initiative framework or another (see *europa.eu.int/comm/research/*).

In general, EU research activities are structured around consecutive four-year, so-called Framework Programmes (FP or FWP) that aim to encourage broad and innovative research in and across many fields. The current and sixth FP (FP6, for the period 2002/2003 to 2006) includes an Information Society Technologies (1ST, *www.cordis.lu/ist/*) 'priority'.

- Priorities for FP research, technology development, and demonstration activities are identified on the basis of a set of common criteria reflecting the major concerns of increasing industrial competitiveness and the quality of life for European citizens in a global information society.

Consistently identifying and following sweb-related technologies developed under EU auspices proves challenging because the overview structure of published information about these projects is not organized with this view in mind.

As an example of identifying sweb-relevant development, consider the following selected FP6 projects under the 1ST category 'semantic-based knowledge systems':

- *Acemedia*: Integrating knowledge, semantics, and content for user-centered intelligent media services.

- *Agentlink II*: A coordination network for agent-based computing.
- *AIM@SHAPE*: Advanced and Innovative Models and Tools for the development of Semantic-based systems for Handling, Acquiring, and Processing Knowledge Embedded in multidimensional digital objects, (*sic*)
- *Alvis*: Superpeer semantic search engine.
- *ASPIC*: Argumentation Service Platform with Integrated Components for logical argumentation in reasoning, decision-making, learning and communication.
- *DIP*: Data, Information, and Process integration with Semantic Web Services.
- *KB20* and *Knowledge Web*: Realizing the semantic web (*sic*).
- *NEWS*: News Engine Web Services to develop News Intelligence Technology for the Semantic Web.
- *REWERSE*: Reasoning on the Web with Rules and Semantics.
- *SEKT*: Semantically-Enable Knowledge Technologies.
- *SIMAC*: Semantic Interaction with Music Audio Contents.

Some, like KB20 and Knowledge Web (see later), are recast continuations of FP5 projects and look likely to have an independent future Web presence in the context of a growing European KM community of active users. The actual direct relevancy for sweb technology as described in this book may not be that large, however. Others may be more relevant, as the acronym expansion suggests, yet exist only for the FP-charter years and not produce any direct follow-up.

But this selection is from only one of several categories where individual projects might have more or less, and sometimes high, sweb relevancy. The full FP6 1ST category list (encompassing almost 400 projects) is here annotated with an indication of potentially relevant areas.

- *Advanced displays*: flexible displays, flexible large-area displays, holographic displays, rendering technology for virtual environments, and arbitrary-shape displays.
- *Applications and services for the mobile user and worker*: context-aware coordination, Web-based and mobile solutions for collaborative work environment, using Ambient Intelligence service Networks, location-based services, harmonized services over heterogeneous mobile, IN and WLAN infrastructures, Mobile Life (ubiquitous new mobile services), secure business applications in ubiquitous environments, wearable computing, and others.
- *Broadband for all*, several related to implementing a fixed wireless infrastructure and enhanced 'anywhere' access.
- *Cognitive systems*: any or all could be highly relevant.
- *Cross-media content for leisure and entertainment*: spans from new forms of collaborative Web-publishing to various interactive activities.
- *eHealth*: several concern sweb-enhancements in diagnostic methods, health care, research, and patient quality-of-life.
- *eInclusion*: several concern ambient intelligence systems in various contexts, disability compensation, partial overlap with eHealth.
- *Embedded systems*: several relevant as enabling technology for intelligent devices.
- *eSafety for road and air transport*: some have possible sweb relevancy.

- *Future and Emerging Technologies*: generally very speculative, but a few possible.
- *General accompanying actions*: a few consider legal or social aspects relevant to sweb.
- *GRID-based systems for solving complex problems*: complements and may enable much sweb functionality.
- *Improving risk management*: most appear as application areas that might use sweb.
- *Micro- andnano systems*: some have relevance for mote and ubiquitous sweb devices.
- *Mobile and wireless systems beyond SG*: some have relevance for ubiquitous sweb network.
- *Multimodal interfaces*: several have high relevance for new human-sweb interface designs and input solutions for various real-world environments.
- *Networked audiovisual systems and home platforms*: some possibly relevant to general interface and sweb-everywhere aspects.
- *Networked businesses and governments*: several are relevant in terms of identity management, security-trust (legitimacy) issues, municipal and *e-gov* services application.
- *Open development platforms for software and services*: relevant when sweb enhances development and deployment.
- *Optical, opto-electronic, and photonic functional components*: mainly about physical-layer infrastructure components.
- *Products and services engineering 2010*: rather abstract attempts to visualize future engineering environments, probably sweb-enhanced.
- *Pushing the limits of CMOS and preparing for post-CMOS*: new integrated-circuit logic-chip technologies.
- *Research networking test-beds*: (nothing defined here in 2005).
- *Semantic-based knowledge systems*: highly relevant, see previous list.
- *Technology-enhanced learning and access to cultural heritage*: application areas enabled and enhanced with various forms of emerging sweb technology.
- *Towards a global dependability and security framework*: projects that directly address higher-level proposed parts of the W3C sweb vision, with focus on identity and authentication (biometrics), cryptography, privacy, e-justice and judicial co-operation, mobile device security, and contracts in a virtual environment.

A further two site selections summarize over 580 individual FP5 projects around common themes:

- Systems and Services for the Citizen;
- New Methods of Work and Electronic Commerce.

All this may give some indication of the diversity of research that is regularly renewed, yet follows a number of broad themes that deliberately focus on the lives and livelihoods of European citizens.

Considerable overlap may also exist with other sections, for example in that certain W3C sweb activities already described are manifest as an FP Priority, at least in part. Given that work on this book began in 2002, in the closing year of FP5, significant changes to the selection of highlighted projects were required when the span of FP6 and results later became apparent.

OntoWeb Network and Knowledge Web

The EU-funded *OntoWeb Network* project (portal site *www.ontoweb.org*) may serve as an example. It promoted ontology-based information exchange for knowledge management and electronic commerce. Concluded in May 2004 (FP5), it was formally superseded by another EU-funded project (FP6) with 4-year charter, *Knowledge Web (knowledgeweb.semanticweb. org)*, which has the main goal to support the transition process of ontology technology from academia to industry.

The 'extended' *OntoWeb* portal site, based on RDF syndication, remains an important resource: *OntoWeb* represented a large and distributed consortium (comprising 127 members).

Where the predecessor explored basic technology for KM, Knowledge Web has its main focus on deliverables and promotional events. An important aspect reflects a current political agenda by striving to extend enabling IT to the new EU members in eastern Europe.

Bit 7.6   *OntoWeb* research and *Knowledge Web* application have a European-specific focus

The work often addresses or assumes infrastructure relevant to European conditions. Published results may therefore seem strange or even irrelevant to a North American context.

Deliverables are offered in the *industrial, research, educational*, and *management* areas. Industry deliverables range from use-case examples and recommendations to reports and demonstrator sites.

An *OntoWeb*-related site was *OntoWeb Edu (qmir.dcs.qmul.ac.uk/ontoweb/)*, provided for the benefit of the sweb community as part of the deliverables. The task of this site was to manage, coordinate, and initiate educational initiatives related to Web standardization activities and topics related to the Semantic Web in general.

The organization also aimed to provide a set of online technologies to support publication, retrieval, and debate of pedagogic material related to the Semantic Web.

The *EducaNext Portal (www.educanext.org)* is a replacement service supporting the creation and sharing of knowledge for higher education. It is the conventional Knowledge Web repository for learning objects in the area of the Semantic Web.

AIC at SRI

The Artificial Intelligence Center (AIC, *www.ai.sri.com*) *at SRI International*, formerly known as *Stanford Research Institute*, has an established nucleus of long-term projects in core areas of artificial intelligence, many of which concern emerging technologies applicable to the Semantic Web and agents. These projects are organized under six primary research programs containing smaller research groups and projects:

- *Perception*, covering the many aspects of machine seeing and processing internal models of the physical environment.

- *Representation and Reasoning*, including the topics of generative planning, reactive planning/control, fuzzy control, evidential reasoning, automated deduction, knowledge browsing and editing, agent architectures, and distributed reasoning.
- *Natural Language*, exploring multimedia/multimodal interfaces, spoken language and dialog systems, and text-understanding systems.
- *Bioinformatics Research Group*, developing biological knowledge bases, pathway tools, and *BioOntologies*.
- *Cognitive Computing Group*.
- *IOMG*.

Currently (April 2005), little is public about the last two, only that they together host the project 'Cognitive Assistant that Learns and Organizes (CALO)', part of DARPA's 'Perceptive Agent that Learns (PAL)' program.

Bit 7.7   AIC research addresses a number of 'classic' AI areas

Fundamental research deals with hardware and software implementations appropriate to autonomous mobile devices (such as robots) that can reason. Much of this research is equally applicable to software agents and distributed devices.

Some (past and published) projects have special relevance to sweb technologies.

Open Connectivity

Open Knowledge Base Connectivity (OKBC, see *www.ai.sri.com/~okbc/*) is a protocol specification for accessing knowledge bases stored in knowledge representation systems (KRS). It figures in several of the sweb-related projects described in this book, and its ontology model can be browsed online at the *Ontolingua Server* (see Chapter 9).

OKBC provides a uniform model of KRSs based on a common conceptualization of classes, individuals, slots, facets, and inheritance. OKBC is defined independently of programming language and consists of a set of operations that provide a generic interface to an underlying KRS. This interface isolates an application from many of the idiosyncrasies of a specific KRS. Tools developed from the OKBC model can access any compliant KRS in a consistent way – for example, graphical browsers, frame editors, analysis tools, and inference tools.

It was part of the DARPA-sponsored program on Rapid Knowledge Formation (RKF) and previously High Performance Knowledge Bases (HPKB). SRI developed technology to support collaborative construction and effective use of distributed large-scale repositories of highly expressive reusable ontologies.

Agent-based Computing

AIC has a number of significant projects that concern agent-based computing (see *www.ai.sri.com/~ortiz/agents*), classified into the following basic groups:

- Collaboration
- Delegated computing
- Automated negotiation
- Distributed generative-planning
- Distributed robotics, which typically concerns autonomous, sensor and actuator equipped small robots that can collaboratively explore real-world environments (agents on wheels, so to speak)
- *SoftBots*, which includes research into user interface, mixed initiative, and distributed information agents
- Massive agent-based systems, which in this context means thousands of interacting agents, or more

Some agent projects are especially relevant to the themes in this book:

- *Autonomous Negotiating Teams* (ANTS) is developing new incremental negotiation and coalition-formation strategies for agents. It emphasizes focused and time-sensitive interactions between agents that range over abstraction spaces.
- *Open Agent Architecture* (OAA) is a framework for building distributed communities of agents, defined as any software processes that meets the OAA conventions (of registering services, sharing common functionality, and using the Interagent Communication Language).
- *Timely Information management in Dynamic collaborative Environments* (TIDE) aims to deliver information to collaborating agents in a dynamic environment by evaluating the relevance of incoming information to the current tasks of the collaborators. Collaborators may discover one another by constantly evaluating the relevance of changing collaborator tasks.
- *Multi-agent Planning Architecture* (MPA) is a framework for integrating diverse technologies into a system capable of solving complex planning problems that cannot be solved by any individual system but require the coordinated efforts of many (including participation of human experts).
- *Smart Workflow for Collection Management* (SWIM) researches information flows between agents, and aims to build a model that describes the communication between the agents in a workflow plan.
- *SenseNet* is developing environment-exploring teams of small autonomous robots, leveraging the work of the OAA project and that of the Perception realm for machine vision processing.

Many projects provided early prototype deliverables that were tested in the field.

- For example, an early implementation of ANTS, demonstrated at a Marine Corps air station in 2001, created weekly and daily flight schedules for 300 air missions, making changes at a minute's notice when necessary. Deployment of similar systems for civilian Air Traffic Control has been discussed.
- A TIDES implementation, also in 2001, reviewed foreign radio broadcasts in real-time, providing operators with information on biological events and diseases. The system automatically transcribed the broadcasts into text and extracted needed information.

Bioscience Communities

A vast amount of data is generated in biological research and therefore many innovative and distributed resources are required to manage these data, and new technologies are needed to share and process the data effectively. It is no surprise therefore to see various segments of bioscience communities involved with sweb technologies.

BioPathways Consortium

The *BioPathways Consortium* (*www.biopathways.org*) is an open group formed to catalyze the emergence and development of computational pathways biology. An important additional mission is 'to support and coordinate the development and use of open technologies, standards and resources for representing, handling, accessing, analyzing, and adding value to pathways information'. This support often takes the form of implementing sweb technology, for example:

- Pathway text-mining
- Ontologies and formalisms
- Pathways database integration
- *BioPAX*

BioPAX was mentioned as an implementation example in the Web Ontology Languages section of Chapter 6.

Gene Ontology Consortium

The *Gene Ontology Consortium* (GO, *www.geneontology.org*) has the goal to produce a controlled vocabulary that can be applied to all organisms, specifically even as structured knowledge of gene and protein roles in cells is accumulating and changing.

GO provides three structured networks of defined terms to describe and organize gene product attributes – ontologies that are seen as important goals in their own right. The three organizing principles of GO are *molecular function, biological process*, and *cellular component*. The ontology collects terms to be used as attributes of gene products by collaborating databases, facilitating uniform queries across them. Controlled vocabularies of terms are structured to allow attribution and querying at different levels of granularity.

However, GO is not intended as a database of gene sequences, nor a catalog of gene products. A number of other caveats and cautions about the role of GO are given in the context of the documentation and definitions (*www.geneontology.org/doc/GO.doc.html*).

Addressing the concern of how functional and other information concerning genes are to be captured, documented gene information is made accessible to biologists or structured in a computable form. Described genes are annotated to nodes in the ontology model, which also provides connections to other annotated genes that have similar biological function, cellular localization, or molecular process.

Using the open sweb languages, the approach implements a general framework for incorporating annotations from many sources into a given reference structure so that other researchers (or agents) can explore all possible relationships, identify commonalities or similarities, and infer new generic characteristics.

8

Application and Tools

Often, the best way to get the feel of a concept is to play around with it. Early prototyping and live demonstration systems provide valuable experiences and insight into the practicalities of the design concepts. Such feedback can be valid even if the prototypes do not necessarily function in the ideal full-scale way, or use the intended, perhaps as yet unimplemented, infrastructure features. We may compare some of this early application work with the mock-up or scale-model principle used in traditional engineering to study, for example, specific usability features.

Functional prototyping is surprisingly common on the Internet where it is not unusual to find very early releases of software available for public trial use – frequently as of the first '0.x' point-versions. Perhaps equally surprising to newcomers is how many of these very early versions are actually usable in practical situations. Part of the reason for this early availability lies in the traditions of open source development, where feedback from many early adopters is a vital factor in guiding further development. Projects are generally well maintained (often at *sourceforge.net*) with published version control and bug tracking, community discussion forums, documentation, and a selection of last-stable versions and daily builds free to download.

Such 'release early and often' policy clearly helps provide relatively polished prototypes even in the low point-versions. Adopting this kind of live-research, cutting-edge software does however have its risks for the user. For example, a design change between v0.3 and v0.4 can introduce incompatibilities that make transition difficult. The casual user may run across features that are not implemented yet, or unresolved bugs. The work is, well... unfinished, after all. That said, much of this early software can still be used profitably – and is often used extensively for practical work.

Chapter 8 at a Glance

This chapter looks at a some application areas of the Semantic Web where prototype tools are already implemented and available for general use. *Web Annotation* introduces the concept of annotating any Web resource in ways that are independent of the resource server and that can be shared.

- *The Annotea Project* describes one of the seminal Web annotations efforts to introduce authoring of metadata on a larger scale.
- *Evaluating Web Annotation* examines the pros and cons of this technology, and why it has not yet become a standard browser feature.

Infrastructure Development looks at tools and methods for ontology deliverables.

- *Develop and Deploy an RDF Infrastructure* examines the situation in terms of why general support for the infrastructure is not yet in place.
- *Building Ontologies* looks at tool-sets for constructing ontologies.

Information Management looks at tools and methods within this application sector.

- *Haystack* introduces a recent all-in-one information client that might revolutionize the way users access and manipulate data objects.

Digital Libraries examines the attempts to imbue existing large information repositories with semantic metadata.

- *Applying RDF Query Solutions* to hack interesting services based on digital libraries is a quick way to leverage information repositories.
- *Project Harmony* investigated key issues encountered when describing complex multi-media resources in digital libraries.
- *DSpace* describes a joint project by MIT Libraries and HP to capture, index, preserve, and distribute the intellectual output of MIT.
- *Simile* is a joint project by HP and MIT library to build persistent digital archive.

Web Syndication presents tools for the small-scale content provider.

- *RSS and Other Content Aggregators* describes syndication services for the Web that might become more useful when augmented by the functionality of the Semantic Web and Web Services.

Metadata Tools provides a brief listing of some common metadata tools.

- *Browsing and Authoring Tools* focuses on the needs of the content creator.
- *Metadata Gathering Tools* looks to the needs of the agent and content consumer.

Web Annotation

Web annotation refers to the process of making comments, notes, explanations, or other types of external remarks that can be 'attached' to any Web document (or a selected part of a document) without actually needing to touch and modify the document itself. Instead, URI associations connect annotations stored on separate servers with the respective documents. Such annotations can be made by anyone at any time.

Bit 8.1   Annotations are physically separate entities from the document

Web annotations in no way affect the content of the original document in this model. To determine that an annotation exists you must resolve the document URL against the annotation server where the annotation is stored.

Anyone who browses a Web-annotated document, and has access to the corresponding annotation server, can elect to load and see the annotations attached to it. One reason might be to see what an associated peer group thinks about the document, another might be to see supplemental pointers to related, updated, or more detailed material. Typically, filters may be applied when viewing annotations to hide or only partially see the full complement of annotations made.

* The ability to make annotations on the Web, in all the concept simplicity of freely adding comments non-destructively to published and changing material, is a significant early step to addressing the lack of metadata in the current Web, as described in Chapter 1.

In addition, because the URI-association is persistent irrespective of the actual Web document content, older annotations can provide a valuable indication of previous content on pages or sites that have been extensively reworked.

Of special interest is the innate ability to annotate resources that are temporarily unavailable or no longer exist, perhaps to provide pointers to alternative resources. Such an annotation is technically an 'orphaned' one, along with any that were made earlier when the resource was still available.

Bit 8.2   An orphaned Web annotation is never lost, only dislocated

Even in the extreme case where the original document is not found, the browser is still (usually) displaying the document URL, and thus also any associated annotations. This feature is due to how annotations build on client and site-external resources.

One obvious course of action for a permanently removed document is to delete orphaned annotations and recreate relevant ones for other pages instead. On the other hand, the capability of a client still to display existing comments for lost Web resources, and for visitors to add new public comments after the fact, is an intriguing way to combat the usual dead-end experience of a '404-not-found' server response.

Annotation in general is a simple enough concept, well-established in people's minds and academic traditions since antiquity. Readers have often added notes in the margin of a page, between the lines, or on a separate page. Sometimes the extent of our knowledge of the writings of some old authority rests entirely on the preserved compiled commentaries of their note-taking students – the original works are long lost. Referential commenting has also successfully been adapted to modern software tools that provide various ways to insert notes physically into digitally stored documents or post to particular shared stores and keep track of who made what comment when and where. So too have other common forms of annotation been adapted: underlining, highlighting, or circling text.

There are varying beliefs of what note-taking capabilities should be offered in an annotation system and how such annotations should be processed and rendered. Different types of annotation mechanisms might therefore be required for private notes compared to notes meant to be shared with other readers, such as discussion-like comments or questions.

Different design choices suggest themselves depending on whether or not the person making the comment has the capability to modify the original document, or whether the document can track changes. Some situations might require the ability to integrate many different forms of annotation material, not just text, perhaps mixing files in varying media types from various sources. In the case of Web content, the situation is commonly that the person wishing to comment can only read the original, so comments must be stored and processed separately.

Implementation of annotation capability on the Web is seen to be a natural extension in the common good. Web-annotation technology thus appears a good place to start when examining the technologies to implement an enhanced Web, especially since the early solutions often leverage the existing infrastructure and do not need to wait on other developments before deployment. The Web community has been exploring various such technologies for some time.

This section introduces a number of prototype annotation solutions with available tools that can be used now, to give a feel for what might come. Although primarily proof-of-concept prototypes, these solutions, or something much like them, could become an integral part of the Semantic Web as client support spreads or the new infrastructure is deployed.

Future development might add communication with the server hosting the annotated document, thus explicitly bringing the content creator into the annotation loop.

The Annotea Project

Annotea (www.w3.org/2001/Annotea/) is an open W3C project (an early part of its Semantic Web effort) devoted to enhancing the W3C collaboration environment with shared annotations. It is classified as a **LEAD** (Live Early Adoption and Demonstration) project, meaning that the prototype environment and tools are publicly available, with about the same caveat as using any test-status software, that functionality and protocols might change and that stored metadata might not persist.

The intent is that *Annotea* uses and helps to advance W3C standards when possible – for example, it uses an RDF-based annotation schema for describing annotations as metadata, and *XPointer* for locating the annotations in the annotated document. *XPointer* implements a fragment identifier for any URI-reference that enables locating, with a fine granularity, a resource within the internal structure of the XML document.

Bit 8.3   Annotations must be sufficiently fine-grained in location

The intuitive way to annotate is to be able to specify an arbitrary, precise location or selection for an annotation within the document. Not all solutions have this capability, yet, but it is interesting that *XPointer* technology can be used to closely simulate it.

- A side-note is in order about W3C URL references that contain dates. An 'old' date does not mean that the content is outdated or not updated since then. The persistent URL is an organizational convention to indicate when the category or resource was originally set up. The resource will itself convey required information about updates or by a later version elsewhere.

How Web Annotations Work

Web annotations are stored as metadata on annotation servers and presented to the user by a client capable of understanding this metadata and of interacting with an annotation server using the HTTP service protocol. Annotation in the *Annotea* model is in fact an independent *Web service*.

Bit 8.4   Web annotations augment existing content through third-party resources

Annotation servers thus define a kind of 'overlay' on the existing document infrastructure of the Web. Such incremental enhancement promotes rapid adoption of a new technology.

It is perfectly feasible to have many partially overlapping annotation spaces independently maintained by different servers – some public and some private. Annotation data may be stored either in a local file system (and thus often called 'local annotations') or on a remote server (then logically enough called 'remote annotations').

Bit 8.5   Even local annotation capability is a significant enhancement

The ability for a user to mix local and public annotations, and selectively publish local ones as public, can have important consequences for collaborative work.

Figure 8.1 provides a simple illustration of the Web annotation principle, showing how two users interact with a Web-published document.

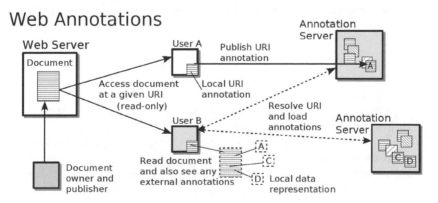

Figure 8.1   Two users interact with a published document in the context of Web annotations. With read access to document and a known relevant annotation server, any user can see a composite, annotated document, and add further public or private comments associated with the same URI

Despite the Web document being read-only from their perspective, remote users can make both local and remote annotations that are URI-associated and point to relevant selections of the document text. Published remote annotations are viewable together with the original document by anyone who can connect to the respective servers. The client software can locally render this information as a composite document, as if the annotations were actually incorporated into the original document.

Bit 8.6   Annotation solutions enhance document representation

Recall the distinction made in Chapter 1 that the Web is about having client software render representations of the results to server queries. Significant enhancement results when the client representation is a composite of results from different sources.

The two modes of creating and accessing annotation metadata, local and remote, are described further in the following sections. Not all early client implementations made this local/remote distinction, however, and some were only for remote annotations.

Local Annotations

Only the 'owner' (according to the system access settings) can see local annotations. By design, this is not only because storage is on the local machine, but also because annotation is associated with a particular user identity.

The annotation metadata is stored on the local system as user-specific data in a special client-maintained directory. Its location and structure may vary between different platforms and client implementations, though typically it will contain RDF- and XML-based files. Local storage does not require any Web server to be installed because the files are managed directly by the client.

In *Annotea*, an index file consists of a list of RDF description containers for annotations made to the same document. These specify the annotation header information and the context information (in *XPointer* syntax). The latter allows the client to render an annotation icon at the proper location and know the span of the selected text.

A simplified and commented version of such a file from a local store is presented here:

```
<?xml version="1.0" ?>
<r:RDF xmlns:r="http://www.w3.org/1999/02/2 2-rdf-syntax-ns#"
xmlns: ... more w3.org/purl.org preamble ... >
  <r:Description>
  <r:type r:resource="http://www.w3.org/2000/10/annotation-
     ns#Annotation" />
  <r:type r:resource="http://www.w3.org/2000/10/annotation-
     Type#Comment" />
  <a:annotates r:resource=" URL-of-Document " />
  <a:context> URL-of-Document #xpointer(start-and-range-speci-
     fication) </a:context>
  <d:title>Annotation of Document Title </d:title>
  <d:creator> Annotation-Owner </d:creator>
```

```
<a:created>2002-11-13T21:43:41</a:created>
<d:date>2002-11-13T21:43:50</d:date>
<a:body r:resource="file:/// local-profile-path /annotations/
    annots284.156.html" />
</r:Description>
<r:Description>
 ...(next container = another annotation to the same document)
</r:Description>
</r:RDF>
```

A consequence of separating the metadata and annotation body in this way is that the latter need only be loaded when the user requests explicitly to see a particular annotation's content. This reduces resource overhead and rendering latency when dealing with remotely stored annotations, as described next.

Referencing local files as the body, and making them HTML documents, means that in principle any form of local file can be included and used as comment material. For the local user, it provides a tool for organizing information.

At any time, a local annotation can be converted to a public and shared one by publishing it to a remote annotation server – 'Post annotation' from the Annotations menu of the client. Once this process is completed, the local annotation is deleted because it is considered to have been moved to the remote server.

Remote Annotations

Remote annotations are accessed through the Web, and as they may be shared in various ways, they are often called *public* annotations. Remote storage involves a minimum of Web server, associated scripts, and a database engine to store and resolve the RDF-format annotations. Annotations are posted and managed using normal HTTP requests from client to server. Appropriate operations are simulated by the client for local storage.

The *Annotea* protocol recognizes five HTTP operations between client and server in an annotation exchange:

- **Post**. The client publishes a new annotation to the annotation server. Both metadata and annotation body are sent. As an alternative to an explicit body, the URI to an existing document is also allowed.
- **Query**. The client sends a query with a document URI to the server and gets backs any associated annotation metadata that is stored. The metadata contains the server URI references to the associated annotation bodies.
- **Download**. The server sends the body data for a specified annotation to the requesting client. The client must know the server URI of the annotation body in order to make the request.
- **Update**. The client modifies an annotation and publishes these modifications back to the server. Again, the client must know the server URI of the associated annotation body.
- **Delete**. The client deletes a specified annotation from the server. In a sense, this operation is a special case of Update. Not everyone, however, agrees that delete should be supported.

Posting information to the server uses the HTTP POST method, while requests use the GET method. Assuming that the server can return an appropriate URI to the client, and the server permissions allow it for an authenticated user, the client can post subsequent modifications using the PUT and DELETE methods. For various reasons, a POST variant for sending updates is also accepted instead of PUT.

As it happens, the design specifies that the comment body be a separate Web document, hence the 'option' of posting an annotation body's document address instead of embedding its contents. The basic referencing syntax is

```
<a:body r:resource=" http://www.domain.tld/mycomment.html "/>
```

Embedding involves declaring a literal segment in the XML post's body container:

```
<a:body>
  <r:Description>
  <h:ContentType>text/html</h:ContentType>
  <h:ContentLength>249</h:ContentLength>
  <h:Body r:parseType="Literal">
    <html xmlns = "http://www.w3.org/1999/xhtml">
    <head>
      <title>More examples</title>
    </head>
    <body>
      <p>For further explorations of this concept, see my
      <a href="http://domain.tld/concept/examples.html">concept
             examples</a>.</p>
    </body>
    </html>
  </h:Body>
  </r:Description>
</a:body>
```

The embedded content is a complete HTML document in its own right, just encapsulated in the XML Body tag, in turn part of the Description container put inside the posting RDF-structure's expanded Body.

Bit 8.7   Annotation bodies can include arbitrary Web resources

The URI-referencing and XML/XHTML-encapsulation of actual annotation body provides an open-ended and useful way to allow any kind of accessible resource to function as an annotation – from simple text to referencing complex Web service functionality.

By design, the posting operation has been made a packaged, atomic process, even for multiple resource components (such as metadata and body) so as to avoid opening any window between the two where another client might try to modify the body based on the metadata alone.

Metadata can be filtered both locally and in the remote server query process. To support the latter, the design adopted a modified version of *Algernon* query language syntax, called *Algae*. A client can pass a subject-predicate-object structure (or RDF triple, as described in Chapter 5) as a query to the server and parse the answer. It is assumed that frequent query types are implemented as click selections in the client GUI, but the full power of Algae is available through the protocol.

Although W3C does offers a public server for trying out annotations for demonstration purposes, users are encouraged to install annotation servers for their own special purposes. For administrators and users with some experience in dealing with Web servers, *Perl* scripts and database engines such as *MySQL*, it is all fairly straightforward and well described in the official 'How-to'.

Posting annotations on a public annotation server will generally entail first creating a user account on that server and accepting its published terms of service. Administrators might restrict annotation posting to particular user groups, depending on the purpose. Viewing already published annotations, as with the demonstration servers, may on the other hand be open to users without requiring registration.

Viewing Annotations

To view (and create) annotations in the *Annotea* model, the client must support the mechanisms and the server protocol. However, support for the technology is poor, despite its long availability as an official W3C LEAD effort. Only the following clients have usable implementations:

- *Amaya*. Full support is limited to this W3C reference browser/editor (*www.w3.org/ amaya*), the most mature prototype implementation. While unlikely ever to be the browser of preference, it remains an essential baseline validation and evaluation tool for Web content.
- *Annozilla*. Implements support within the framework of the open source Mozilla browser (see *annozilla.mozdev.org*). At v0.5 at the time of writing, and compatible with Mozilla 1.6, it leverages Mozilla's built-in handling of RDF to process annotation metadata.
- *Snufkin*. Supports Annotea with MS Internet Explorer 5 or better. This solution is at very early prototype status (v0.25 at time of writing) and builds on a third-party ActiveX component called ScriptX.

Some W3C work was done on *JavaScript* interfaces using '*bookmarklets*', but the technology is no longer actively supported. The Snufkin prototype (*www.jibbering.com/ snufkin/*) appears to have remained unchanged at v0.25 for years.

SWAD-E Annotea Tools (*www.w3.org/2001/sw/Europe/200209/annodemo/readme.html*) is a small library of simple tools written in *Ruby* and available for incorporation or use as model code.

An illustration of practical Web annotation in *Amaya* is shown in Figure 8.2.

The client supports fine-grained annotation by specific location in the document text. The user may also locate by block selection, and by referencing entire document (which is the basic resolution of Web URLs). In the latter case, the pencil icon is placed at the document top.

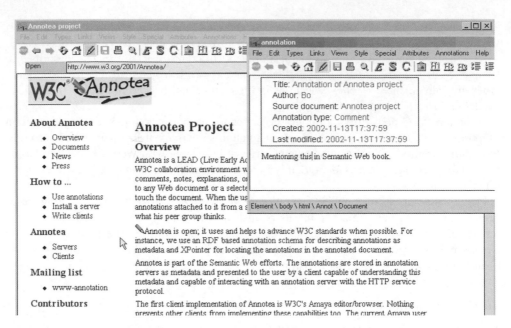

Figure 8.2 Creating (or viewing or editing) annotations in the Amaya client. When annotations are loaded, they are indicated by the pencil icon at the proper location. Right-clicking (or double-clicking) on the icon opens the annotation window, while left-clicking highlights the selection

Configuring the URL of a remote annotation server, the picture becomes more interesting, because then the user can view the annotations of others to the same page. Figure 8.3 shows a recent example of the same Web page, but now including the public shared annotations.

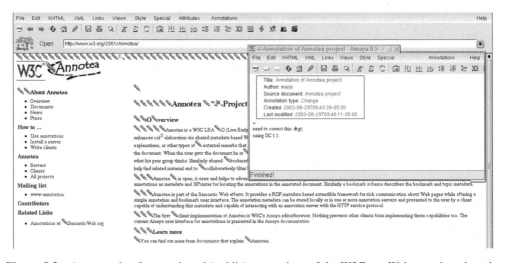

Figure 8.3 An example of many shared (public) annotations of the W3C top Web page based on the W3C demonstration server's annotation store. Note that rendered annotation icon may vary depending on declared type

Several things are evident from this capture. One is an indication of the sheer number of public annotations that can accumulate over time; another is how the client's rendered icon insertion, although a useful visual aid, can disrupt the intended flow of the original content. Fortunately, a local filter component allows the user to hide selectively annotations from a summary list sorted by author, type, or server.

Some technical problems with Web annotations remain:

- *Client overhead*, which comes from extra processing, many connections with the annotation server(s), and rendering. A client is often slow to render an annotated page due to the many icon insertions that force redraws of the original page layout.
- *Latency issues*, which are a function of sending requests to the database. For local annotations, the client simply accesses the local file system, but resolving remote annotations can introduce significant delays for the user. We might, however, expect annotation services within an intranet to be less delay prone, comparable in latency with a strictly local annotation store.
- *Server load*, which would become significant if every page browsed by an annotation-aware client pulled metadata from a single server. However, if annotation became ubiquitous on the Web, it seems a reasonable assumption that annotation storage would move in the direction of massively distributed and adaptive storage, which would alleviate the problem considerably.

Dealing with local annotations is far less problematic in all these respects.

Evaluating Web Annotation

In the *Annotea* model, the major benefit is the ability for anyone to create and publish comments attached to any Web-browsed resource. Such annotations can be for private use (at least in clients that support it) or for sharing with others.

It is in the nature of the current Web that those browsing generally have no control over page content, so virtual attachments using URI-associations and the fine-grained location resolution provided by *XPointer* have great utility.

It is also a valuable functionality feature to be able to use and mix local (and user-specific) annotation stores with an arbitrary and configurable number of remote stores. *Amaya* provides the feature to convert local annotations to remote ones on a specified server, in addition to publishing directly. Presumably, many or most annotation services will implement some sort of member authentication in that context, perhaps even for viewing, so remote annotation might be much more restricted than the current open demonstration systems.

We may foresee several levels of annotation system access, following examples set in Knowledge Management and Technology monitoring:

- public open servers (such as *Yahoo!*);
- communities of interests servers (access by registration and moderators);
- corporate servers (intranet-like, access by role or organization, perhaps);
- individual private repository (owner-only access, local or remote server).

There will be more levels, perhaps, and everything in-between.

Downsides exist as well, mainly due to the same lack of control, but also dependent on how the annotations are published.

Similar to the familiar 'document not found' frustration, context-bound annotations can be rendered incomprehensible by the removal of a resource from a given URI. Annotations risk becoming misplaced or 'unstuck' in the document due to inserted, modified, or deleted content.

The fine-grained localization provided by *XPointer* is not foolproof if content is heavily revised in the same document. Such internal placement is syntactical by nature, based on the embedded structural tags. In most cases, markup is still at the mercy of the creator and the use of semantically impoverished HTML.

Annotation dislocations seem inevitable, and it is probably a good thing to allow users to move or relocate existing annotations. However, as noted previously, the existence of persistent metadata independent of the original document is also a good thing. Tied to the original URI, it can provide valuable information about an unavailable document, or of previous content in cases where the site structure has changed. Equally valuable is the ability to continue to annotate even unavailable resource location on the Web, perhaps to provide pointers to alternatives.

Bit 8.8   Persistent independent metadata enhances content availability

Collaborative efforts to annotate unavailable or changing Web resources can greatly enhance the user experience and overall perceived quality of the Web.

So, although the annotation tool might initially be seen as somewhat limited in the current mostly-HTML environment, where resource types can disappear and content is unpredictably revised by the resource owner or is generated dynamically, it can still provide surprising new functionality. Annotation otherwise works best with archive resources with stable URI. A greater deployment of semantically more valid markup, such as well-defined XML, might reduce the dislocation problems within documents significantly.

Why still only Prototype?

For a technology with such promising potential that has been around for so long (in Internet terms), it may seem odd that Web annotation is not already a well-established feature of all browser clients. There are several reasons for this turn of events.

In a PC Web browser market that for years has been 90–95% dominated by a single client implementation, *MS Internet Explorer*, it should come as no surprise that a lack of support in it for any new technology is a serious obstacle to deployment. The situation is similar to the lack of deployment of a standard technology for browser-based micropayment systems.

Bit 8.9   Deployment is a key factor in acceptance and use, though not the only one

If Web annotation became just another menu entry in the major browsers, then the Web could easily and quickly have become awash in annotations and annotation services.

Another reason is that *Amaya* is never likely to become a browser for the masses, and it is the only client to date with full support for *Annotea*. The main contender to MSIE is *Mozilla*, and it is in fact of late changing the browser landscape. A more mature *Annozilla* may therefore yet reach a threshold penetration into the user base, but it is too early to say.

Issues and Liabilities

These days, no implementation overview is complete without at least a brief consideration of social and legal issues affected by the proposed usage and potential misuse of a given technology.

One issue arises from the nature of composite representations rendered in such a way that the user might perceive it as being the 'original'. Informed opinion diverges as to whether such external enhancement constitutes at minimum a 'virtual infringement' of the document owner's legal copyright or governance of content.

The question is far from easy to resolve, especially in light of the established fact that any representation of Web content is an arbitrary rendering by the client software – a rendering that can in its 'original' form already be 'unacceptably far' from the specific intentions of the creator. Where lies the liability there? How far is unacceptable?

An important aspect in this context is the fact that the document owner has *no control* over the annotation process – in fact, he or she might easily be totally unaware of the annotations. Or if aware, might object strongly to the associations.

• This situation has some similarity to the issue of 'deep linking' (that is, hyperlinks pointing to specific resources within other sites). Once not even considered a potential problem but instead a natural process inherent to the nature of the Web, resource referencing with hyperlinks has periodically become the matter of litigation, as some site owners wish to forbid such linkage without express written (or perhaps even paid-for) permission.

Even the well-established tradition of academic reference citations has fallen prey to liability concerns when the originals reside on the Web, and some universities now publish cautionary guidelines strongly discouraging online citations because of the potential lawsuit risks. The fear of litigation thus cripples the innate utility of hyperlinks to online references.

When anyone can publish arbitrary comments to a document on the public Web in such a way that other site visitors might see the annotations as if embedded in the original content, some serious concerns do arise. Such material added from a site-external source might, for example, incorrectly be perceived as endorsed by the site owner.

At least two landmark lawsuits in this category were filed in 2002 for commercial infringement through third-party pop-up advertising in client software – major publishers and hotel chains, respectively, sued *The Gator Corp* with the charge that its *adware* pop-up component violated trademark/copyright laws, confused users, and hurt business revenue. The outcome of such cases can have serious implications for other 'content-mingling' technologies, like Web Annotation, despite the significant differences in context, purpose, and user participation.

- A basic argument in these cases is that the Copyright Act protects the right of copyright owners to display their work as they wish, without alteration by another. Therefore, the risk exists that annotation systems might, despite their utility, become consigned to at best closed contexts, such as within corporate intranets, because the threat of litigation drives its deployment away from the public Web.

Legal arguments might even extend constraints to make difficult the creation and use of generic third-party metadata for published Web content. Only time will tell which way the legal framework will evolve in these matters. The indicators are by turns hopeful and distressing.

Infrastructure Development

Broadening the scope of this discussion, we move from annotations to the general field of developing a semantic infrastructure on the Web. As implied in earlier discussions, one of the core sweb technologies is RDF, and it is at the heart of creating a sweb infrastructure of information.

Develop and Deploy an RDF Infrastructure

The W3C RDF specification has been around since 1997, and the discussed technology of Web annotation is an early example of its deployment in a practical way.

Although RDF has been adopted in a number of important applications (such as *Mozilla, Open Directory, Adobe,* and *RSS 1.0*), people often ask developers why no 'killer application' has emerged for RDF as yet. However, it is questionable whether 'killer app' is the right way to think about the situation – the point was made in Chapter 1 that in the context of the Web, the Web itself is the killer application.

Nevertheless, it remains true that relatively little RDF data is 'out there on the public Web' in the same way that HTML content is 'out there'. The failing, if one can call it that, must at least in part lie with the lack of metadata authoring tools – or perhaps more specifically, the lack of embedded RDF support in the popular Web authoring tools.

For example, had a widely used Web tool such as *MS FrontPage* generated and published usable RDF metadata as a matter of course, it seems a foregone conclusion that the Web would very rapidly have gained RDF infrastructure. MS FP did spread its interpretation of CSS far and wide, albeit sadly broken by defaulting to absolute font sizes and other unfortunate styling.

- The situation is similar to that for other interesting enhancements to the Web, where the standards and technology may exist, and the prototype applications show the potential, but the consensus adoption in user clients has not occurred. As the clients cannot then in general be assumed to support the technology, few content providers spend the extra effort and cost to use it; and because few sites use the technology, developers for the popular clients feel no urgency to spend effort implementing the support.

It is a classic barrier to new technologies. Clearly, the lack of general ontologies, recognized and usable for simple annotations (such as bookmarks or ranking) and for searching, to name two common and user-near applications, is a major reason for this impasse.

Building Ontologies

The process of building ontologies is a slow and effort-intensive one, so good tools to construct and manage ontologies are vital. Once built, a generic ontology should be reusable, in whole or in parts. Ontologies may need to be merged, extended, and updated. They also need to be browsed in an easy way, and tested.

This section examines a few of the major tool-sets available now. Toolkits come in all shapes and sizes, so to speak. To be really usable, a toolkit must not only address the ontology development process, but the complete ontology life-cycle: identifying, designing, acquiring, mining, importing, merging, modifying, versioning, coherence checking, consensus maintenance, and so on.

Protégé

Protégé (developed and hosted at the *Knowledge Systems Laboratory, Stanford University*, see *protege.stanford.edu*) is a Java-based (and thus cross-platform) tool that allows the user to construct a domain ontology, customize knowledge-acquisition forms, and enter domain knowledge. At the time of last review, the software was at a mature v3.

System developers and domain experts use it to develop Knowledge Base Systems (KBS), and to design applications for problem-solving and decision-making in a particular domain. It supports import and export of RDF Schema structures.

In addition to its role as an ontology editing tool, *Protégé* functions as a platform that can be extended with graphical widgets (for tables, diagrams, and animation components) to access other KBS-embedded applications. Other applications (in particular within the integrated environment) can also use it as a library to access and display knowledge bases.

Functionality is based on Java applets. To run any of these applets requires Sun's Java 2 Plug-in (part of the Java 2 JRE). This plug-in supplies the correct version of Java for the user browser to use with the selected Protégé applet. The *Protégé OWL Plug-in* provides support for directly editing Semantic Web ontologies.

Figure 8.4 shows a sample screen capture suggesting how it browses the structures.

Development in Protégé facilitates conformance to the OKBC protocol for accessing knowledge bases stored in Knowledge Representation Systems (KRS). The tool integrates the full range of ontology development processes:

- Modeling an ontology of classes describing a particular subject. This ontology defines the set of concepts and their relationships.
- Creating a knowledge-acquisition tool for collecting knowledge. This tool is designed to be domain-specific, allowing domain experts to enter their knowledge of the area easily and naturally.
- Entering specific instances of data and creating a knowledge base. The resulting KB can be used with problem-solving methods to answer questions and solve problems regarding the domain.
- Executing applications: the end product created when using the knowledge base to solve end-user problems employing appropriate methods.

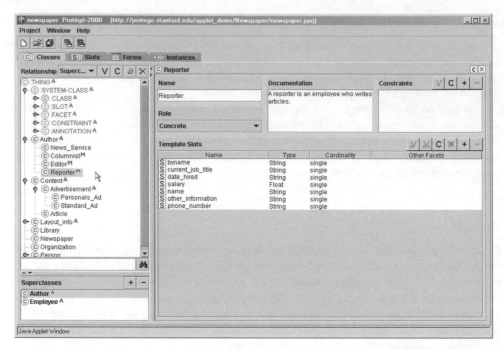

Figure 8.4 Browsing the 'newspaper' example ontology in Protégé using the browser Java-plug-in interface. Tabs indicate the integration of the tool – tasks supported range from model building to designing collection forms and methods

The tool environment is designed to allow developers to re-use domain ontologies and problem-solving methods, thereby shortening the time needed for development and program maintenance. Several applications can use the same domain ontology to solve different problems, and the same problem-solving method can be used with different ontologies.

Protégé is used extensively in clinical medicine and the biomedical sciences. In fact, the tool is declared a 'national resource' for biomedical ontologies and knowledge bases supported by the *U.S. National Library of Medicine.* However, it can be used in any field where the concept model fits a class hierarchy.

A number of developed ontologies are collected at the *Protégé Ontologies Library* (*protege.stanford.edu/ontologies/ontologies.html*). Some examples that might seem intelligible from their short description are given here:

- **Biological Processes**, a knowledge model of biological processes and functions, both graphical for human comprehension, and machine-interpretable to allow reasoning.
- **CEDEX**, a base ontology for exchange and distributed use of ecological data.
- **DCR/DCM**, a Dublin Core Representation of DCM.
- **GandrKB** (Gene annotation data representation), a knowledge base for integrative modeling and access to annotation data.
- **Gene Ontology** (GO), knowledge acquisition, consistency checking, and concurrency control.

- **Geographic Information Metadata**, ISO 19115 ontology representing geographic information.
- **Learner**, an ontology used for personalization in eLearning systems.
- **Personal Computer – Do It Yourself** (PC-DIY), an ontology with essential concepts about the personal computer and frequently asked questions about DIY.
- **Resource-Event-Agent Enterprise** (REA), an ontology used to model economic aspects of e-business frameworks and enterprise information systems.
- **Science Ontology**, an ontology describing research-related information.
- **Semantic Translation** (ST), an ontology that supports capturing knowledge about discovering and describing exact relationships between corresponding concepts from different ontologies.
- **Software Ontology**, an ontology for storing information about software projects, software metrics, and other software related information.
- **Suggested Upper Merged Ontology** (SUMO), an ontology with the goal of promoting data interoperability, information search and retrieval, automated inferencing, and natural language processing.
- **Universal Standard Products and Services Classification** (UNSPSC), a coding system to classify both products and services for use throughout the global marketplace.

A growing collection of OWL ontologies are also available from the site (*protege. stanford.edu/plugins/owl/owl-library/index.html*).

Chimaera

Another important and useful ontology tool-set system hosted at KSL is *Chimaera* (see *ksl.stanford.edu/software/chimaera/*). It supports users in creating and maintaining distributed ontologies on the Web. The system accepts multiple input format (generally OKBC-compliant forms, but also increasingly other emerging standards such as RDF and DAML). Import and export of files in both DAML and OWL format are possible.

Users can also merge multiple ontologies, even very large ones, and diagnose individual or multiple ontologies. Other supported tasks include loading knowledge bases in differing formats, reorganizing taxonomies, resolving name conflicts, browsing ontologies, and editing terms. The tool makes management of large ontologies much easier.

Chimaera was built on top of the *Ontolingua Distributed Collaborative Ontology Environment*, and is therefore one of the services available from the *Ontolingua Server* (see Chapter 9) with access to the server's shared ontology library.

Web-based merging and diagnostic browser environments for ontologies are typical of areas that will only become more critical over time, as ontologies become central components in many applications, such as e-commerce, search, configuration, and content management.

We can develop the reasoning for each capability aspect:

- *Merging capability* is vital when multiple terminologies must be used and viewed as one consistent ontology. An e-commerce company might need to merge different vendor and network terminologies, for example. Another critical area is when distributed team members need to assimilate and integrate different, perhaps incomplete ontologies that are to work together as a seamless whole.

- *Diagnosis capability* is critical when ontologies are obtained from diverse sources. A number of 'standard' vocabularies might be combined that use variant naming conventions, or that make different assumptions about design, representation, or reasoning. Multidimensional diagnosis can focus attention on likely modification requirements before use in a particular environment. Log generation and interaction support assists in fixing problems identified in the various syntactic and semantic checks.

The need for these kinds of automated creation, test, and maintenance environments for ontology work grows as ontologies become larger, more distributed, and more persistent.

KSL provides a quick online demo on the Web, and a fully functional version after registration (*www-ksl-svc.stanford.edu/*). Other services available include *Ontolingua, CML,* and *Webster.*

OntoBroker

The *OntoBroker* project (*ontobroker.semanticweb.org*) was an early attempt to annotate and wrap Web documents. The aim was to provide a generic answering service for individual agents. The service supported:

- clients (or agents) that query for knowledge;
- providers that want to enhance the accessibility of their Web documents.

The initial project, which ran until about 2000, was successful enough that it was transformed into a commercial Web-service venture in Germany, *Ontoprise* (*www.ontoprise. de*). It includes an RDF inference engine that during development was known as the Simple Logic-based RDF Interpreter (*SiLRI,* later renamed *Triple*).

Bit 8.10   *Knowledge* is the capacity to act in a context

This *Ontoprise*-site quote, attributed to Dr. Karl-Erik Sveiby (often described as one of the 'founding fathers' of Knowledge Management), sums up a fundamental view of much ontology work in the context of KBS, KMS and CRM solutions.

The enterprise-mature services and products offered are:

- *OntoEdit*, a modeling and administration framework for Ontologies and ontology-based solutions.
- *OntoBroker*, the leading ontology-based inference engine for semantic middleware.
- *SemanticMiner*, a ready-to-use platform for KMS, including ontology-based knowledge retrieval, skill management, competitive intelligence, and integration with *MS Office* components.
- *OntoOffice*, an integration agent component that automatically, during user input in applications (*MS Office*), retrieves context-appropriate information from the enterprise KBS and makes it available to the user.

The offerings are characterized as being 'Semantic Information Integration in the next generation of Enterprise Application Integration' with Ontology-based product and services solutions for knowledge management, configuration management, and intelligent dialog and customer relations management.

Kaon

KAON (the *KArlsruhe ONtology,* and associated Semantic Web tool suite, at *kaon. semanticweb.org*) is another stable open-source ontology management infrastructure targeting business applications, also developed in Germany. An important focus of KAON is on integrating traditional technologies for ontology management and application with those used in business applications, such as relational databases.

The system includes a comprehensive Java-based tool suite that enables easy ontology creation and management, as well as construction of ontology-based applications. KAON offers many modules, such as API and RDF API, Query Enables, Engineering server, RDF server, portal, OI-modeller, text-to-onto, ontology registry, RDF crawler, and application server.

The project site caters to four distinct categories: users, developers, researchers, and partners. The last represents an outreach effort to assist business in implementing and deploying various sweb applications. KAON offers experience in data modeling, sweb technologies, semantic-driven applications, and business analysis methods for sweb. A selection of ontologies (modified OWL-S) are also given.

Documentation and published papers cover important areas such as conceptual models, semantic-driven applications (and application servers), semantic Web management, and user-driven ontology evolution.

Information Management

Ontologies as such are both interesting in themselves and as practical deliverables. So too are the tool-sets. However, we must look further, to the application areas for ontologies, in order to assess the real importance and utility of ontology work.

As an example in the field of information management, a recent prototype is profiled that promises to redefine the way users interact with information in general – whatever the transport or media, local or distributed – simply by using an extensible RDF model to represent information, metadata, and functionality.

Haystack

Haystack (*haystack.lcs.mit.edu*), billed as 'the universal information client' of the future, is a prototype information manager client that explores the use of artificial intelligence techniques to analyze unstructured information and provide more accurate retrieval. Another research area is to model, manage, and display user data in more natural and useful ways.

Work with information, not programs.   Haystack motto

The system is designed to improve the way people manage all the information they work with on a day-to-day basis. The *Haystack* concept exhibits a number of improvements over current information management approaches, profiling itself as a significant departure from traditional notions. Core features aim to break down application barriers when handling data:

- **Genericity**, with a single, uniform interface to manipulate e-mail, instant messages, addresses, Web pages, documents, news, bibliographies, annotations, music, images, and more. The client incorporates and exposes all types of information in a single, coherent manner.
- **Flexibility**, by allowing the user to incorporate arbitrary data types and object attributes on equal footing with the built-in ones. The user can extensively customize categorization and retrieval.
- **Objects-oriented**, with a strict user focus on data and related functionality. Any operation can be invoked at any time on any data object for which it makes sense. These operations are usually invoked with a right-click context menu on the object or selection, instead of invoking different applications.

Operations are module based, so that new ones can be downloaded and immediately integrated into all relevant contexts. They are information objects like everything else in the system, and can therefore be manipulated in the same way. The extensibility of the data model is directly due to the RDF model, where resources and properties can be arbitrarily extended using URI pointers to further resources.

The RDF-based client software runs in Java SDK v1.4 or later. The prototype versions remain firmly in the 'play with it' proof-of-concept stages. Although claimed to be robust enough, the design team makes no guarantees about either interface or data model stability – later releases might prove totally incompatible in critical ways because core formats are not yet finalized.

The prototype also makes rather heavy demands on the platform resources (*MS Windows* or *Linux*) – high-end GHz P4 computers are recommended. In part because of its reliance on the underlying JVM, users experience it as slow. Several representative screen captures of different contexts of the current version are given at the site (*haystack.lcs.mit.edu/ screenshots.html*).

Haystack may represent the wave of the future in terms of a new architecture for client software – extensible and adaptive to the Semantic Web. The release of the Semantic Web Browser component, announced in May 2004, indicates the direction of development. An ongoing refactoring of the Haystack code base aims to make it more modular, and promises to give users the ability to configure their installations and customize functionality, size, and complexity.

Digital Libraries

An important emergent field for both Web services in general, and the application of RDF structures and metadata management in particular, is that of digital libraries. In many respects, early digital library efforts to define metadata exchange paved the way for later generic Internet solutions.

In past years, efforts to create digital archives on the Web have tended to focus on single-medium formats with an atomic access model for specified items. Instrumental in achieving a relative success in this area was the development of metadata standards, such as Dublin

Core or MPEG-7. The former is a metadata framework for describing simple text or image resources, the latter is one for describing audio-visual resources.

The situation in utilizing such archives hitherto is rather similar to searching the Web in general, in that the querying party must in advance decide which medium to explore and be able to deal explicitly with the retrieved media formats.

However, the full potential of digital libraries lies in their ability to store and deliver far more complex multimedia resources, seamlessly combining query results composed of text, image, audio, and video components into a single presentation. Since the relationships between such components are complex (including a full range of temporal, spatial, structural, and semantic information), any descriptions of a multimedia resource must account for these relationships.

Bit 8.11   Digital libraries should be medium-agnostic services

Achieving transparency with respect to information storage formats requires powerful metadata structures that allow software agents to process and convert the query results into formats and representational structures with which the recipient can deal.

Ideally, we would like to see a convergence of current digital libraries, museums, and other archives towards generalized memory organizations – digital repositories capable of responding to user or agent queries in concert. This goal requires a corresponding convergence of the enabling technologies necessary to support such storage, retrieval, and delivery functionality.

In the past few years, several large scale projects have tackled practical implementation in a systematic way. One massive archival effort is the *National Digital Information Infrastructure and Preservation Program* (NDIIPP, *www.digitalpreservation.gov*) led by the U.S. Library of Congress. Since 2001, it has been developing a standard way for institutions to preserve LoC digital archives.

In many respects, the Web itself is a prototype digital library, albeit arbitrary and chaotic, subject to the whims of its many content authors and server administrators. In an attempt at independent preservation, digital librarian Brewster Kahle started the *Internet Archive* (*www.archive.org*) and its associated search service, the *Way Back Machine*. The latter enables viewing of at least some Web content that has subsequently disappeared or been altered. The archive is mildly distributed (mirrored), and currently available from three sites.

A more recent effort to provide a systematic media library on the Web is the BBC Archive. The BBC has maintained a searchable online archive since 1997 of all its Web news stories (see *news.bbc.co.uk/hi/english/static/advquery/advquery.htm*). The BBC Motion Gallery (*www.bbcmotiongallery.com*), opened in 2004, extends the concept by providing direct Web access moving image clips from the BBC and CBS News archives. The BBC portion available online spans over 300,000 hours of film and 70 years of history, with a million more hours still offline.

Launched in April 2005, the BBC Creative Archive initiative (*creativearchive.bbc.co.uk*) is to give free (U.K.) Web access to download clips of BBC factual programmes for non-commercial use. The ambition is to pioneer a new approach to public access rights in the digital age, closely based on the U.S. Creative Commons licensing. The hope is that it would eventually include AV archival material from most U.K. broadcasters, organizations, and creative individuals. The British Film Institute is one early sign-on to the pilot project which should enter full deployment in 2006.

Systematic and metadata-described repositories are still in early development, as are technologies to make it all accessible without requiring specific browser plug-ins to a proprietary format. The following sections describe a few such prototype sweb efforts.

Applying RDF Query Solutions

One of the easier ways to hack interesting services based on digital libraries is, for example, to leverage the *Dublin Core RDF* model already applied to much stored material. *RDF Query* gives significant interoperability with little client-side investment, with a view to combining local and remote information.

Such a solution can also accommodate custom schemas to map known though perhaps informally Web-published data into RDF XML (and DC schema), suitable for subsequent processing to augment the available RDF resources. Both centralized and grassroot efforts are finding new ways to build useful services based on RDF-published data.

Social (and legal) constraints on reusing such 'public' data will probably prove more of a problem than any technical aspects. Discussions of this aspect are mostly deferred to the closing chapters. Nevertheless, we may note that the same RDF technology can be implemented at the resource end to constrain access to particular verifiable and acceptable users (Web Access). Such users may be screened for particular 'credentials' relevant to the data provenance (perhaps colleagues, professional categories, special interest groups, or just paying members).

With the access can come annotation functionality, as described earlier. In other words, not only are the external data collections available for local use, but local users may share annotations on the material with other users elsewhere, including the resource owners. Hence the library resource might grow with more interleaved contributions.

We also see a trend towards the *Semantic Portal* model, where data harvested from individual sites are collected and 'recycled' in the form of indexing and correlation services.

Project Harmony

The *Harmony Project* (found at *www.metadata.net/harmony/*) was an international colla-boration funded by the Distributed Systems Technology Centre (DSTC, *www.dstc.edu.au*), Joint Information Systems Committee (JISC, *www.jisc.ac.uk*), and National Science Foun-dation (NSF, *www.nsf.gov*), which ran for three years (from July 1999 until June 2002).

The goal of the Harmony Project was to investigate key issues encountered when describing complex multimedia resources in digital libraries, the results (published on the site) applied to later projects elsewhere. The project's approach covered four areas:

- **Standards**. A collaboration was started with metadata communities to develop and refine developing metadata standards that describe multimedia components.
- **Conceptual Model**. The project devised a conceptual model for interoperability among community-specific metadata vocabularies, able to represent the complex structural and semantic relationships that might be encountered in multimedia resources.

- **Expression**. An investigation was made into mechanisms for expressing such a conceptual model, including technologies under development in the W3C (that is, XML, RDF, and associated schema mechanisms).
- **Mapping**. Mechanisms were developed to map between community-specific vocabularies using the chosen conceptual model.

The project presented the results as the ABC model, along with pointers to some practical prototype systems that demonstrate proof-of-concept. The ABC model is based in XML (the syntax) and RDF (the ontology) – a useful discursive overview is 'The ABC Ontology and Model' by Carl Lagoze and Jane Hunter, available in a summary version (at *jodi.ecs.soton. ac.uk/Articles/v02/i02/Lagoze/*), with a further link from there to the full text in PDF format.

The early ABC model was refined in collaboration with the *CIMI Consortium* (*www. cimi.org*), an international association of cultural heritage institutions and organizations working together to bring rich cultural information to the widest possible audience. From 1994 through 2003, CIMI ran Project CHIO (Cultural Heritage Information Online), an SGML-based approach to describe and share museum and library resources digitally.

Application to metadata descriptions of complex objects provided by CIMI museums and libraries resulted in a metadata model with more logically grounded time and entity semantics. Based on the refined model, a metadata repository of RDF descriptions and new search interface proved capable of more sophisticated queries than previous less-expressive and object-centric metadata models.

Although CIMI itself ceased active work in December 2003 due to insufficient funding, several aspects lived on in addition to the published Web resources:

- *Handscape* (*www.cimi.org/whitesite/index.html*), active until mid-2004, explored the means for providing mobile access in a museum environment using existing hand-held devices, such as mobile phones, to access the museum database and guide visitors.
- MDA (*www.mda.org.uk*), an organization to support the management and use of collections, is also the owner and developer of the SPECTRUM international museum data standard.
- CIMI XML Schema (*www.cimi.org/wg/xml_spectrum/index.html*), intended to describe museum objects (and based on SPECTRUM) and an interchange format of OAI (Open Archives Initiative) metadata harvesting, is currently maintained by MDA.

Prototype Tools

A number of prototype tools for the ABC ontology model emerged during the work at various institutions. While for the most part 'unsupported' and intended only for testing purposes, they did demonstrate how to work with ABC metadata in practice. Consequently, these tools provided valuable experience for anyone contemplating working with some implementation of RDF schema for metadata administration.

One such tool was the *Cornell ABC Metadata Model Constructor* by David Lin at the Cornell Computer Science Department (*www.cs.cornell.edu*). The Constructor (demo and prototype download at *www.metadata.net/harmony/constructor/ABC_Constructor.htm*) is a pure Java implementation, portable to any Java-capable platform, that allows the user to construct, store, and experiment with ABC models visually. Apart from the Java RTE runtime, the tool also assumes the *JenaAPI* relational-database back-end to manage the RDF data.

This package is freely available from *HPL Semweb* (see *www.hpl.hp.com/semweb/ jena-top.html*).

The Constructor tool can dynamically extend the ontology in the base RDF schema to more domain-specific vocabularies, or load other prepared vocabularies such as qualified or unqualified Dublin Core.

DSTC demonstration tools encompass a number of online search and browse interfaces to multimedia archives. They showcase different application contexts for the ABC model and include a test ABC database of some 400 images contributed from four museums, the *SMIL Lecture and Presentation Archive*, and the *From Lunchroom to Boardroom MP3 Oral History Archive*.

The DSTC prototypes also include *MetaNet* (*sunspot.dstc.edu.au:8888/Metanet/Top. html*), which is an online English dictionary of '-nyms' for metadata terms. A selectable list of 'core' metadata words (for example, agent) can be expanded into a table of *synonyms* (equivalent terms), *hyponyms* (narrower terms), and *hypo-hyponyms* (the narrowest terms). The objective is to enable semantic mappings between synonymous metadata terms from different vocabularies.

The *Institute for Learning and Research Technology* (ILRT, *www.ilrt.bristol.ac.uk*) provides another selection of prototype tools and research.

Schematron, for example, throws out the regular grammar approach used by most implementations to specify RDF schema constraints and instead applies a rule-based system that uses *XPath* expressions to define assertions that are applied to documents. Its unique focus is on *validating* schemas rather than just defining them – a user-centric approach that allows useful feedback messages to be associated with each assertion as it is entered.

Creator Rick Jelliffe makes the critique that alternatives to the uncritically accepted grammar-based ontologies are rarely considered, despite the observation that some constraints are difficult or impossible to model using regular grammars. Commonly cited examples are co-occurrence constraints (if an element has attribute A, it must also have attribute B) and context-sensitive content models (if an element has a parent X, then it must have an attribute Y). In short, he says:

> *If we know XML documents need to be graphs, why are we working as if they are trees? Why do we have schema languages that enforce the treeness of the syntax rather than provide the layer to free us from it?*

A comparison of six schema languages (at *www.cobase.cs.ucla.edu/tech-docs/dongwon/ ucla-200008.html*) highlights how far *Schematron* differs in its design. Jeliffe maintains that the rule-based systems are more expressive. A balanced advocacy discussion is best summarized as the feeling that grammars are better that rule-based systems for some things, while rule-based systems are better than grammars for other things. In 2004, the *Schematron* language specification was published as a draft ISO standard.

Another interesting ILRT tool is *RDFViz* (*www.rdfviz.org*). The online demo can generate graphic images of RDF data in DOT, SVG, and 3D VRML views.

The *Rudolf Squish* implementation (*swordfish.rdfweb.org/rdfquery/*) is a simple RDF query engine written in Java. The site maintains a number of working examples of RDF query applications, along with the resources to build more. The expressed aim is to present practical and interesting applications for the Semantic Web, exploring ways to make them real, such as with co-depiction photo metadata queries.

One such example is **FOAF** (Friends of a Friend) described in Chapter 10 in the context of sweb technology for the masses. The FOAF project is also exploring social implications and anti-spam measures. The system provides a way to represent a harvesting-opaque 'hashed' e-mail address. People can be reliably identified without openly having to reveal their e-mail address. The FOAF *whitelists experiment* takes this concept a step further by exploring the use of FOAF for sharing lists of non-spammer mailboxes, to aid in implementing collaborative mail filtering tools.

DSpace

One major digital archive project worth mentioning is *DSpace* (*www.dspace.org*), a joint project in 2002 by *MIT Libraries* and *Hewlett-Packard* to capture, index, preserve, and distribute the intellectual output of the *Massachusetts Institute of Technology*. The release software is now freely available as open source (at *sourceforge.net/projects/dspace/*). Research institutions worldwide are free to customize and extend the system to fit their own requirements.

Designed to accommodate the multidisciplinary and organizational needs of a large institution, the system is organized into 'Communities' and 'Collections'. Each of these divisions retains its identity within the repository and may have customized definitions of policy and workflow.

With more than 10,000 pieces of digital content produced each year, it was a vast collaborative undertaking to digitize MIT's educational resources and make them accessible through a single interface. MIT supported the development and adoption of this technology, and of federation with other institutions. The experiences are presented as a case study (*dspace.org/implement/case-study.pdf*).

DSpace enables institutions to:

- *capture* and *describe* digital works using a submission workflow module;
- *distribute* an institution's digital works over the Web through a search and retrieval system;
- *preserve* digital works over the long term – as a sustainable, scalable digital repository.

The multimedia aspect of archiving can accommodate storage and retrieval of articles, preprints, working papers, technical reports, conference papers, books, theses, data sets, computer programs, and visual simulations and models. Bundling video and audio bit streams into discrete items allows lectures and other temporal material to be captured and described to fit the archive.

The broader vision is of a *DSpace* federation of many systems that can make available the collective intellectual resources of the world's leading research institutions. MIT's implementation of *DSpace*, which is closely tied to other significant MIT digital initiatives such as MIT *OpenCourseWare* (OCW), is in this view but a small prototype and preview of a global repository of learning.

Bit 8.12   *DSpace* encourages wide deployment and federation

In principle anyone wishing to share published content can set up a *DSpace* server and thus be ensured of interoperability in a federated network of *DSpace* providers.

The Implementation

DSpace used a qualified version of the Dublin Core schema for metadata, based on the DC *Libraries Application Profile* (LAP), but adapted to fit the specific needs of the project. This selection is understandable as the requirements of generic digital libraries and MIT publishing naturally coincide in great measure.

The way data are organized is intended to reflect the structure of the organization using the system. Communities in a *DSpace* site, typically a university campus, correspond to laboratories, research centers, or departments. Groupings of related content within a Community make up the Collections.

The basic archival element of the archive is the *Item*, which may be further subdivided into bitstream bundles. Each bitstream usually corresponds to an ordinary computer file. For example, the text and image files that make up a single Web document are organized as a bundle belonging to the indexed document item (specifically, as the Dublin Core metadata record) in the repository.

Figure 8.5 shows the production system deployed by MIT Libraries (at *libraries.mit.edu/dspace*).

The single public Web interface allows browsing or searching within any or all of the defined Communities and Collections. A visitor can also subscribe to e-mail notification when items are published within a particular area of interest.

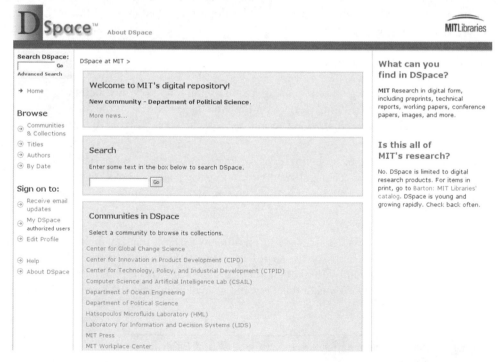

Figure 8.5 Top Web page to MIT Libraries, the currently deployed DSpace home. While limited to already digital research 'products', the repository is constantly growing

Table 8.1 Conceptual view of the DSpace-adapted Dublin Core metadata model. The actual qualifier terms have been recast into a more readable format in this table

Element	Qualifiers	Notes
Contributor	Advisor, Author, Editor, Illustrator, Other	A person, organization, or service responsible for the content of the resource. Possibly unspecified.
Coverage	Spatial, Temporal	Characteristics of the content.
Creator	N/A	Only used for harvested metadata, not when creating.
Date	Accessioned, Available, Copyright, Created, Issued, Submitted	Accessioned means when DSpace took possession of the content.
Identifier	Govdoc, ISBN, ISSN, SICI, ISMN, Other, URI	See Glossary entry for **Identifier**.
Description	Abstract, Provenance, Sponsorship, Statement of responsibility, Table of contents, URI	Provenance refers to the history of custody of the item since its creation, including any changes successive custodians made to it.
Format	Extent, Medium, MIME type	Size, duration, storage, or type.
Language	ISO	Unqualified is for non-ISO values to specify content language.
Relation	Is format of, Is part of, Is part of series, Has part, Is version of, Has version, Is based on, Is referenced by, Requires, Replaces, Is replaced by, URI	Specifies the relationship of the document with other related documents, such as versions, compilations, derivative works, larger contexts, etc.
Rights	URI	Terms governing use and reproduction of the content.
Source	URI	Only used for harvested metadata, not when creating.
Subject	Classification, DDC, LCC, LCSH, MESH, Other	Index term, typically a formal content classification system.
Title	Alternative	Title statement or title proper. Alternative is for variant form of title proper appearing in item, such as for a translation.
Type	N/A	Nature or genre of content.

The design goal of being a sustainable, scalable digital repository (capable of holding the more than 10,000 pieces of digital content produced by MIT faculty and researchers each year) places heavy demands on efficient searching and notification features.

The metadata structure for *DSpace* can be illustrative of how a reasonably small and simple structure can meet very diverse and demanding requirements. Table 8.1 outlines the core terms and their qualifiers, used to describe each archived item in the RDF metadata.

We may note the heavy linkage to existing standards (institutional, national and international) for systematically identifying and classifying published intellectual works. Note also the reference to 'harvesting' item metadata from other sources.

From an architectural point of view, *DSpace* can be described as three layers:

- **Application**. This top layer rests on the *DSpace* public API, and for example supports the Web user interface, metadata provision, and other services. Other Web and envisioned Federation services emanate from this API as well.

- **Business logic**. In the middle layer, we find system functionality, with administration, browsing, search, recording, and other management bits. It communicates by way of the public API to service the Application layer, and by way of the Storage API to access the stored content.
- **Storage layer**. The entire edifice rests on this bottom layer, representing the physical storage of the information and its metadata. Storage is virtualized and managed using various technologies, but central is a RDMS wrapper system that currently builds on *PostgreSQL* to answer queries.

Preservation Issues

Preservation services are potentially an important aspect of *DSpace* because of the long-term storage intention. Therefore, it is vital also to capture the specific formats and format descriptions of the submitted files.

The bitstream concept is designed to address this requirement, using either an implicit or explicit reference to how the file content can be interpreted. Typically, and when possible, the reference is in the form of a link to some explicit standard specification, otherwise it is linked implicitly to a particular application. Such formats can thus be more specific than MIME-type.

Support for a particular document format is an important issue when considering preservation services. In this context, the question is how long into the future a hosting institution is likely to be able to preserve and present content of a given format – something that should be considered more often in general, not just in this specific context.

Bit 8.13   Simple storage integrity is not the same as content preservation

Binary data is meaningless without context and a way to reconstruct the intended presentation. Stored documents and media files are heavily dependent on known representation formats.

Storing bits for 100 years is easier than preserving content for 10. It does us no good to store things for 100 years if format drift means our grandchildren can't read them.
        Clay Shirky, professor at New York University and consultant to the Library of Congress

Each file submitted to *DSpace* is assigned to one of the following categories:

- **Supported** formats presume published open standards.
- **Known** formats are recognized, but no guarantee of support is possible, usually because of the proprietary nature of the format.
- **Unsupported** formats are unrecognized and merely listed as unknown using the generic 'application/octet-stream' classification.

Note that the category 'supported' implies the ability to make the content usable in the future, using whatever combination of techniques (such as migration, emulation, and so on) is appropriate given the context of the retrieval need. It assumes required markup-parsing tools can be built from published specifications.

Bit 8.14   Proprietary formats can never be fully supported by any archival system

Although documents stored in closed formats might optionally be viewable or convertible in third-party tools, there is a great risk that some information will be lost or misinterpreted. In practice, not even the format owner guarantees full support in the long term, a problem encountered when migrating documents between software versions.

Proprietary formats for which specifications are not publicly available cannot be supported in *DSpace*, although the files may still be preserved. In cases where those formats are native to tools supported by MIT Information Systems, guidance is available on converting files into open formats that are fully supported.

However, some 'popular' proprietary formats might in practice seem well supported, even if never classified as better than 'known', as it assumes enough documentation can be gathered to capture how the formats work. Such file specifications, descriptions, and code samples are made available in the *DSpace Format Reference Collection.*

In general, *MIT Libraries DSpace* makes the following assertive claims concerning format support in its archives:

- Everything put in *DSpace* will be retrievable.
- We will recognize as many file formats as possible.
- We will support as many known file formats as possible.

The first is seen as the most important in terms of archive preservation.

There are two main approaches to practical digital archiving: *emulation* and *migration*. Capturing format specifications allow both, and also on-the-fly conversion into current application formats. Preserving the original format and converting only retrieved representations has the great advantage over migration that no information is lost even when an applied format translation is imperfect. A later and better conversion can still be applied to the original. Each migration will however permanently lose some information.

Removal of archived material is handled in two ways:

- An item might be 'withdrawn', meaning hidden from view – the user is presented with a tombstone icon, perhaps with an explanation of why the material is no longer available. The item is, however, still preserved in the archive and might be reinstated at some later time.
- Alternatively, an item might be 'expunged', meaning it is completely removed from the archive. The hosting institution would need some policy concerning removal.

Simile

Semantic Interoperability of Metadata and Information in unLike Environments is the long name for the *Simile* joint project by W3C, HP, MIT Libraries, and MIT CSAIL to build a persistent digital archive (*simile.mit.edu*).

The project seeks to enhance general interoperability among digital assets, schemas, metadata, and services across distributed stores of information – individual, community, and institutional. It also intends to provide useful end-user services based on such stores. *Simile* leverages and extends *DSpace*, enhancing its support for arbitrary schemas and metadata, using RDF and other sweb technologies. The project also aims to implement a digital asset dissemination architecture based on established Web standards, in places called '*DSpace II*'.

The effort seeks to focus on well-defined, real-world use cases in the library domain, complemented by parallel work to deploy *DSpace* at a number of leading research libraries. The desire is to demonstrate compellingly the utility and readiness of sweb tools and techniques in a visible and global community.

Candidate use cases where *Simile* might be implemented include annotations and mining unstructured information. Other significant areas include history systems, registries, image support, authority control, and distributed collections. Some examples of candidate prototypes are in the list that follows:

- Investigate use of multiple schemas to describe data, and interoperation between multiple schemas;
- Prototype dissemination kernel and architecture;
- Examine distribution mechanisms;
- Mirroring *DSpace* relational database to RDF;
- Displaying, editing, and navigating RDF;
- *RDF Diff*, or comparing outputs;
- Semantic Web processing models;
- History system navigator;
- Schema registry and submission process;
- Event-based workflow survey and recommendations;
- Archives of *Simile* data.

The project site offers service demonstrations, data collections, ontologies, and a number of papers and other resources. Deliverables are in three categories: *Data Acquisition, Data Exploration* and *Metadata Engine*.

Web Syndication

Some might wonder why Web syndication is included, albeit briefly, in a book about the Semantic Web. Well, one reason is that aggregation is often an aspect of syndication, and both of these processes require metadata information to succeed in what they attempt to do for the end user. And as shown, RDF is involved.

Another reason is that the functionality represented by syndication/aggregation on the Web can stand as an example of useful services on a deployed Semantic Web infrastructure. These services might then be augmented with even more automatic gathering, processing and filtering than is possible over the current Web.

A practical application has already evolved in the form of the *semblog,* a SWAD-related development mentioned in Chapter 7. In Chapter 9, some examples of deployed applications of this nature are described.

RSS and Other Content Aggregators

RSS, which originally stood for *RDF Site Summary,* is a portal content language. It was introduced in 1999 by *Netscape* as a simple XML-based channel description framework to gather content site snapshots to attract more users to its portal. A by-product was headline syndication over the Web in general.

Today, the term RSS (often reinterpreted as *Rich Site Summary*) is used to refer to several different but related things:

- a lightweight syndication format;
- a content syndication system;
- a metadata syndication framework.

In its brief existence, RSS has undergone only one revision, yet has been adopted as *one of the most widely used Web site XML applications.* The popularity and utility of the RSS format has found uses in many more scenarios than originally anticipated by its creators, even escaping the Web altogether into desktop applications.

A diverse infrastructure of different registries and feed sources has evolved, catering to different interests and preferences in *gathering* (and possibly *processing* and *repackaging*) summary information from the many content providers. However, RSS has in this development also segued away from its RDF metadata origins, instead dealing more with actual content syndication than with metadata summaries.

Although a 500-character constraint on the description field in the revised RSS format provides enough room for a blurb or abstract, it still limits the ability of RSS to carry deeper content. Considerable debate eventually erupted over the precise role of RSS in syndication applications.

Opinions fall into three basic camps in this matter of content syndication using RSS:

- support for content syndication in the RSS core;
- use of RSS for metadata and of *scriptingNews* for content syndication;
- modularization of lightweight content syndication support in RSS.

The paragraph-based content format *scriptingNews* has a focus on Web writing, which over time has lent some elements to the newer RSS specification (such as the item-level description element).

But as RSS continues to be redesigned and re-purposed, the need for an enhanced metadata framework also grows. In the meantime existing item-level elements are being overloaded with metadata and markup, even RDF-like elements for metadata inserted *ad hoc*. Such extensions cause increasing problems for both *syndicators* and *aggregators* in dealing with variant streams.

Proposed solutions to these and future RSS metadata needs have primarily centered around the inclusion of more optional metadata elements in the RSS core, essentially in the grander scheme putting the RDF back into RSS, and a greater modularization based on XML namespaces.

On the other hand, if RSS cannot accommodate the provision of support in the different directions required by different developers, it will probably fade in favor of more special purpose formats.

Many people find aggregator/RSS and portal sites such as *O'Reilly's Meerkat* (see Chapter 9) or *NewsIsFree* (*www.newsisfree.com*) invaluable time savers. Subscribing to such services provides summary (or headline) overviews of information that otherwise must be manually visited at many sites or culled from long newsletters and mailing lists.

Metadata Tools

In constructing toolkits for building arbitrary metadata systems on the Web, several key usability features have been identified – for example, the availability of assisted selection, automated metadata creation, and portal generation.

A brief summary by category concludes this chapter. See these lists as starting points for further investigation. Some toolkits are discussed in broader contexts elsewhere in the book to highlight particular environments or ontologies. The tools selected here are deemed representative mainly because they are referenced in community lists by the people most likely to have reason to use them.

Browsing and Authoring Tools

This category includes tools for creating and viewing metadata, usually in RDF.

- *IsaViz* (*www.w3.org/2001/11/IsaViz/*) is a visual browsing and authoring tool (or actually, environment) developed and maintained by the W3C for RDF models. Users can browse and author RDF models represented as graphs. V2.1 was released in October 2004.
- *Metabrowser* (*metabrowser.spirit.net.au*) is a server-client application pair. The browser client can show metadata and content of Web pages simultaneously. The commercial version also allows metadata to be edited and created. The server provides a full range of site-search features for visitors.
- *xdirectory* (*www.esprit-is.com*) offers a fully configurable, browser-based environment for creating and publishing web-based information directories. It is a commercial product.
- *Protégé* (discussed in detail earlier) allows domain experts to build knowledge-based systems by creating and modifying reusable ontologies and problem-solving methods. It has support for editing RDF schema and instance data knowledge bases.

Metadata Gathering Tools

In many cases, the primary goal is not to create the metadata, but to harvest or derive it from already published Web content and generate catalogs. A selection of toolkit applications follow:

- *Scorpion* (*www.oclc.org/research/software/default.htm*) is a project of the OCLC Office of Research exploring the indexing and cataloging of electronic resources, with automatic subject recognition based on well-known schemes like DDC (Dewey Decimal) or LCC (Library of Congress).
- *Mantis* (*orc.rsch.oclc.org:6464/toolkit.html*) is an older toolkit for building Web-based cataloging systems with arbitrary metadata definitions and interfaces administration.

A user can create a template without knowing XML. Mantis includes conversion and an integrated environment.

- *DC-dot* (*www.ukoln.ac.uk/metadata/dcdot/*) extracts and validates Dublin Core metadata (DCM) from HTML resources and Web-published MS Office files (and can perform conversion). The site offers a Web form to generate and edit metadata online, generated for a document at any specified URL, optionally as tagged RDF.
- *MKDoc* (*www.mkdoc.org*) is a GPL CMS that can produce both HTML and RDF (as qualified DCM) for every document on a Web site. It also supports DCM via the generation of RSS 1.0 syndication feeds.
- *Nordic URN generator* (*www.lib.helsinki.fi/meta/*) is a Web service integrated into the DC metadata creator to allocate URN identifiers for resources in the Nordic countries, according to the French NBN library specifications.
- *IllumiNet Corpus* (*www.illuminet.se*) is a commercial search engine service written in Java that can index content, context, and metadata in documents on the network or on the local file system and output XML and RSS results. It is used as a 'distributed relational database' for document-oriented solutions with metadata as indexed keys.
- *Klarity* (*www.klarity.com.au*) includes a 'feature parser' (also as developer SDK) that can automatically generate metadata for HTML pages based on concepts found in the text. Other products deal with summaries and metrics of '*aboutness*'.
- *HotMETA* (*www.dstc.edu.au/Products/metaSuite/HotMeta.html*) is a commercial entry-level Java product for managing DC (or similar) metadata. It consists of *Broker* (a Web-enabled metadata repository and query engine), *Gatherer* (a Web crawler and document indexer), and *MetaEdit* (a grapical metadata editor for creating, validating, and maintaining metadata in the repository and stand-alone files).

As can be seen, the ambition level varies considerably, but there are many solutions to automate converting and augmenting existing Web content with metadata.

9

Examples of Deployed Systems

There is at present no distinct line between prototyping and deployment of Semantic Web systems – it is all development testing so far. Therefore, the decision to here shift to a new chapter about 'deployed' systems, with some apparent overlap of previous project descriptions, may seem arbitrary.

Why, for example, is not *Annotea* discussed here? It could have been, but it served better as an introduction in Chapter 8 to 'tools'. Also, Web annotation as an application has not yet found as widespread acceptance as it deserves (many due to a lack of user awareness and to the absence of general client support) so it does not really qualify as 'deployed'.

When does a prototype system become so pervasive in practical applications that one can legitimately speak of a deployed system? This transition is discernible only in hindsight, when the full usage pattern becomes clearer. However, it is in the nature of these open-source systems that an exact user base cannot easily be determined, nor the degree to which they start to support other systems, such as agents.

Most 'deployed' examples are essentially Web-accessible RDF-structures and the services associated with them. In itself, RDF packaging is a significant step forward, giving a large potential for software agents, but like any infrastructure it lacks dramatic public appeal. Nothing can yet aspire to the 'killer application' level of public awareness.

Most likely, the transition from Web to Semantic Web will not be a dramatic, watershed moment, but rather a gradual, almost unnoticed accumulation of more services and nifty features based on various aspects of emerging sweb technology. One day, if we stop to think about it, we will just notice that the Web is different – the infrastructure was upgraded, the agents are at work, and convenience is greatly enhanced from the way we remember it.

Consensus incremental change is like that. It sneaks up on you. It can change the world while you are not looking, and you often end up wondering how you ever managed without it all.

Chapter 9 at a Glance

This chapter examines a number of Semantic Web application areas where some aspect of the technology is deployed and usable today. *Application Examples* is a largely random sampler of small and large 'deployments' that in some way relate to SW technologies.

The Semantic Web: Crafting Infrastructure for Agency   Bo Leuf
© 2006 John Wiley & Sons, Ltd

- *Retsina Semantic Web Calendar Agent* represents a study of a prototype personal assistant agent, an example of what might become pervasive PA functionality in only a few years.
- *MusicBrainz and Freedb* shows how metadata and RDF exchange is already present in common user contexts on the Web, quietly powering convenience features.
- *Semantic Portals and Search* describes practical deployment of semantic technology for the masses, in the form of metadata-based portals and search functionality. A related type of deployment is the semantic weblog with a view to better syndication.
- *WordNet* is an example of a semantic dictionary database, available both online and to download, which forms a basis for numerous sweb projects and applications.
- *SUMO Ontology* promotes data interoperability, information search and retrieval, auto-mated inferencing, and natural language processing.
- *Open Directory Project* is the largest and most comprehensive human-edited directory of the Web, free to use for anyone, and available as RDF-dumps from the Web.
- *Ontolingua* provides a distributed collaborative environment to browse, create, edit, modify, and use ontologies. The section describes several related projects.

Industry Adoption assesses the scale of deployment and the adoption of frameworks surrounding sweb technologies.

- *Adobe XMP* represents a large-scale corporate adoption of RDF standards and embedding support in all major publishing-tool products.
- *Sun Global Knowledge Engineering (GKE)* is developing an infrastructure to manage distributed processing of semantic structures.

Implemented Web Agents examines the current availability of agent technology in general.

- *Agent Environments* discusses agents that directly interact with humans in the workspace or at home.
- *Intelligent Agent Platforms* examines mainly the FIPA-compliant specifications and implementations that underlie industrial adoption.

Application Examples

Practical applications of sweb core technologies are many and varied, more than one might suspect, but not always advertised as being such. Scratch the surface of any system for 'managing information' these days and the chances are good that at least some components of it represent the direct application of some specific sweb technologies.

Managing information is more and more seen as managing information metadata; the stored information needs to be adequately described. In general, a significant driving motivation to publish and share open metadata databases is to confer greater interoperability and reuse of existing data. Greater access and interoperability in turn promotes the discovery of new uses though relational links with other database resources.

Bit 9.1   Published and shared databases promote metadata consistency
Shared databases promote a publish-once mentality. This can focus reuse and error checking, and the development of relational links to other useful resources.

Since users and applications can refer to the same public metadata, the risk of introducing new errors in published information is greatly reduced. The tendency is then to have closer ties between the published metadata and the sources, which also promotes accuracy, since the people who update the data are the ones who have the most interest in them being correct.

Deployed systems reflect not just the technology, but also the social and commercial issues they reveal in their design and implementation. Sweb systems in general show a clear slant towards open access, distributed and individual responsibility for own data, and collaborative efforts. Also discernible is the remarkable convergence of design based on self-interest (seen in commercial efforts) and design based on the public good (as promoted by the W3C and various other organizations). In both cases, open and collaborative efforts give the greater returns.

Overall, the desire is to promote simplicity and transparency in usage. The complex systems should appear simple, very simple, to the casual user. Often known as the *'Perl Philosophy'* in its canonic formulation (but by Alan Kay, of *Smalltalk* and *Object Oriented Programming* fame), the 'simple things should be simple and complex things should be possible' maxim is a powerful design factor, and an ideal worth striving for.

Bit 9.2   Sweb technology should inform and empower, not distract or intimidate

The ultimate goal of many efforts is, after all, the ability to delegate most tasks to automatic agents. Ideally, such delegation should require as little detailed instruction as possible from the user, relying instead on logic rules and inference.

A dominating proportion of the following examples deal directly with human language, which is perhaps to be expected. In this early stage of the Semantic Web, most of the development focus is still on implementing the capability for software to reason around human-readable text (and ultimately also speech and gesture) in relation to objects and events in the real world.

Increasingly, however, applications will reference 'raw data' packaged as RDF assertions, leaving human-targeted representations (as text, speech, visual, or other media) until the final user presentation.

Early fundamental application examples typically build on using XML and RDF, and one or more of the sweb protocols, to communicate. For example, leveraging *RDF Database Access Protocol* results in several practical use cases:

- *Web Page Access Control Lists (ACL)*, where a Web server queries the RDF database to authenticate access to a given URI-defined resource before continuing processing. A server-side Web application (itself access controlled) allows modification of the access control information. (Implemented for the W3C Web site.)
- *Web Page Annotations (Annotea,* described in Chapter 8), which represents an 'enhanced' Web browser capable of communicating with a database of 'annotations' about pages or portions of pages. The browser allows annotations to be added, and automatically indicates the presence of annotations. (Implemented in W3C *Amaya*).
- *Semantic Web Browser*, which provides a user interface to view and alter the properties of objects as recorded in RDF databases. The user can manage databases, and attach validation and inference processors.

One problem in any overview of this kind of leading edge technology, and the applications based on it, is that individual implementations – in fact, entire families of implementations – can suddenly fall by the wayside due to significant shifts in the underlying concepts, or changes in priorities or funding. An active development topic can for any number of reasons unexpectedly become simply an historical footnote.

Retsina Semantic Web Calendar Agent

The *Retsina Semantic Web Calendar Agent* (*RCal,* home at *www.daml.ri.cmu.edu/Cal/*) was an early working example of RDF markup of 'events' so that they can be browsed on the Web. The project is seen as a practical example of sweb agency to solve common tasks, namely automatically coordinating distributed schedules and event postings.

Client software was developed as an add-on for *MS Windows 2000* or *XP* systems running *MS Outlook 2000* or *Outlook XP.* The resource URL indicates the origins of the project in the DAML phase of XML schema and ontology development for the Semantic Web.

RCal can use other existing ontologies, such as the *Hybrid iCal-like RDF Schema* or the *Dublin Core* ontology. It may also link to user contact information described on a Web home page.

Basic Functionality

Retsina functionality is implemented in several parts. Some of these parts reside on distributed Web services, or sometimes just in the prototype as a single server instance. Some are components that reside on the user system complementing its personal information manager (PIM).

The *Calendar Agent* (itself also called *RCal*) supports distributed meeting scheduling through a number of autonomous processes:

- It requests and compares available meeting slots from each attendee's personal calendar agent.
- It proposes optimal meeting slots for all attendees and returns this information to the others.
- Each agent compares proposed meeting requests against its own user's schedule, as defined within the PIM client's agenda component.
- Confirmed slots that each user could attend are sent back to the proposer agent, which then determines the best meeting slot.
- Replies with the final meeting schedule are distributed to the participants and automatically inserted as 'tentative' in their respective schedules.

A *Calendar Parser* can work in a synergistic way with the Calendar Agent by providing the agent with schedules posted on the Semantic Web and identifying the contact details of the attendees at the posted meetings.

Schedules can be browsed using a Web-page interface, or imported into the client. The *Electronic Secretary* is a Web-based interface to a distributed *Meeting Scheduling Agent,* which can negotiate meeting requests with *RCal* agents.

Why Another Calendar Agent?

Although calendar and scheduling agents are by no means new, sweb agency provides a different approach: more automated, open-ended, and efficient. It adds ontology models for reasoning and inference. Additionally, the RDF approach is not tied to proprietary functionality and exchange formats of any client.

Other systems that manage meeting requests on behalf of their users are only effective as far they incorporate an up-to-date and accurate model of user preferences and current activities. The difficulty in these locally managed systems lies in their reliance on explicit user input of activities and anticipated appointments – they cannot automatically alert users to irregular or unanticipated events of interest that might be announced in some public space.

Therefore, to provide assistance and reduce the load on their user, traditional calendar-agent automation paradoxically requires significant efforts from the user to maintain a valid schedule model.

Bit 9.3   Useful scheduling agents must have the capability to reason

People are more comfortable giving and tweaking general guidance rules or examples than having to input lists of specifics and spell out exhaustive detail.

RCal simplifies data acquisition to importing posted schedule markup from the Semantic Web directly into the user's calendar. The agent need only be directed to a schedule that the agent can reason about and automatically add into the calendar. Users no longer need to type in accurately each event that they would like to attend. Other services can also be dynamically offered to the user, as deemed relevant to the content of the browsed schedules.

MusicBrainz and Freedb

A very popular application of published metadata is found in online repositories of audio CD listings. The application area also illustrates issues that arise when such metadata may be viewed as infringing protected intellectual property.

In a music CD, human-readable artist and track information is provided only on the printed jewel-box cover or physical label – neither is readable by the player for playback display. Each CD does however contain identification data that can be compiled into an 'almost-unique' lookup index to appropriate metadata published in a database.

For example, *MusicBrainz* (*www.musicbrainz.org*) is a community music meta-database that attempts to create collaboratively a comprehensive music information site using RDF. The database can be searched manually on the Web, and is accessible through player clients that incorporate the query interface.

Such an online database confers the advantage of consistent metadata for selected CDs. Users can browse and search the catalog to examine what music different bands have published and how artists relate to each other, and use the information to discover new music. The application example is chosen mainly because it provides an easily accessible illustration that many readers will find familiar.

The roots of the project are in a software program developed many years ago to play audio CDs on UNIX systems and display track information from a stored list. Over time, the Internet community created a large index file with information about thousands of music CDs. The creation in 1996 of the *Internet Compact Disc Database* (*cddb.com* – site now defunct) changed the situation by applying a client-server database model instead of a single flat file. Despite entry duplication and a lack of error checking, the service became hugely popular.

Originally a free service with an open database built by the user community, CDDB was subsequently bought by *Escient* (*www.escient.com*), who made the service proprietary for its own home entertainment equipment and services. Based on the core 'Media Recognition and Metadata Delivery' service, the company (see *http://www.escient.com/entertainmentservices. html*) provides customers with an integrated movie/music guide and player that automatically recognizes and displays DVD/CD cover art, album and song titles, artist information, and more.

The product interface gives access to in-depth information, behind-the-scenes material, explore favorite artist/actor, biographies, discographies, filmographies and influences, and naturally opt in to 'personalized buying experiences'.

In the change to a commercial offering, third-party clients were closed off, both legally and by re-engineering the protocol. The former CDDB user community instead reorganized around five new open projects, of which two survived: *MusicBrainz* and *Freedb*.

Freedb (*www.freedb.org*) rebuilt a database along the same lines as the former CCDB, and currently comprises over one and a half million CD album listings. It operates under GPL (as did the original CDDB).

Unlike *Freedb, MusicBrainz* did not attempt to be a drop-in replacement for CDDB; instead it is based on a relational database and a number of features more advanced than the original CDDB concept. Its licensing model was Open Content, but a change was made to the *Creative Commons* model due to the nature of the database – the core listings as public domain.

- See Creative Commons Public Domain (*www.creativecommons.org/licenses/publicdomain*), and derived or contributor data under the Creative Commons *'Attribution-NonCommercial-ShareAlike'* License (*creativecommons.org/licenses/by-nc-sa/2.0/*).

Bit 9.4   Published and shared meta-databases must consider licensing models

---

Since databases of music CD listing metadata so directly impinge on perceived digital content rights, aggressively asserted by the music industry, proper consideration of the status of data licensing was seen as critical. Hence the extra detail in this section.

One stated goal of the collaborative repositories is to allow the Internet community to discover new music without any of the bias introduced by marketing departments of the recording industry. Independent music metadata, and its ability to identify uniquely music, also enables unambiguous communication about music.

At the time of writing, *MusicBrainz* remains the smaller offering, comprising about 212,000 album listings (or about 2.6 million tracks, though not all tracks need correspond to albums). However, it may have the greater potential. The goal is to build an extensive music encyclopaedia, owned and constructed by the users, distributed across several servers, and verified virtually error free.

In the design/vision phase as a third-generation project is *SuperBrainz*, where the core concepts are applied to other open community meta-databases, such as for book titles, DVDs, movies, encyclopaedias, and so on. The design choice of RDF is then a fortunate one, because it promises the potential of extensive agent processing as client software becomes more capable. In fact, due to the popularity of this kind of Web-based service, it is bound to promote implementation of greater client capability.

Semantic Portals and Syndication

The perceived benefit from practical application of sweb technologies, mainly XML, was initially seen as more precise search and data mining. While worth pursuing, as indicated by the discussion in Chapter 1, these are not however application areas one sees mentioned much.

Unfortunately, general XML-based searching requires an investment in data preparation that virtually nobody authoring content is willing to make. Part of the problem is that even now, few tools make metadata creation easy or transparent enough for routine online publishing.

Still, these areas are where 'sweb for the masses' is likely to have significant impact. Characteristic is that large volumes of material are published in distributed sites, and portal sites (semantic portals, or *'semportals'*) collect and update abstracts or links based on metadata.

SWED Semportal

The *Semantic Web Environmental Directory* (SWED, *www.swed.org.uk*) is a prototype of a meta-directory of environmental organizations and projects. It has the goal to develop a realistically maintainable and easy to use directory about environmental organizations and projects throughout the UK. It was developed as part of the SWAD-Europe initiative.

Where most directories centralize the storage, management, and ownership of the information, in SWED the organizations and projects themselves hold and maintain their own information. SWED instead 'harvests' information published on individual Web sites and uses it to create the directory. Using this directory, searching users may then be directed to individual sites.

The SWED directory provides one or more views of the distributed and self-published data. However, others can also harvest and collate the information and provide different views. The approach can enrich the aggregate information by adding further information which may in turn be harvested.

Meerkat WS

The *O'Reilly Network Meerkat Open Wire Service* site (*www.oreillynet.com/meerkat/*) is one of a select few useful, public Web services used by thousands of other sites daily. The public browsing interface is a Web-based syndicated content reader. It is based on the three foundation technologies of 'second-generation' WS (REST design, as described in Chapter 5): HTTP, RSS (in the XML-standard version), and URI addressing as its 'API'. Figure 9.1 shows the main public Web interface.

Figure 9.1 The public Web interface to the Meerkat WS. Thousands of sites use a common core Web (sweb) technology to automatically contribute to this popular content aggregator

However, it was not entirely clear before REST philosophy was popularized that the *Meerkat* HTTP/XML-based interface qualified as a complete WS, and not just a set of clever interfaces coupled to content aggregation. In addition, the fact that content aggregation is mainly a one-way transfer simplifies the design parameters enough so that basic HTTP suffices.

Its simplicity and reliance on core technologies means the service is powerful, robust, and scalable. Its function is dependent on the fact that it can integrate information from hundreds of other contributing sites that also use RSS over HTTP. It can be seen as a precursor to more formal Semantic Portals.

> **Bit 9.5   Many emerging WS applications will arise *ad hoc* out of the core technologies**
>
> When many sites automatically share an XML vocabulary, a protocol, and a URI namespace due to the common infrastructure, new interoperative services can arise organically, without premeditated design across the sites.

The success of *Meerkat* strongly suggests that common XML vocabularies may be the most important enabler of large-scale, distributed Web services.

Figure 9.2 The discrete checkbox, labeled 'Code fragments only', just under the search form to the upper left on this Web page, is in fact the visible sign of a rare sweb-enhanced search function

Safari Search

A deployed example of a metadata-based search can be found in the *O'Reilly Network Safari Bookshelf* (*safari.oreilly.com*). Simple queries can be formulated to be constrained to particular semantic contexts in the text body of the online books. Search results in such cases may include only those books that reference the search pattern in text typeset as code examples. The functionality relies on systematic use of known typesetting metatags.

For user convenience, the common constraint of 'code fragments only' is implemented as a checkbox in the search form, as illustrated in Figure 9.2.

An explicit HTTP-GET version includes textual context in the result list:

```
http://safari.oreilly.com/JVXSL.asp?x=1l&srchText=
    (CODE+NET::ldap)
```

A simple implementation, yet it involves rigorous data preparation behind the scenes. Just this issue is at the heart of the problem with deployment of sweb technology; common Web publishing is far more casual, and far less tolerant of such requirements.

In the database world, sweb life is easier. The trend is for online database engines to at least support XML-query, perhaps as an *XPath* implementation, as an alternative input in addition to the normal RPC mode. The commercial ones are already fully hybrid, but the OSS ones are moving the same way.

We may also note the Apache XML Project's *Apache Xindice* native-XML database ('*zeen-dee-chay*', *xml.apache.org/xindice/* and *wiki.apache.org/xindice*). As a flexible *ad-hoc* indexer and searcher, *Xindice* enables you to index all elements or attributes with wildcard parameters, and to add and drop indexes on the fly. It makes management of XML data easier, and is recently an add-on to most any application server, which in the Apache context usually means *Tomcat*.

Semblogging

XML database management gets more interesting due to a parallel development in RSS 2, where in addition to XML metadata, the content is properly namespace-declared as XHTML-body.

The upshot is that content authors who want to start sprinkling semantic cues into their weblog sites can now do so in a way that does not break existing RSS applications. The first benefit is undoubtedly self-interest, enabling a precise search of the own archive. However, it also painlessly extends the same capability to the readers by way of RSS.

The *Semantic Blogging Demonstrator (/jena.hpl.hp.com:3030/blojsom-hp/blog/)* explores the practical use of sweb technologies augmenting 'the blogging paradigm'. It is intentionally applied to the domain of bibliography management to show how sweb technologies can 'push blogging from a communal diary browsing experience to a rich information sharing activity'. Functionally, SBD is divided into *semantic view, semantic navigation* and *semantic query*.

The composite in Figure 9.3 shows two overlapped views of a weblog posting:

- Normal view is visually unremarkable except for a few extra link 'controls'.
- Table view is a first step into semantic browsing with numerous new features.
- See also the paper 'Semantic Blogging: Spreading the Semantic Web Meme' by Steve Cayzer (found at *idealliance.org/papers/dx_xmle04/papers/03-05-03/03-05-03.html*).

WordNet

During the 1990s, *Princeton WordNet (www.cogsci.princeton.edu/~wn/)* developed a comprehensive lexical database for English, where nouns, verbs, adjectives, and adverbs are

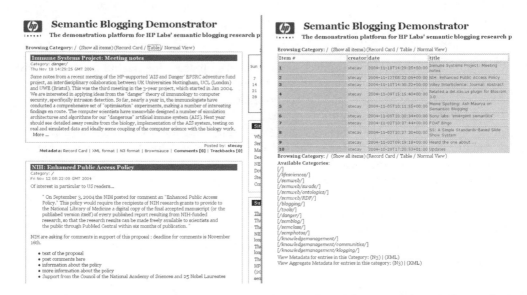

Figure 9.3 Example of a semantic blog: 'normal' view on left partially overlain by 'table' (metadata) view right of the top article. Not shown is the 'record card' view, which is a more structured text layout. Extra view/navigation/query functionality results from adding semantic metadata and markup to the basic structure and also piping it through the RSS

organized into synonym sets, each representing one underlying lexical concept. Different relations link the synonym sets.

The design is inspired by current psycholinguistic theories of human lexical memory and it exposes many semantic connections in the English language. Ontology work, such as in OWL, and lexical/liguistic applications commonly leverage *WordNet* resources and derivative databases.

Online Reference

Of specific relevance to the Semantic Web context is that the database is available online, as shown in Figure 9.4. The core can also be downloaded and integrated into various application contexts.

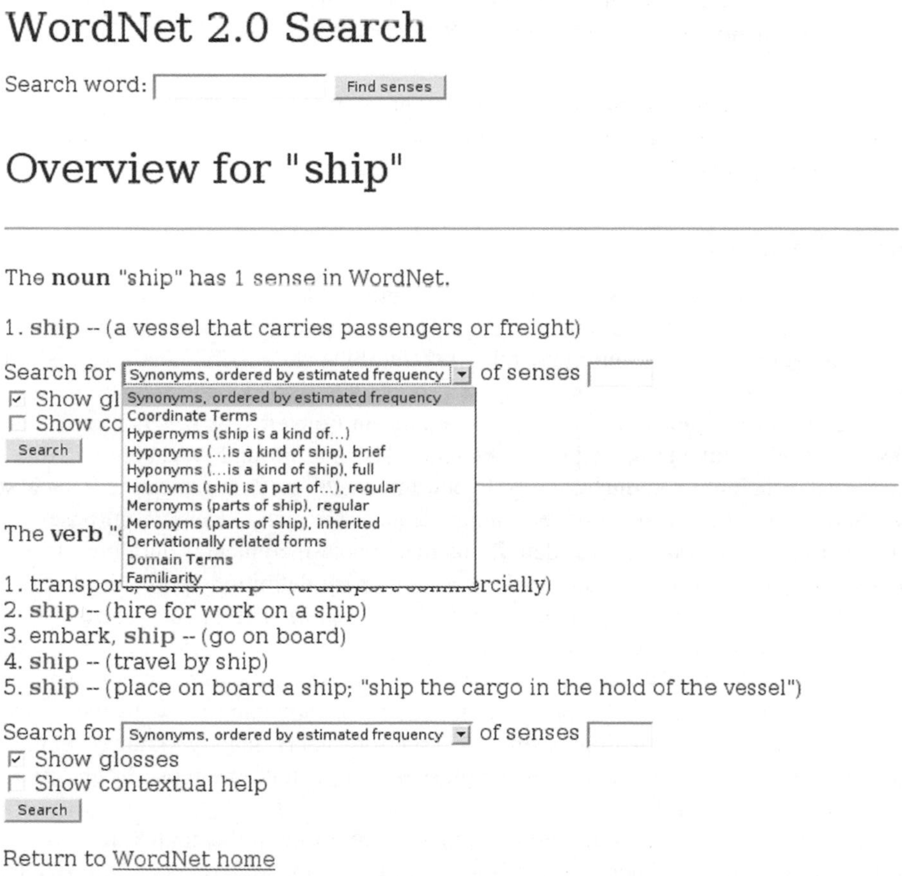

WordNet 2.0 Search

Search word: [   ]   [Find senses]

Overview for "ship"

The **noun** "ship" has 1 sense in WordNet.

1. ship -- (a vessel that carries passengers or freight)

Search for [Synonyms, ordered by estimated frequency ▼] of senses [   ]
☑ Show gl | Synonyms, ordered by estimated frequency
☐ Show co | Coordinate Terms
[Search] | Hypernyms (ship is a kind of...)
           | Hyponyms (...is a kind of ship), brief
           | Hyponyms (...is a kind of ship), full
           | Holonyms (ship is a part of...), regular
           | Meronyms (parts of ship), regular
The **verb** " | Meronyms (parts of ship), inherited
           | Derivationally related forms
           | Domain Terms
1. transpor | Familiarity ~~~~~~~~~~~~~~~~~~~~rcially)
2. ship -- (hire for work on a ship)
3. embark, ship -- (go on board)
4. ship -- (travel by ship)
5. ship -- (place on board a ship; "ship the cargo in the hold of the vessel")

Search for [Synonyms, ordered by estimated frequency ▼] of senses [   ]
☑ Show glosses
☐ Show contextual help
[Search]

Return to WordNet home

Figure 9.4   The WordNet online query page, the Web front-end for the database, showing some of the relational dimensions that can be explored

Theory

The initial idea was to provide an aid for use in searching dictionaries conceptually, rather than merely alphabetically. As the work progressed, the underlying psycholinguistic theories were found insufficiently developed, or based on artificially small word selections. It was decided to expose such hypotheses to the full range of the common vocabulary.

WordNet divides the lexicon into five categories: *nouns*, *verbs*, *adjectives*, *adverbs*, and *function words*. The relatively small set of English function words is in fact omitted from the database on the assumption that they are in effect a side-effect of the syntactic processing of language. The database attempts to organize the lexicon in terms of word meanings (or *concepts*) rather than forms, making it more like a thesaurus than a traditional dictionary – an approach that led to some interesting insights concerning suitable lexical structures:

- Nouns are organized in lexical memory as topical hierarchies.
- Verbs are organized by a variety of entailment relations.
- Adjectives and Adverbs are organized as N-dimensional hyperspace structures.

Each of these lexical structures thus reflects a different way of categorizing experience. However, to impose a single organizing principle on all syntactic categories would badly misrepresent the psychological complexity of lexical knowledge.

Representing Meaning

It turns out that the practical structure cannot ignore the lexical relationships entirely, but these remain subordinate to the semantic ones. The most common rendered view remains the '*synset*', or set of synonyms, but many other relationships are possible and potentially more valuable.

Storing 'meaning' values for words poses significant problems, as it is not known how to represent meaning outside the minds of humans. The interim and workable solution is to store formal definitions. Agents using rule-and-inference systems can process such structures, thereby emulating a form of reasoning adequate for many semantic purposes.

At a deeper level, 'constructive' definitions that enable incremental building of meaning also pose significant problems (and traditional dictionary definitions probably do not meet the constructive requirements either), but *differential definitions* that can distinguish meanings are adequate for the desired semantic mapping.

A defining or distinguishing gloss in *WordNet* contexts, therefore, is *not* intended for constructing a new lexical concept by someone not already familiar with the word in a particular sense. The assumption is that the user (or agent) already 'knows' English (in whatever sense is applicable) and uses the gloss to differentiate the sense in question from others with which it could be confused.

The most important relation, therefore, is that of *synonymy* in the contextual sense – that is, semantic similarity. General analysis of conceptual models supports this choice by suggesting that what it important about any given symbolic representation is not so much how it relates to what is represented, but instead how it relates to other representations.

Bit 9-6   Practical representations of relations are functionally sufficient

Although constructive models of semantic meaning are beyond current technology, it turns out that differential models are tractable. More to the point, differential models may in fact provide a better functional approach to lay the semantic foundations, if representations of similarity relations underpin our own conceptual/cognitive ability.

Related Resources

The *Princeton WordNet* resource, recently at v2.0, has over the years served as a focus for a number of interesting experiments in semantic-related processing. Background material is published as the 'Five Papers' resource, available from the Web site and in printed form.

- An updated version of these papers is in the book, *WordNet: An Electronic Lexical Database* (MIT Press: *mitpress.mit.edu/book-home.tcl?isbn=026206197X*).

Of specific relevance to the Semantic Web context is that Sergey Melnik and Stefan Decker released an RDF representation of *WordNet*, including an RDF Schema of the terms used to represent it. The main resource page for the effort is found at *SemanticWeb.org* (*www.semanticweb.org/library/*), and the downloadable representation and schema mean that the database is available as a metadata resource using standard RDF parsing.

International efforts such as *EuroWordNet* (EWN, *www.globalwordnet.org*) provide multilingual databases with *wordnets* for several European languages (for example, Dutch, Italian, Spanish, German, French, Czech, and Estonian). Each represents a unique language internal system of *lexicalizations*. In addition, all are linked to an *Inter-Lingual-Index* (ILI), which maps from the words in one language to similar words in any other language.

The EWN ILI also gives access to a shared top-ontology of 63 semantic distinctions, a common semantic framework for all the languages. Many institutes and research groups are using the EWN specification to develop similar *wordnets* in other languages (European and non-European). Known examples are Swedish, Norwegian, Danish, Greek, Portuguese, Basque, Catalan, Romanian, Lithuan, Russian, Bulgarian, and Slovenic.

In this broader context, the *Global WordNet Association* (*www.globalwordnet.org*) is a free, public and non-commercial organization that provides a platform for discussing, sharing, and connecting *wordnets* for all languages in the world. The aim is to stimulate further building of *wordnets*, further standardization, interlinking and development of tools, and dissemination of information.

Lexical FreeNet

One public *WordNet* application is *Lexical FreeNet* (*www.lexfn.com*), a finite relation expression network that allows the visitor to explore a large number of different kinds of relationships between any two words that occur in the database. The results are often intriguing, and sometimes surprising or mystifying. The site's front page with its selections is shown in Figure 9.5.

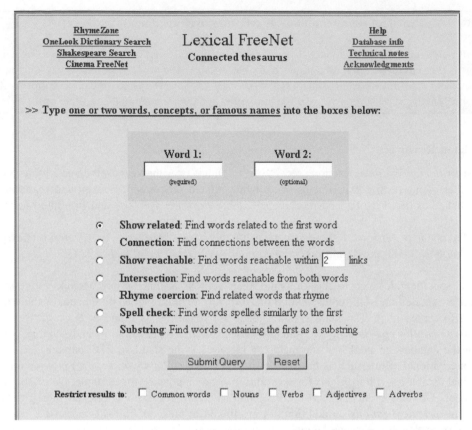

Figure 9.5   The Lexical FreeNet top page, where the visitor can explore various relationships between any two words that occur in the WordNet database. The results are intriguing and often surprising

The lower half of the same page, seen in Figure 9.6, allows the user to select different kinds of relationships through individual filters.

The shown interface options, leveraging and exposing conceptual relations defined within the *WordNet* model and formulating them in a human-readable way, provide an illuminating insight into the large number of different dimensions of meaning relations in human languages.

Most of the given examples are used without conscious thought during free associations or when playing with words. They illustrate the scope of interpretation issues confronting attempts to emulate in software the kind of associative semantic reasoning people take for granted. They also suggest the complexities involved when attempting to encode a comprehensive database of such relationships, one requirement for implementing intelligent sweb agent software.

SUMO Ontology

The SUMO project (*Suggested Upper Merged Ontology*, see *www.ontologyportal.org*, *ontology.teknowledge.com*, and *suo.ieee.org*) started as part of the *IEEE Standard Upper*

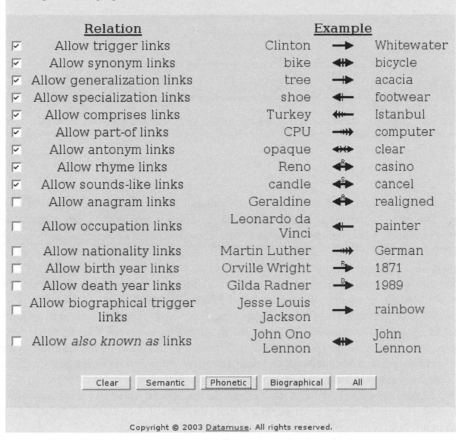

>> Optionally, you can select which links to allow in your query:

Relation	Example		
☑ Allow trigger links	Clinton	➡	Whitewater
☑ Allow synonym links	bike	⬌	bicycle
☑ Allow generalization links	tree	➡	acacia
☑ Allow specialization links	shoe	⬅	footwear
☑ Allow comprises links	Turkey	⬅	Istanbul
☑ Allow part-of links	CPU	➡	computer
☑ Allow antonym links	opaque	⬌	clear
☑ Allow rhyme links	Reno	⬌ᴿ	casino
☑ Allow sounds-like links	candle	⬌ˢ	cancel
☐ Allow anagram links	Geraldine	⬌ᴬ	realigned
☐ Allow occupation links	Leonardo da Vinci	⬅	painter
☐ Allow nationality links	Martin Luther	➡	German
☐ Allow birth year links	Orville Wright	➡ᴮ	1871
☐ Allow death year links	Gilda Radner	➡ᴰ	1989
☐ Allow biographical trigger links	Jesse Louis Jackson	➡	rainbow
☐ Allow *also known as* links	John Ono Lennon	⬌	John Lennon

| Clear | Semantic | Phonetic | Biographical | All |

Figure 9.6 The bottom half of the Lexical FreeNet page, where many different kinds of relations can be toggled. The buttons add grouped selections. The examples are illustrative of the kinds of associative semantic reasoning people take for granted

Ontology Working Group. The goal was a standard upper ontology to promote data inter-operability, information search and retrieval, automated inferencing, and natural language processing.

SUMO is an *open-standard resource* that can be re-used for both academic and commercial purposes without fee. The extensions are available under *GnuGPL*. Since 2004, SUMO and its domain ontologies form the largest formal public ontology currently in existence, and the only one mapped to all of the *WordNet* lexicon. It comprises about 20,000 terms and 60,000 axioms, such as commonly used in search, linguistics, and reasoning.

An ontology, in the sense used in computer contexts (as explained earlier in Chapter 6), consists of a set of concepts, relations, and axioms that formalize a field or interest. An *upper ontology* is limited to concepts that are *meta, generic, abstract, or philosophical.*

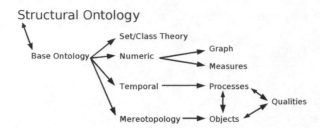

Figure 9.7   A simple graph that illustrates the relationships (or dependencies, as indicated by the arrow heads) between sub-ontologies in a modular ontology such as SUMO. Each sub-ontology is self-contained

The point is to make the ontology general enough to address a broad range of domain areas. Therefore, concepts specific to particular domains are not included in an upper ontology as such. However, an upper ontology does provide a well-defined structure upon which can be constructed ontologies for specific domains, such as medicine, finance, or engineering.

Domain-specific ontologies are defined at the SUMO site for a number of different areas, as downloadable examples. Currently listed (2005) are ontologies of Communications, Countries and Regions, Distributed Computing, Economy, Finance, Engineering Components, Geography, Government, Military, North American Industrial Classification System, People, Physical Elements (periodic table), Transnational Issues, Transportation, Viruses, and World Airports. Additional ontologies of terrorism and weapons of mass destruction are available on special request.

The SUMO model is modular, and encompasses sets of sub-ontologies that are self-contained. Figure 9.7 shows a simple graph that illustrates this hierarchical subdivision and the dependency relationships between the different sub-ontologies.

Most readers are unlikely to be familiar with the term *mereotopology*. It refers to a formal theory of 'part and whole' that uses topological means to derive ontological laws pertaining to the boundaries and interiors of wholes, to relations of contact and connectedness, to the concepts of surface, point, neighbourhood, and so on.

- This theory has a number of advantages for ontological purposes over standard treatments of topology in set-theoretic terms. A detailed presentation of the subject is the 1996 paper '*Mereotopology*: A Theory of Parts and Boundaries' by Barry Smith (which can be found at *ontology.buffalo.edu/smith/articles/mereotopology.htm*).

As practical implementations, the download section also offers the *WordNet* mappings, a Mid-Level ontology (MILO), along with several tools:

- an ontology browser for SUMO concepts or related language terms and visual editing system;
- an inference and ontology management system;
- a number of conversion and translation utilities;
- language generation templates, from SUMO to English and other languages.

The site collects many papers that describe SUO and SUMO efforts published over the past few years, including the core one: 'Towards a Standard Upper Ontology' by Niles and Pease (2001).

Comparisons with Cyc and OpenCyc

A fairly comprehensive upper-level ontology did in fact already exist when SUMO was started, but several factors made it relevant to proceed with a new effort regardless. A critical issue was the desire to have a fully open ontology as a standards candidate.

The existing upper-level ontology, *Cyc,* developed over 15 years by the company *Cycorp* (*www.cyc.com*), was at the time mostly proprietary. Consequently, the contents of the ontology had not been subject to extensive peer review. The *Cyc* ontology (billed as 'the world's largest and most complete general knowledge base and commonsense reasoning engine') has nonetheless been used in a wide range of applications.

Perhaps as a response to the SUMO effort, *Cycorp* released an open-source version of its ontology, under Lesser GPL, called *OpenCyc* (*www.opencyc.org*). This version can be used as the basis of a wide variety of intelligent applications, but it comprises only a smaller part of the original KB. A larger subset, known as *ResearchCyc*, is offered as free license for use by 'qualified' parties.

The company motivates the mix of proprietary, licensed, and open versions as a means to resolve contradictory goals: an open yet controlled core to discourage divergence in the KB, and proprietary components to encourage adoption by business and enterprise wary of the forced full disclosure aspects of open-source licensing.

OpenCyc, though limited in scope, is still considered adequate for implementing, for example:

- speech understanding
- database integration
- rapid development of an ontology in a vertical area
- e-mail prioritizing, routing, automated summary generation, and annotating functions

However, SUMO is an attractive alternative – both as a fully open KB and ontology, and as the working paper of an IEEE-sponsored open-source standards effort.

Users of SUMO, say the developers, can be more confident that it will eventually be embraced by a large class of users, even though the proprietary *Cyc* might initially appear attractive as the *de facto* industry standard. Also, SUMO was constructed with reference to very pragmatic principles, and any distinctions of strictly philosophical interest were removed, resulting in a KB that should be simpler to use than *Cyc*.

Open Directory Project

The *Open Directory Project* (ODP, *www.dmoz.org*) is the largest and most comprehensive human-edited directory of the Web, free to use for anyone. The DMOZ alias is an acronym for Directory Mozilla, which reflects ODP's loose association with and inspiration by the open-source Mozilla browser project. Figure 9.8 shows a recent screen capture of the main site's top directory page.

Figure 9.8 The largest Web directory, DMOZ ODP, a free and open collaboration of the Web community, and forming the core of most search-portal directories. It is based on an RDF-like KB

The database is constructed and maintained by a vast, global community of volunteer editors, and it powers the core directory services for the Web's largest and most popular search engines and portals, such as *Netscape Search, AOL Search, Google, Lycos, HotBot, DirectHit*, and hundreds of others. For historical reasons, *Netscape Communication Corporation* hosts and administers ODP as a non-commercial entity. A social contract with the Web community promises to keep it a free, open, and self-governing resource.

Of special interest to Semantic Web efforts is that the ODP provides RDF-like dumps of the directory content (from *rdf.dmoz.org/rdf/*). Typical dumps run around several hundred MB and can be difficult to process and import properly in some clients.

The conceptual potential remains promising, however. Just as Web robots today collect lexical data about Web pages, future 'bots might collect and process metadata, delivering ready-to-insert and up-to-date RDF-format results to the directory.

Ontolingua

Hosted by the *Knowledge Systems Laboratory* (KSL) at Stanford University, *Ontolingua* (more formally, the *Ontolingua Distributed Collaborative Ontology Environment, www. stanford.edu/software/ontolingua/*) provides a distributed collaborative environment to browse, create, edit, modify, and use ontologies. The resource is mirrored at four other sites for maximum availability.

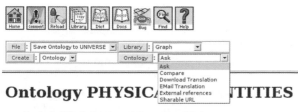

Ontology PHYSIC⌐ ⌐TITIES

- **Last modified:** *Sunday, 12 September 2004*
- **Generality:** High
- **Maturity:** Moderate
- **I/O Syntax:** Case Insensitive
- **Private by default:** No
- **Source code:** physical-quantities.lisp

Ontology documentation:

In engineering analysis, physical quantities such as the length of a beam or the velocity of a body are routinely modeled by variables in equations with numbers as values. While human engineers can i▄▄▄

Summary of Physical-Quantities:

Physical-Quantities includes the following ontologies:

```
Abstract-Algebra
     Frame-Ontology
          Kif-Relations
               Kif-Sets
                    Frame-Ontology ...
```

Figure 9.9 Browsing a sample public ontology online in the Ontolingua Server. The toolset allows interactive session work on shared or private ontologies using a stock Web browser as interface to the provided tools

The *Ontolingua Server* (access alias as *ontolingua.stanford.edu*) gives interactive public access to a large-scale repository of reusable ontologies, graded by generality and maturity (and showing any dependency relationships), along with several supported ontology working environments or toolset services (as described in Chapter 8).

The help system includes comprehensive guided tours on how to use the repository and the tools. Any number of users can connect from around the world and work on shared ontology libraries, managed by group ownership and individual sessions. Figure 9.9 shows a capture from a sample browser session.

In addition to the HTTP user and client interface, the system also provides direct access to the libraries on the server using NGFP (New Generic Frame Protocol) through special client-side software based on the protocol specifications.

The range of projects based on such ontologies can give some indication of the kinds of practical Semantic Web areas that are being studied and eventually deployed. Most projects initially occupy a middle ground between research and deployment. Some examples are briefly presented in the following sections.

CommerceNet

The *CommerceNet Consortium* (*www.commerce.net*) started in 1994 as an *Ontolingua* project with the overall objective to demonstrate the efficiencies and added capabilities afforded by making semantically-structured product and data catalogs accessible on the Web.

The idea was that potential customers be able to locate products based on descriptions of their specifications (not just keywords or part numbers) and compare products across multiple catalogs. A generic product ontology (that includes formalized structures for agreements, documentation, and support) can be found on the Ontology Server, along with some more specialized vendor models.

Several pilot projects involved member-company catalogs of test and measurement equipment, semiconductor components, and computer workstations – for example:

- *Integrated Catalog Service Project*, part of a strategic initiative to create a global Business-to-Business (B2B) service. It enables sellers to publish product-catalog data only once across a federation of marketplaces, and buyers to browse and search customized views across a wide range of sellers. The two critical characteristics of the underlying technology are: highly structured data (which enables powerful search capabilities), and delivery as a service rather than a software application (significantly lowers adoption barriers and total costs).
- *Social Security Administration SCIT Proof-of-Concept Project*, established with the U. S. Social Security Administration (SSA). It developed Secured Customer Interaction Technologies (SCIT) to demonstrate the customer interaction technologies that have been used successfully by industry to ensure secure access, data protection, and information privacy in interlinking and data snaring between Customer Relationship Management (CRM) and legacy systems.

CommerceNet was also involved with the Next Generation Internet (NGI) Grant Program, established with the State of California, to foster the creation of new high-skill jobs by accelerating the commercialization of business applications for the NGI. A varying but overall high degree of Semantic Web technology adoption is involved, mainly in the context of developing the associated Web services.

More recent ongoing studies focused on developing and promoting *Business Service Networks* (**BSN**), which are Internet business communities where companies collaborate in real time through loosely coupled business services. Participants register business services (such as for placing and accepting orders or payments) that others can discover and incorporate into their own business processes with a few clicks of a mouse. Companies can build on each other's services, create new services, and link them into industry-transforming, network-centric business models.

The following pilot projects are illustrative of issues covered:

- *Device.Net* examined and tested edge-device connectivity solutions, expressing the awareness that pervasive distributed computing will play an increasingly important role in future networks. The practical focus was on the health-care sector, defining methods and channels for connected devices (that is, any physical object with software).
- *GlobalTrade.Net* addressed automated payment and settlement solutions in B2B transactions. Typically, companies investigate (and often order) products and services online, but they usually go offline to make payments, reintroducing the inefficiencies of traditional paper-based commerce. The goal was to create a 'conditional payment service' proof-of-concept pilot to identify and test a potential B2B trusted-payments solution.

- *Health.Net* had the goal to create a regional (and ultimately national) health-care network to improve health-care. Initially, the project leveraged and updated existing local networks into successively greater regional and national contexts. The overall project goals were to improve quality of care by facilitating the timely exchange of electronic data, to achieve cost savings associated with administrative processes, to reduce financial exposure by facilitating certain processes (related to eligibility inquiry, for example), and to assist organizations in meeting regulatory requirements (such as HIPAA).
- *Source.Net* intended to produce an evolved sourcing model in the high-technology sector. A vendor-neutral, Web-services based technology infrastructure delivers fast and inexpensive methods for inter-company collaboration, which can be applied to core business functions across industry segments. A driving motivation was the surprising slow adoption of online methods in the high-tech sector – for example, most sourcing activity (80%) by mid-sized manufacturing companies still consists of exchanging human-produced fax messages.
- *Supplier.Net* focused on content management issues related to Small and Medium Enterprise (SME) supplier adoption. Working with ONCE (*www.connect-once.com*), *CommerceNet* proposed a project to make use of enabling WS-technology to leverage the content concept of 'correct on capture', enabling SMEs to adopt suppliers in a cost effective way.

Most of these projects resulted in deployed Web services, based on varying amounts of sweb components (mainly ontologies and RDF).

Bit 9.7   Web Services meet a great need for B2B interoperability

It is perhaps surprising that business in general has been so slow to adopt WS and BSN. One explanation might be the pervasive use of Windows platforms, and hence the inclination to wait for .NET solutions to be offered.

Major BSN deployment is so far mainly seen in the Java application environments. The need for greater interoperability, and for intelligent, trusted services, can be seen from U.S. corporate e-commerce statistics from the first years of the 21st century:

- only 12% of trading partners present products online,
- only 33% of their products are offered online;
- only 20% of products are represented by accurate, *transactable* content.

Other cited problems include that companies evidently pay scant attention to massive expenditures on in-house or proprietary services, and that vendors and buyers tend to have conflicting needs and requirements.

The Enterprise Project

Enterprise (developed by *Artificial Intelligence Applications Institute*, University of Edinburgh, *www.aiai.ed.ac.uk/~entprise/*) represented the U.K. government's major initiative to promote the use of knowledge-based systems in enterprise modeling. It was aimed at providing a method and computer toolset to capture and analyze aspects of a business, enabling users to identify and compare options for meeting specified business requirements.

Bit 9.8   European sweb initiatives for business seem largely unknown in the U.S.

Perhaps the ignorance is the result of U.S. business rarely looking for or considering solutions developed outside the U.S. Perhaps it is also that the European solutions tend to cater more specifically to the European business environment.

At the core is an ontology developed in a collaborative effort to provide a framework for enterprise modeling. (The ontology can be browsed on the *Ontolingua Server*, described earlier.)

The toolset was implemented using an agent-based architecture to integrate off-the-shelf tools in a plug-and-play style, and included the capability to build processing agents for the ontology-based system. The approach of the Enterprise project addressed key problems of communication, process consistency, impacts of change, IT systems, and responsiveness.

Several end-user organizations were involved and enabled the evaluation of the toolset in the context of real business applications: *Lloyd's Register*, *Unilever*, *IBM UK*, and *Pilkington Optronics*. The benefits of the project were then delivered to the wider business community by the business partners themselves. Other key public deliverables included the ontology and several demonstrators.

InterMed Collaboratory and GLIF

InterMed started in 1994 as a collaborative project in Medical Informatics research among different research sites (hospitals and university institutions, see *camis.stanford.edu/projects/ intermed-web/*) to develop a formal ontology for a medical vocabulary.

Bit 9.9   The health-care sector has been an early adopter of Sweb technology

The potential benefits and cost savings were recognized early in a sector experiencing great pressure to become more effective while cutting costs.

A subgroup of the project later developed *Guideline Interchange Language* to model, represent and execute clinical guidelines formally. These computer-readable formalized guidelines can be used in clinical decision-support applications. The specified *GuideLine Interchange Format* (GLIF, see *www.glif.org*) enables sharing of agent-processed clinical guidelines across different medical institutions and system platforms. GLIF should facilitate the contextual adaptation of a guideline to the local setting and integrate it with the electronic medical record systems.

The goals were to be precise, non-ambiguous, human-readable, computable, and platform independent. Therefore, GLIF is a formal representation that models medical data and guidelines at three levels of abstraction:

- conceptual flowchart, which is easy to author and comprehend;
- computable specification, which can be verified for logical consistency and completeness;
- implementable specification, which can be incorporated into particular institutional information systems.

Besides defining an ontology for representing guidelines, GLIF included a medical ontology for representing medical data and concepts. The medical ontology is designed to facilitate the mappings from the GLIF representation to different electronic patient record systems.

The project also developed tools for guideline authoring and execution, and implemented a guideline server, from which GLIF-encoded guidelines could be browsed through the Internet, downloaded, and locally adapted. Published papers cover both collaborative principles and implementation studies. Several tutorials aim to help others model to the guidelines for shared clinical data.

Although the project's academic funding ended in 2003, the intent was to continue research and development, mostly through the *HL7 Clinical Guidelines Special Interest Group* (*www.hl7.org*). HL7 is an ANSI-accredited Standards Developing Organization operating in the health-care arena. Its name (Level 7) associates to the OSI communication model's highest, or seventh, application layer at which GLIF functions. Some HL7-related developments are:

- *Trial Banks*, an attempt to develop a formal specification of the clinical trials domain and to enable knowledge sharing among databases of clinical trials. Traditionally published clinical test results are hard to find, interpret, and synthesize.
- *Accounting Information System*, the basis for a decision aid developed to help auditors select key controls when analyzing corporate accounting.
- *Network-based Information Broker*, develops key technologies to enable vendors and buyers to build and maintain network-based information brokers capable of retrieving online information about services and products from multiple vendor catalogs and databases.

Industry Adoption

Mainstream industry has in many areas embraced interoperability technology to streamline their business-to-business transactions. Many of the emerging technologies in the Semantic Web can solve such problems as a matter of course, and prime industry for future steps to deploy more intelligent services.

For example, electric utility organizations have long needed to exchange system modeling information with one another. The reasons are many, including security analysis, load simulation purposes, and lately regulatory requirements. Therefore, RDF was adopted in the U.S. electric power industry for exchanging power system models between system operators. Since a few years back the industry body (NERC) requires utilities to use RDF together with schema called EPRI CIM in order to comply with interoperability regulations (see *www.langdale.com.au/XMLCIM.html*).

The paper industry also saw an urgent need for common communication standards. *PapiNet* (see *www.papinet.org*) develops global transaction standards for the paper supply chain. The 22-message standards suite enables trading partners to communicate every aspect of the supply chain in a globally uniform fashion using XML.

Finally, the *HR-XML Consortium* (*www.hr-xml.org*) promotes the development and promotion of standardized XML vocabularies for human resources.

These initiatives all address enterprise interoperability and remain largely invisible outside the groups involved, although their ultimate results are sure to be felt even by the end consumer of the products and services. Other adopted sweb-related solutions are deployed much closer to the user, as is shown in the next section.

Adobe XMP

The *eXtensible Metadata Platform* (XMP) is the *Adobe* (*www.adobe.com*) description format for *Network Publishing*, profiled as 'an electronic labeling system' for files and their components.

Nothing less than a large-scale corporate adoption of core RDF standards, XMP implements RDF deep into all *Adobe* applications and enterprise solutions. It especially targets the author-centric electronic publishing for which *Adobe* is best known (not only PDF, but also images and video).

Adobe calls XMP the first major implementation of the ideas behind the Semantic Web, fully compliant with the specification and procedures developed by the W3C. It promotes XMP as a standardized and cost-effective means for supporting the creation, processing, and interchange of document metadata across publishing workflows.

XMP-enabled applications can, for instance, populate information automatically into the value fields in databases, respond to software agents, or interface with intelligent manufacturing lines. The goal is to apply unified yet extensible metadata support within an entire media infrastructure, across many development and publishing steps, where the output of one application may be embedded in complex ways into that of another.

For developers, XMP means a cross-product metadata toolkit that can leverage RDF/XML to enable more effective management of digital resources. From *Adobe's* perspective, it is all about content creation and a corporate investment to enable XMP users to broadcast their content across the boundaries of different uses and systems.

Given the popularity of many of *Adobe's* e-publishing solutions, such pervasive embedding of RDF metadata and interfaces is set to have a profound effect on how published data can get to the Semantic Web and become machine accessible. It is difficult to search and process PDF and multimedia products published in current formats.

It is important to note that the greatest impact of XMP might well be for published photographic, illustration, animated sequences, and video content.

Bit 9.10   Interoperability across multiple platforms is the key

With XMP, *Adobe* is staking out a middle ground for vendors where proprietary native formats can contain embedded metadata defined according to open standards so that knowledge of the native format is not required to access the marked metadata.

The metadata is stored as RDF embedded in the application-native formats, as XMP packets with XML processing instruction markers to allow finding it without knowing the file format. The general framework specification and an open source implementation are available to anyone. Since the native formats of the various publishing applications are binary and opaque to third-party inspection, the specified packet format is required to safely embed the open XML-based metadata. Therefore, the metadata is framed by a special header and trailer sections, designed to be easily located by third-party scanning tools.

Persistent Labels

The XMP concept is explained through the analogy of product labels in production flow – part human readable, part machine-readable data. In a similar way, the embedded RDF in any data item created using XMP tools would enable attribution, description, automated tracking, and archival metadata.

Bit 9.11   Physical (RFID) labels and virtual labels seem likely to converge

Such a convergence stems from the fact that increasingly we create virtual models to describe and control the real-world processes. A closer correspondence and dynamic linking/tracking (through URIs and sensors) of 'smart tags' will blur the separation between the physical objects and their representations.

Editing and publishing applications in this model can retrieve, for example, photographic images from a Web server repository (or store them, and the created document) based on the metadata labels. Such labels can in addition provide automated auditing trails for accounting issues (who gets paid how much for the use of the image), usage analysis (which images are most/least used), end usage (where has image A been used and how), and a host of other purposes.

The decision to go with an open, extensible standard such as RDF for embedding the metadata rested on several factors, among them a consideration of the relative merits of three different development models.

Table 9.1 summarizes the evaluation matrix, which can apply equally well to most situations where the choice lies between using proprietary formats and open standards.

The leverage that deployed open standards give was said to be decisive. The extensible aspect was seen as critical to XMP success because a characteristic of proprietary formats is that they are constrained to the relatively sparse set of distinguishing features that a small group of in-the-loop developers determine at a particular time. Well-crafted extensible

Table 9.1.   Relative merits of different development models for XMP

Property	Proprietary	Semi-closed	Open W3C
Accessible to developers	No	Yes	Yes
Company controls format	Yes	Yes	No
Leverage Web-developer work	No	No	Yes
Decentralization benefits	No	No	Yes

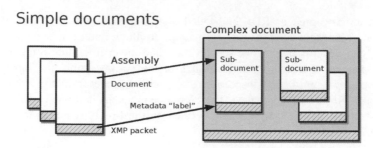

Figure 9.10 How metadata labels are preserved when documents are incorporated as subcomponents in an assembled, higher-level document

formats that are open have a dynamic ability to adapt to changing situations because anyone can add new features at any time.

Therefore, *Adobe* bootstraps XMP with a core set of general XMP schemas to get the content creator up and running in common situations, but notes that any schema may be used as long as it conforms to the specifications. Such schemas are purely human-readable specifications of more opaque elements. Domain-specific schemas may be defined within XMP packets. (These characteristics are intrinsic to RDF.)

Respect for Subcomponent Compartmentalization

An important point is that XMP framework respects an operational reality in the publishing environment: *compartmentalization.*

When a document is assembled from subcomponent documents, each of which contains metadata labels, the sub-document organization and labels are preserved in the higher-level containing document. Figure 9.10 illustrates this nesting principle.

The notion of a sub-document is a flexible one, and the status can be assigned to a simple block of information (such as a photograph) or a complex one (a photograph along with its caption and credit). Complex nesting is supported, as is the concept of context, so that the same document might have different kinds and degrees of labels for different circumstances of use.

In general terms, if any specific element in a document can be identified, a label can be attached to it. This identification can apply to workflow aspects, and recursively to other labels already in the document.

XMP and Databases

A significant aspect of XMP is how it supports the use of traditional databases. A developer can implement correspondences in the XMP packet to existing fields in stored database records. During processing metadata labels can then leverage the application's *WebDAV* features to update the database online with tracking information on each record.

> *We realize that the real value in the metadata will come from interoperation across multiple software systems. We are at the beginning of a long process to provide ubiquitous and useful metadata.*   Adobe

The expectation is that XMP will rapidly result in many millions of Dublin Core records in RDF/XML as the new XMP versions of familiar products deploy and leverage the workflow advantage that Adobe is implementing as the core technology in all Adobe applications.

XMP is both public and extensible, accessible to users and developers of content creation applications, content management systems, database publishing systems, Web-integrated production systems, and document repositories. The existing wide adoption of the current Adobe publishing products with proprietary formats suggests that embedded XMP will surely have profound impact on the industry.

Sun Global Knowledge Engineering (GKE)

Sun Microsystems (*www.sun.com*) made an early commitment to aggregate knowledge across the corporation, and thus to develop the required sweb-like structures to manage and process this distributed data.

Sun's situation is typical of many large corporations, spanning diverse application areas and needs. Many sources of data in numerous formats exist across the organization, and many users require access to the data: customer care, pre-emptive care, system integration, Web sites, etc.

RDF was selected to realize full-scale, distributed knowledge aggregation, implement the business rules, and mediate access control as a function of information status and person role.

- The technology is called *Global Knowledge Engineering* (GKE) and an overview description is published in a Knowledge Management technical whitepaper (see *sg.sun. com/events/presentation/files/kmasia2002/Sun.KnowledgeMngmnt_FINAL.pdf*).

The GKE infrastructure includes the following components:

- The *swoRDFish* metadata initiative, an RDF-based component to enable the efficient navigation, delivery, and personalization of knowledge. It includes a controlled vocabulary, organizational classifications and business rules, along with a core set of industry-standard metadata tags.
- A *content management system* based on Java technology. This component provides flexible and configurable workflows; enforces categorization, cataloging, and tagging of content; and enables automated maintenance and versioning of critical knowledge assets.

Sun markets the *Sun Open Net Environment* (Sun ONE), based on GKE infrastructure, to enable enterprises to develop and deploy *Services on Demand* rapidly – delivering Web-based services to employees, customers, partners, suppliers, and other members of the corporate community.

Of prime concern in GKE and Sun ONE is support for existing, legacy formats, while encouraging the use of open standards like XML and SOAP, and integrated Java technologies. The goal is to move enterprise CMS and KMS in the direction of increasing interoperability for data, applications, reports, and transactions.

Implemented Web Agents

An implementation aspect not explicitly mentioned so far concerns the agent software – much referenced in texts on ontology and semantic processing (as in, 'an agent can . . .'). The technology is discussed in Chapter 4, but when it comes down to lists of actual software, casual inspection in the field often finds mainly traditional user interfaces, tools, and utilities.

A spate of publications fuelled the interest in agents, both among developers and the Web community in general, and one finds considerable evidence of field trials documented in the latter half of the 1990s. But many of these Web sites appear frozen in the era, and have not been updated for years. Program successfully concluded. So, where are the intelligent agents?

Part of the explanation is that the agent concept was overly hyped due to a poor understanding of the preconditions for true agency software. Therefore, much of the practical work during 1995 – 2000, by necessity, dealt with other semantic components that must first provide the infrastructure – semantic markup, RDF, ontology, KBS+KMS – the information ecology in which agent software is to live and function.

Another aspect is that the early phases of commercialization of a new technology tend to be less visible on the public Web, and any published information around may not explicitly mention the same terms in the eventual product release, usually targeting enterprise in any case.

Finally, deployment might have been in a closed environment, not accessible from or especially described on the public Web. One approach then is to consider the frameworks and platforms used to design and implement the software, and to infer deployments from any forward references from there, as is done in a later section.

To gain some perspective on agents for agent-user interaction, a valuable resource is UMBC *AgentWeb* (*agents.umbc.edu*). The site provides comprehensive information about software agents and agent communication languages, overview papers, and lists of actual implementations and research projects. Although numerous implementations are perhaps more prototype than fully deployed systems, the UMBC list spans an interesting range and points to examples of working agent environments, grouped by area and with approximate dating.

Agent Environments

By environment, we mean agents that directly interact with humans in the workspace or at home. The area is often known by its acronyms HCI (Human Computer Interaction / User Interface) and IE (Intelligent Environment). Two example projects are:

- *HAL: The Next Generation Intelligent Room* (2000, *www.ai.mit.edu/projects/hal/*). HAL was developed as a highly interactive environment that uses embedded computation to observe and participate in the normal, everyday events occurring in the world around it. As the name suggests, HAL was an offshoot of the MIT AI Lab's Intelligent Room, which was more of an adaptive environment.
- *Agent-based Intelligent Reactive Environments* (AIRE, *www.ai.mit.edu/projects/aire/*), which is the current focus project for MIT AI research and supplants HAL. AIRE is dedicated to examining how to design pervasive computing systems and applications for people. The main focus is on IEs – human spaces augmented with basic perceptual sensing, speech recognition, and distributed agent logic.

MIT, long a leading actor in the field, has an umbrella *Project Oxygen*, with the ambitious goal of entirely overturning the decades-long legacy of machine-centric computing. The vision is well-summarized on the overview page (*oxygen.lcs.mit.edu/Overview.html*) and is rapidly developing prototype solutions:

In the future, computation will be human-centric. It will be freely available everywhere, like batteries and power sockets, or oxygen in the air we breathe. It will enter the human world, handling our goals and needs and helping us to do more while doing less. We will not need to carry our own devices around with us. Instead, configurable generic devices, either handheld or embedded in the environment, will bring computation to us, whenever we need it and wherever we might be. As we interact with these 'anonymous' devices, they will adopt our information personalities. They will respect our desires for privacy and security. We won't have to type, click, or learn new computer jargon. Instead, we'll communicate naturally, using speech and gestures that describe our intent ('send this to Hari' or 'print that picture on the nearest color printer'), and leave it to the computer to carry out our will. (MIT Project Oxygen)

The project gathers new and innovative technology for the Semantic Web under several broad application areas:

- *Device Technologies*, which is further subdivided into Intelligent Spaces and Mobile Devices (with a focus on multifunctional hand-held interfaces).
- *Network Technologies*, which form the support infrastructure (examples include *Cricket*, an indoor analog to GPS, *Intentional Naming System* for resource exploration, *Self-Certifying* and *Cooperative* file systems, and trusted-proxy connectivity).
- *Software Technologies*, including but not limited to agents (for example, architecture that allows software to adapt to changes in user location and needs, and that ensures continuity of service).
- *Perceptual Technologies*, in particular Speech and Vision *(Multimodal, Multilingual, SpeechBuilder)*, and systems that automatically track and understand inherently) human ways to communicate (such as gestures and whiteboard sketching).
- *User Technologies,* which includes the three development categories of Knowledge Access, Automation, and Collaboration (includes Haystack and other sweb-support software).

Some of these application areas overlap and are performed in collaboration with other efforts elsewhere, such as with W3C's SWAD (see Chapter 7), often using early live prototypes to effectuate the collaboration process.

Agentcities

Agentcities (*www.agentcities.org*) is a global, collaborative effort to construct an open network of online systems hosting diverse agent based services. The ultimate aim is to create complex services by enabling dynamic, intelligent, and autonomous composition of individual agent services. Such composition addresses changing requirements to achieve user and business goals.

The *Agentcities Network* (accessed as *www.agentcities.net*) was launched with 14 distributed nodes in late 2001, and it has grown steadily since then with new platforms worldwide (as a rule between 100 and 200 registered as active at any time). It is a completely open network – anybody wishing to deploy a platform, agents, or services may do so simply by registering the platform with the network. Member status is polled automatically to determine whether the service is reported as 'active' in the directory.

The network consists of software systems connected to the public Internet, and each system hosts agent systems capable of communicating with the outside world. These agents may then host various services. Standard mechanisms are used throughout for interaction protocols, agent languages, content expressions, domain ontologies, and message transport protocols.

Accessing the network (using any browser) lets the user browse:

- *Platform Directory,* which provides an overview of platform status;
- *Agent Directory,* which lists reachable agents;
- *Service Directory,* which lists available services.

The prototype services comprise a decidedly mixed and uncertain selection, but can include anything from concert bookings to restaurant finders, weather reports to auction collaboration, or cinema finders to hotel bookings.

Intelligent Agent Platforms

The Foundation for Intelligent Physical Agents (FIPA, *www.fipa.org*) was formed in 1996 (registered in Geneva, Switzerland) to produce software standards for heterogeneous and interacting agents and agent-based systems. It promotes technologies and interoperability specifications to facilitate the end-to-end interworking of intelligent agent systems in modern commercial and industrial settings. In addition, it explores development of intelligent or cognitive agents – software systems that may have the potential for reasoning about themselves or about other systems that they encounter.

Thus the term '**FIPA-compliant**' agents, which one may encounter in many agent development contexts, for example in *Agentcities.* Such compliance stems from the following base specifications:

- *FIPA Abstract Architecture* specifications deal with the abstract entities that are required to build agent services and an agent environment. Included are specifications on domains and policies, and guidelines for instantiation and interoperability.
- *FIPA Agent Communication* specifications deal with Agent Communication Language (ACL) messages, message exchange interaction protocols, speech act theory-based communicative acts, and content language representations. Ontology and ontology services are covered.
- *FIPA Interaction Protocols* ('IPs') specifications deal with pre-agreed message exchange protocols for ACL messages. The specifications include *query, response, Contract Net, auction, broker, recruit, subscribe,* and *proposal* interactions.

We may note the inclusion of boilerplate in the specification to warn that use of the technologies described in the specifications may infringe patents, copyrights, or other intellectual property rights of FIPA members and non-members. Unlike the recent W3C policy of trying to ensure license-free technologies for a standard and guaranteed open infrastructure, FIPA makes no such reservations.

On the other hand, FIPA seeks interoperability between existing and sometimes proprietary agent-related technologies. FIPA compliance is a way for vendors to maximize agent utility and participate in a context such as the Semantic Web by conforming to an abstract design model. Compliance specifications have gone through several version iterations, and the concept is still evolving, so that earlier implementations based on FIPA-97, for example, are today considered obsolete.

FIPA application specifications describe example application areas in which FIPA-compliant agents can be deployed. They represent ontology and service descriptions specifications for a particular domain, 'experimental' unless noted otherwise:

- *Nomadic application support* (formal standard), to facilitate the adaptation of information flows to the greatly varying capabilities and requirements of nomadic computing (that is, to mobile, hand-held and other devices).
- *Agent software integration*, to facilitate interoperation between different kinds of agents and agent services.
- *Personal travel assistance*, to provide assistance in the pre-trip planning phase of user trips, as well as during the on-trip execution phase.
- *Audio-visual entertainment and broadcasting*, to implement information filtering and retrieval in digital broadcast data streams; user selection is based on the semantic and syntactic content.
- *Network management and provisioning* to use agents that represent the interests of the different actors on a VPN (user, service provider, and network provider).
- *Personal assistant,* to implement software agents that act semi-autonomously for and on behalf of users, also providing user-system services to other users and PAs on demand.
- *Message buffering service,* to provide explicit FIPA-message buffering when a particular agent or agent platform cannot be reached.
- *Quality of Service* (formal standard), defines an ontology for representing the Quality of Service of the FIPA Message Transport Service.

Perusing these many and detailed specifications gives significant insight into the state-of-the-art and visions for the respective application areas.

Deployment

FIPA does attempt to track something of actual deployment, though such efforts can never fully map adoption of what in large parts is open source technology without formal registration requirements, and in other parts deployment in non-public intranets.

A number of other commercial ventures are also referenced, directly or indirectly, though the amount of information on most of these sites is the minimum necessary for the purpose of determining degree and type of agent involvement.

Other referenced compliant platforms are not public, instead usually implementing internal network resources. However, a number of major companies and institutions around the world are mentioned by name.

The general impression from the FIPA roster is that some very large autonomous agent networks have been successfully deployed in the field for a number of years, though public awareness of the fact has been minimal. Telecom and military applications for internal support systems appear to dominate.

Looking for an Agent?

Agentland (*www.agentland.com*) was the first international portal for intelligent agents and 'bots – a self-styled one-stop-shop for intelligent software agents run by *Cybion* (*www. cybion.fr*). Since 1996, *Cybion* had been a specialist in information gathering, using many different kinds of intelligent agents. It decided to make collected agent-related resources available to the public, creating a community portal in the process.

The *Agentland* site provides popularized information about the world of agents, plus a selection of agents from both established software companies and independent developers. Probably the most useful aspect of the site is the category-sorted listing of the thousands of different agent implementations that are available for user and network applications.

The range of software 'agents' included is very broad, so it helps to know in advance more about the sweb view of agents to winnow the lists.

Bit 9.12   A sweb agent is an extension of the user, not of the system.
Numerous so-called 'agents' are in reality mainly automation or feature-concatenation tools. They do little to further user intentions or manage delegated negotiation tasks, for instance.

An early caveat was that the English-language version of the site can appear neglected in places. However, this support has improved judged by later visits. Updates to and activity in the French-language version (*www.agentland.fr*) still seem more current and lively, but of course require a knowledge of French to follow.

Part III
Future Potential

10

The Next Steps

In this last part of the book, we leave the realm of models, technology overview, and prototyping projects, to instead speculate on the future directions and visions that the concept of the Semantic Web suggests. Adopting the perspective of 'the Semantic Web is not a destination, it is a journey', we here attempt to extrapolate possible itineraries. Some guiding questions are:

- Where might large-scale deployment of sweb technology lead?
- What are the social issues?
- What are the technological issues yet to be faced?

The first sections of this chapter discuss possible answers, along with an overview of the critique sometimes levelled at the visions. We can also compare user paradigms in terms of consequences:

- The current user experience is to specify *how* (explicit protocol prefix, possibly also special client software) and *where* (the URL).
- The next paradigm is just *what* (the URI, but do not care how or from where).

Many things change when users no longer need to locate the content according to an arbitrary locator address, but only identify it uniquely – or possibly just close enough so that agents can retrieve a selection of semantically close matches.

Chapter 10 at a Glance

This chapter speculates about the immediate future potential for sweb solutions and the implications for users. *What Would It Be Like*? looks at some of the aspects close to people's lives.

- *Success on the Web* explores the success stories of the Web today and shows that these can easily be enhanced in useful ways in the SW.

The Semantic Web: Crafting Infrastructure for Agency   Bo Leuf
© 2006 John Wiley & Sons, Ltd

- *Medical Monitoring* depicts two visions of sweb-empowered health-care: one clinical, the other proactive – both possible using existing infrastructure with sweb extensions.
- *Smart Maintenance* paints a similar picture of proactive awareness, but now with a focus on mechanical devices.

And So It Begins examines not just the glowing positive expectations, but also the caveats, cautions, and criticisms of the Semantic Web concepts.

- *Meta-Critique* discusses the negative view that people simply will not be able to use Semantic Web features usefully, but also gives a reasoned rebuttal. The discussion also includes content issues such as relevancy and trust.
- *Where Are We Now*? marks the 'X' of where sweb development seems to be right now, in terms of the core technologies for the infrastructure.
- *Intellectual Property Issues* is a short review of an area where current contentious claims threaten the potential of the Semantic Web.

The Road Goes Ever On looks at some of the future goals of ontology representations.

- *Reusable Ontologies* represents an aim that will become the great enabler of using published knowledge. Public data conversions into RDF are required to make available many existing large Web repositories.
- *Device Independence* examines how devices should communicate their capabilities in a standardized way.

What Would It Be Like?

Some research projects focus specifically on an exploration of the question:

What would it be like if machines could read what we say in our Web homepages?

This focus assumes that personal and family content on the Web is the information space most interesting yet hardest to navigate effectively. As with many projects, such visions are myopically fixed on the screen-and-PC model of Web usage – that is, browsing Web pages. But let us start there, nonetheless, as it seems to be what many people think of first.

It is patently true that billions of accessible Web pages already exist, and the numbers are increasing all the time. Even if Web browsing will never captivate more than a relative minority of these people (though large in absolute number), browsing functionality is of high importance for them.

Yet browsing is much more than just visiting and reading Web pages. Even constrained to the artificial constraint of seeing the PC screen as the sole interface to the Web, a rich field of functionality and interaction already thrives.

Success on the Web

The Web of today provides examples of both successes (quickly taken for granted) and failures (quickly forgotten). Some of the success stories already build on early aspects of sweb technology, and incidentally indicate activities already proven to be what people want to do on the Web:

- *Shopping*, perhaps no better illustrated than by *Amazon.com*, the online global book supplier in the process of becoming a mini-mall, and by *eBay*, the leader in online auctions. Less obvious is the multitude of more local outlets that emulate them, and the price-comparison sites that then spring up. Yet in the long run, this gradual virtualization of local community businesses is more important for the Web as a whole.
- *Travel Planning*, where itinerary planning, price comparison, and ticket booking services paradoxically both provided new possibilities and new frustrations. The more complex conveniences of a human travel agency were then cut out of the transaction loop. Sweb services can reintroduce much of this integrated functionality, and more.
- *Gaming*, where sweb technology is sometimes applied in the process of enabling new and more involving gaming environments.
- *Support*, where users can now often order, configure, monitor, and manage their own services (for example, telecom customer configuration pages, or for that matter, online banking).
- *Community*, where forums and special-interest pages (such as wedding-gift planners) have quickly become indispensable parts of many people's lives. New aspects of social events, entertainment, and contacts with local authorities continue to move online.

With sweb and agent technology, it is possible to enhance in numerous ways each already proven online Web activity area, without having to look further than implementing more convenience under sufficiently secure conditions.

- Automated bidding monitoring or offer notifications on auction or 'classified ad' sites.
- More advanced price comparisons using business logic to close deals on satisfied conditions.
- Referral and recommendation mechanisms based on the opinions of 'trusted' others.
- Avatar gaming with 'autopilot' mode for those periods when the player is too busy to participate in person, yet when pausing would be detrimental to the game role.
- Automated social events planning, which also includes proposals and planning negotiations between group PIMs.
- Profile-adapted summaries of news and events from many sources for insertion into local agent/agenda database for later reading.
- Notification of changes to selected Web sites, ideally filtered for degree of likely interest.

In addition to these simple enhancements, we have radically new services and possibilities, and many as yet unimagined. Some might be superficial and passing trends, others might become as firmly fixed in everyday life as e-mail, or using a search engine as portal.

Popularized sweb

One of the ways to make sweb technology ubiquitous is to implement functionality that most people want to use, especially if existing support is either hard to use or non-existent.

Managing contacts in a local database, for example, works reasonably well for people with whom one is regularly in touch, say by e-mail. In simpler times, this management method was sufficient, and the trickle of unsolicited mail was mostly from legitimate senders.

These days, hardly anyone looks at mail not from an already known sender, so that it is very hard to establish new contacts, even by those who have someone or something in common with the intended recipient. Unknown mail drowns in and is flushed out with the junk.

What seems lacking is some form of informal matching of common interests, or a trusted referral mechanism that could facilitate new contacts. Since the Web is all about making connections between things, *RDFWeb* (*www.rdfweb.org*) began by providing some basic machinery to help people tell the Web about the connections between the things that matter to them.

Bit 10.1   Sweb solutions should leverage people's own assessment of important detail

In practice, self-selection of descriptive items promotes the formation of connections along the lines of special-interest groups. Automating such connection analysis can be a useful feature.

Linked items, typically Web pages published as machine-understandable documents (in XML, RDF, and XHTML), are harvested by Web-indexing programs that merge the resulting information to form large databases.

In such a database, short, stylized factual sentences can be used to characterize a Web of relationships between people, places, organizations, and documents. These statements summarize information distributed across various Web pages created by the listed individuals, and the direct or indirect links to the home pages of countless other friends-of-friends-of-friends.

Exploiting various features of RDF technology and related tools, such as digital signature, Web-browsing clients can then retrieve structured information that machines can process and act upon.

Suitable goals are summed up in the following capability list:

- Find documents in the Web based on their properties and inter-relationships.
- Find information about people based on their publications, employment details, group membership, and declared interests.
- Share annotations, ratings, bookmarks, and arbitrary useful data fragments using some common infrastructure.
- Create a Web search system that is more like a proper database, though distributed, decentralized, and content-neutral – and less like a lucky dip.

If successful, this feature-set should provide the sorts of functionality that are currently only the proprietary offering of centralized services, plus as yet unavailable functionality.

FOAF Connectivity

The *Friend of a Friend* (**FOAF**, *www.xml.com/pub/a/2004/02/04/foaf.html*) vocabulary provides the basic enabler of a contact facilitation service, namely an XML namespace to define RDF properties useful for contact applications.

The FOAF project was founded by *Dan Brickley* and *Libby Miller*, and it is quietly being adopted in ever larger contexts where automated referral mechanisms are needed without any heavy requirements on formal trust metrics. In other words, it is well suited to expanding social connectivity in electronic media such as chat and e-mail.

The vocabulary is developed in an open Wiki forum to promote the concept of inclusiveness. It leverages *WordNet* (described in Chapter 9) to provide the base nouns in the system. The core of any such vocabulary must here be 'identity' and fortunately it is easily to leverage the unique Web URI of a mailbox – the assumption being that while a person might have more than one, knowledge of an e-mail address is an unambiguous property that in principle allows anyone to contact that person.

Further advantages of using one's mailbox URI as self-chosen identifier is that it is not dependent on any centralized registry service, and that it allows a form of *a priori* filtering. It is quite common to reserve knowledge of one mailbox only for trusted contacts and high-priority mail, while another may be openly published but then collect much junk. (Issues of identity and its governance for authentication purposes are discussed in the section on Trust in Chapter 3.)

Anyone can generate FOAF descriptors using the '*FOAF-a-matic*' service (a client-side Java script and form is available at *www.ldodds.com/foaf/foaf-a-matic.html*) and paste the result into a published Web page. The information is not much different than a simple entry in a contacts list or *vCard* file. Web-browsing FOAF clients know how to extract such information. In this way, existing infrastructure is leveraged to gain new functionality.

People have long used community portals, Web forums, newsgroups, distribution lists (or Yahoo groups), chat, IM, and so on to form social groups. The difference is that these were all separate contact channels. Sweb technology has the potential of integrating all of them so that the user does not need to choose consciously one or another client on the desktop, or remember in which client a particular contact is bookmarked.

One issue in all these traditional channels is *how* to establish contact with like-minded people, and how to filter out those we do not ourselves want to contact us (especially in this age of junk e-mail). It is about community building, and in all communities, one does need to allow and assist new entrants.

Bit 10.2   If we must filter incoming information, we must also implement referrals

Despite the increasing flood of unwanted messages, functional social contacts require that a certain minimum of unsolicited contact attempts be accepted, at least once. Intelligent filtering and trusted referrals is one way of implementing a sane solution.

FOAF adds the automated referral concept by way of inferred relationships.

- If I enjoy chatting with A, and A chats with B, and B knows C, then I might have interests in common with both B and C as well. In addition, B and C may reasonably assume that A will vouch for me, and are thus likely to allow a first contact through their filters – a rule that could easily be automated.

Figure 10.1 illustrates this chain of connectivity.

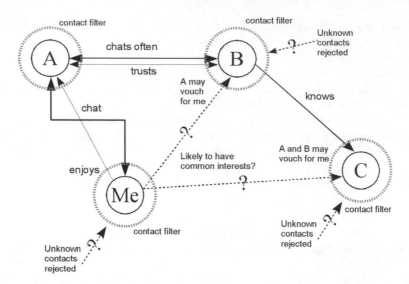

Figure 10.1 The referral aspect of FOAF, where previous contacts can vouch for the establishment of new ones through 'friends-of-a-friend'. Without referrals, new contacts become exceedingly difficult to initiate in an environment with heavy filtering, such as is the case with e-mail today

In FOAF, the simplest referral relation is 'knows', which points to the name and e-mail identity of another person that you assert you know. A FOAF client might then correlate your assertion with the FOAF description of that person and consider it truthful if your name is in that 'knows' list.

The correlation process, merging different assertion lists, can therefore deal with the useful situation that any source can in fact specify relations between arbitrary people. These independent assertions become trusted only to the extent that the respective targets directly or indirectly confirm a return link. It seems reasonable to expect the inference logic to apply some form of weighting.

- For example, a FOAF somewhere on the Web (D) asserts that C knows B. On its own, the statement is only hearsay, but it does at least imply that D knows or 'knows of' both B and C. From B's FOAF, referenced from D, we learn that B knows C, which gives D's assertion greater weight. FOAF lists browsed elsewhere might provide enough valid corroboration for a client to infer that the assertion's weighted trust value is sufficient for its purposes.

Independent descriptions may additionally be augmented from the other FOAF relationships, providing links to trustworthy information, regardless of the state of the initial assertion.

- Suppose C's FOAF refers to a photo album with a labeled portrait likeness. Some client browsing D's FOAF can then merge links to access this additional information about C, known to be trustworthy since it originates from C.

The interesting aspect of FOAF-aware clients is that referrals can be made largely automatic, with the client software following published chains of trust in the FOAF network.

The results of such queries can then be applied to the filtering components to modify the pass criteria, dynamically. The target user would still retain the option of overriding such automated decisions.

- Finding an identifiable photo of D in C's FOAF-linked album might be considered sufficient grounds for C's blocking filters to allow unsolicited first contact from D. For that matter, corresponding FOAF-based processing should probably by default allow D an initial pass to B and A as well, assuming the relations in the previous illustration.

FOAF is designed as a largely static declaration to be published as any other Web document. A proposed extension is *MeNowDocument* (*schema.peoplesdns.com/menow/*) to handle a subset of personal data that frequently or dynamically change. Proposed application contexts include:

- *Blogging* (blog/moblog/glog), providing current mood or activity indicators, implicit links to friends' blogs, and so on.
- *Project collaboration*, providing work and activity status, open queries, or transient notes also easily accessible for other participants.
- *Personal or professional life archival*, such as an always up-to-date CV available on demand.
- *Instant messaging and IRC*, perhaps as an adjunct to presence status; easy access to background personal information.
- *Forums and interactive social networks*, augmenting the usual terse person descriptors.
- *Online gaming* (extra extensions proposed), which could add new parallel channels to the in-game chat.
- *Real-world gaming and group dynamics*, which could become advanced social experiments using mobile devices.
- *Military and government operations* (bioinformatics – proposed), not detailed but several possibilities for team dynamics come to mind, akin to 'Mission Impossible' perhaps.
- *Dating and relationships* (proposed), again not detailed but the obvious feature is easily browsed information about possible common interests, scheduling, and so on.

Agents perusing FOAF and *MeNow* could combine all the features of a centralized directory service with near real-time updates by everyone concerned, all served from a fully decentralized and informal network of people 'who know each other'.

- Proof-of-concept 'bots have been deployed that can interactively answer questions about other participants in a chat room without disrupting the shared flow of the chat; information is pulled from the respective FOAF files but subject to the discretion of each file's owner.

Clearly there are security issues involved here as well. How accessible and public does anyone really want their life and annotated moods? Actually, most people seem quite comfortable with a remarkably large degree of even intimate information online – *as long as they are themselves in control of it*! Many personal weblogs demonstrate this attitude, as do home-made 'adult' Web sites for the more visually inclined.

And that is the thing, distributed p2p architecture does give the individual the basics of control, both of access and of ensuring that the information is valid and relevant. In the best-case scenario, sweb technology can enable people to define and redefine their chosen online identity and persona at will, being securely anonymous in one context and intimately open in another.

Other issues do need investigation, mainly about information policy. Trusted information should be kept from leaking out into untrusted contexts, which implies the ability to partition data according to assigned trust contexts. Provenance mechanisms may be needed in order to correct factual errors and provide accountability. Off-the-record communications should be possible, in analogy with how anonymous sources can be protected in the real world.

Finally, FOAF aggregators will necessarily function in a global context, yet privacy and data-protection laws vary considerably at the national level. This area of potential conflict needs to be resolved so as not to risk needless litigation risks.

Medical Monitoring

Leaving aside the browsing and filtering contexts, let us instead take a visionary example from the medical application domain that can potentially affect anyone, irrespective of interests. It can be illustrated in anecdotal form as follows.

Ms. Jo Smith has a diagnosed but latent heart condition. It is not serious by any means, but is still some cause for concern. Jo's family has a history of sudden cardiovascular deterioration at relatively young ages, which if not caught early and treated in time, can easily prove fatal. Her doctor suggests a new smart continuous-monitoring technology, and sets her up with the required wireless devices – small as buttons – on her person. Jo is instructed to grant the monitoring system permission to access the Web through her normal connectivity devices, such as 3G always-on cellular, Web-aware PIM, and home and work systems, because their unaided transmission range is limited.

After a short trial run to determine baseline parameters against the clinic's reception loop, and test remote adaptive connectivity during the rest of the day, Jo is free to resume her regular life at home and at work.

One day at work, many months later, Jo finds to her surprise a cellular mail to check her agenda. Her personal planner at home has juggled appointments and set up an urgent reminder for a visit to the clinic the following day, and synchronized with her work planner to now await her confirmation. It seems the monitoring devices, long invisible companions, have detected early warning signs of the onset of her heart condition, and therefore taken the initiative to set in motion a chain of networked requests to arrange a complete check-up as soon as possible.

The next day, her doctor has good news after reviewing the data. The half-expected onset of her condition had been detected long before she would have noticed anything herself, and therefore early treatment and medication can stabilize the condition with almost full heart capacity retained. Months of 24/7 data provide an excellent profile of heart activity in all her day-to-day activities, as well as in a few unusual stress situations, such as once when she was almost run over by a car.

The potential invalidity and the high-risk surgery required by a later detection have been avoided, making the initial investment in the monitoring system well worthwhile. The overall savings mean that her health insurance coverage takes a far smaller hit than would otherwise have been the case. But most importantly, she can continue with her activities much as before.

She continues to carry monitoring devices, some embedded in worn accessories. The monitoring is no longer as intensive, and less processed data are sent to the clinic since her heart profile is established. Some new functions have been added, such as to track medication and provide reminders for taking her pills, and to schedule maintenance for the devices themselves.

This kind of capability already exists as prototype systems, but stock clients do not yet support it. Neither, as yet, does the wireless communications infrastructure in rural, or even all urban, areas.

Pro-active Monitoring

Another example, already field tested by *Intel* researchers, concerns pro-active health care for elderly people. Of special concern is the case where old people develop Alzheimer's and similar conditions that tend to incapacitate them over time.

Intel's specific focus is on the mesh of connected devices that can monitor activities in the home and interface with smarter systems to provide interactive aid. Intel has documented these explorations (*www.intel.com/research/prohealth/cs-aging_in_place.htm*), and as white-paper (*ftp://download.intel.com/research/prohealth/proactivepdf.pdf*).

For example, a 'mote' sensor network is deployed to sense the locations of people and objects in the home – video cams, motion sensors, switches of various kinds, and object-mounted RFID tags tracked by a system of wireless transmitters.

The home network ties all this together with interaction devices (for instance, multiple touch pads, PCs, PDA units, and tablet devices) and enhanced home appliances (such as TV, clock radio, telephone, and even light switches).

Typically, Alzheimer's patients forget how to use the newest technologies, so unaided must rely on the more familiar interfaces of older technology to deal with prompts and reminders from the system. The point is that the system is capable of basic interaction using any proximate device.

The home system processes the raw sensor data using sophisticated self-learning 'lesser AI' technologies to generate meaningful 'trending' information. Trending means making inference interpretations from sensor data about probable activities that people in the home are doing.

It is hoped that developed trending analysis of simple daily tasks might be able to detect the onset of conditions like Alzheimer's years before traditional diagnosis methods can, and thus enable early stabilizing treatment to at least delay the deterioration of functionality in the patient.

Ambient display technologies can also be used to provide reassuring feedback to remote locations, such as for medical staff and concerned family members. Such feedback can be as subtle and non-intrusive as at-a-glance positive-presence indicators on picture frames in relatives' homes to signal that the monitored person is home and conditions are normal.

The envisioned system has many advantages over prevalent personal-alarm solutions where the subject must actively seek help using a carried actuator or the telephone, for example.

Smart Maintenance

Already, many of our more complex gadgets have embedded self-diagnostic systems, usually activated at start-up. Some provide continuous or recurring monitoring of system status. PCs have several levels of start-up testing and temperature/fan monitoring, hard disk devices have continuous monitoring to detect the onset of failure, and many more have optional self-test functionality. Modern cars have electronic systems to monitor and log a host of sensor readings, so that maintenance sessions at garages start with a readout of these values to an analyzer tool.

The next step in such monitoring is also *proactive*, in that adding connectivity to these monitoring and self-testing systems enable them to initiate requests for corrective maintenance *before* a detected problem becomes crippling.

As with the first visionary example, such device initiatives could schedule appointments after background queries to planner software at all affected parties.

We can easily extrapolate how such proactive monitoring can be extended into areas of less-critical maintenance, monitoring of consumables, and general availability of less-often used resources. The changes may not seem dramatic or be visible, but they will be profound.

Some day, perhaps, even our potted flowers will be able to ask for water when dry, assisted by sensors, speech synthesizers, and ubiquitous connectivity in the home network. It will not really be the flowers speaking, but the illusion will be convincing.

- Like all technology, proactive monitoring can also be misapplied for other purposes not necessarily in the user/consumer interest. Simple examples already among us are printer ink cartridges and batteries that are device monitored to disallow or penalize the use of non-approved replacements. Nor is everyone comfortable with cars that refuse to start when the sensors detect alcohol fumes, presumably from the driver's breath but perhaps not.

And So It Begins

All the previous material throughout the book may seem to imply that the road to (and the acceptance of) the Semantic Web is an inevitable process – that we stand on the very threshold of a utopian future which will revolutionize human activity as much or more than the explosion of the Internet and the Web has done already.

> *This process [of implementing and deploying sweb technologies] will ultimately lead to an extremely knowledgeable system that features various specialized reasoning services. These services will support us in nearly all aspects of our daily life – making access to information as pervasive, and necessary, as access to electricity is today.*
> Next Web Generation (*informatik.uibk.ac.at/injweb/*)

Technical problems might delay the process, to be sure, and it might not just yet be clear what the best models and solutions are, but the end result is out there, attainable.

Metadata rules, agents cruise, right?

Maybe. Critical voices warn of buying into the envisioned meta-utopia, *'cause it ain't necessarily so.*

> **Bit 10.3   The vision is never the reality; it is just a vision**
>
> The biggest problem with an envisioned future is that the vision is always too narrow. It neglects subtle (and not so subtle) consequences of the envisioned change. In addition, it can never capture the effects of other parallel changes, or of social acceptance.

Some people critique the ubiquity aspect, saying (rightly) that there is a world of people (in fact, a numerical majority) who are totally uninterested in all things related to the Web and computers, many of them even functionally illiterate for one or another reason, and most of them preoccupied with other, for them more pressing issues.

These things are all true, although a later section notes that ubiquitous computing is far more than just more people surfing the Web on PCs or hand-held devices.

Some critique the concept as a matter of principle, perhaps from the viewpoint that the Web is good enough as it is. Others say the implementation will be just too much work given the mind-boggling amount of content already published – billions of documents in legacy HTML, markup already obsolete by the standards of HTML v4 and highly unlikely to be updated any time soon, even to just passable XHTML, let alone XML. Then we have all the Web-published documents in other, less amenable formats: *MS Word doc, Adobe PDF, TeX, PostScript, plain text*, and worse.

However, Tim Berners-Lee pointed out in correspondence with the author that the issue of updating existing Web content to XML is not a critical one in the current vision of the Semantic Web – or more pointedly, that it is an obsolete issue:

> *This is not what the Semantic Web is waiting for. For most enterprise software folks, the Semantic Web is about data integration. For example, getting information from the stock control available in the catalog, in a web across the company. In these cases, the information already exists in machine-readable but not semantic Web form; in databases or XML, but not in RDF. The Semantic Web is about what happens as more and more stuff gets into RDF.*

As noted in the core chapters of this book, 'getting stuff into RDF' is indeed happening, and rapidly in some fields. The consequences sometimes include new intriguing applications based on the availability of the data in this format. Often, in the main, they just mean a more efficient business process somewhere, mostly out of sight, providing new services that most people quickly take for granted.

Meta-Critique

Another line of critique looks at the way people are and behave – with a rather pessimistic view, as it happens. The criticism is then less about the technology, because technical problems ultimately do have technological solutions, but more about how the technology is, or will be, used.

The meta-critic argues that people, being the way people habitually are, can never implement and use the Semantic Web concept consistently or accurately enough for it to become a useful reality – even assuming that they want to.

A world of exhaustive, reliable metadata would be a utopia. It's also a pipe-dream, founded on self-delusion, nerd hubris and hysterically inflated market opportunities.

Cory Doctorow (*www.well.com/~doctorow/metacrap.htm*)

Metadata is fairly useless. Users don't enter it. The SW project will fail since users around the globe will not work to enter descriptive data of any kind when there is no user benefit.

Urs Hölzle, *Google* (Search Engine Day, 2003)

More tempered criticism boils down to doubts about the quality of the metadata plugged into the system. It is the old garbage-in-garbage-out adage yet again – your processed and inferred data are no good if your input data are bad.

Why would the data be bad? Sadly, real world data collection and metadata empowerment can suffer from a number of possible problems.

- The well of information can be poisoned at the source by bad data disseminated for various selfish or malicious reasons.

Let's face it, people lie, often – for the worst and best of reasons. The Web is testimony to this caveat, because it is possible to find at least one Web site that advocates just about any position you can imagine, not to mention all the ones you cannot. Misinformation and outright lies are self-published with ease and are soon quoted elsewhere. Urban legends run rampant around the globe.

- People never get round to it – that is, to doing the markup that would provide the metadata to complement the published human-readable material.

Actually, a lot of people never even get around to finishing or updating the published material itself, let alone start thinking of metadata. Some call it pure laziness, though perhaps such a label is unfair, if only because the current tools do not make metadata markup a natural and transparent part of the process of publishing. It is true, however, that documentation is always imperfect and never finished – but does that mean it is pointless to document things?

- Who decides what to publish, how, and for what purpose?

Everyone has an agenda, whether it is to share and communicate, convert readers to a viewpoint, or sell something. Should declared purpose be part of the metadata, and if so, in what way? In many ways, it must be better to let those who wish to publish to do so directly and thus disclose their real agendas openly, rather than have to go through intermediaries and proxies, all with their own hidden agendas that both filter and obscure that of the authors?

- Who cares? Nobody can read or comprehend it all anyway.

The vast majority of the Internet's millions of self-publishing users are to varying degrees illiterate, according to the critics, who point to the massive incidence of mistakes in spelling, grammar, and elementary punctuation. Undeniably, such mistakes are of serious concern, because we know that software is not as good at recognizing them and inferring the intended meaning as human readers are. To this basic level of textual accuracy then, comes the issue of correctly categorizing the content according to the many and complex hierarchies of metadata by the great untrained masses.

- Self-evaluation of data (such as required when categorizing) is always skewed, sometimes absurdly so.

The assumption that there is a single 'correct' way to build a knowledge schema (an all-encompassing ontology) is assuredly incorrect. In fact, it can prove insurmountable to attain consensus in even very limited domains due to conflicting interests and agendas by those involved in the process of deciding the attributes and designing the metadata hierarchy. Publisher reviews and editing can improve the material, but then again, so can open Web annotations, for instance.

Bit 10.4   The chosen schema will influence or skew the results

If the metadata hierarchy omits certain categories, for example, then the corresponding entities will fall through the net, so to speak.

This issue is not a new one; it has been around ever since humans started creating categories for concepts. In its most subtle form, it shows in the dependencies on particular human languages and assumed knowledge structures even to formulate certain concepts.

- There is no consensus agreement on what is fact. How then can we evaluate assertions?

When all is said and done, people will still disagree. The old joke about diplomacy and negotiation states that talks have a successful conclusion only if everybody gets something to dislike intensely. Since we all describe things differently (*What, for example, is the temperature of 'hot' tea?*), and believe the descriptions are valid from our own personal viewpoints, any attempt to set up a universal schema will make most everyone unhappy.

All these objections are overly provocative, to be sure, but they do point out some of the problems involved when dealing with conceptual spaces and how to categorize knowledge so that software can process it. Perhaps the only way to deal with it is to allow plurality and dissent.

Bit 10.5   Metadata can describe rough approximations, not exact detail

Any inference machine to process human knowledge must take approximations into consideration and be able to deal gracefully with apparent inconsistencies and contradictions in both data and results. It must deal with uncertainties and subjective views.

Web Usage in e-government

One application area of sweb technology mentioned earlier is *e-government* – that is, to make various functions of government and authorities Web-accessible for the citizens.

Recent critique of some early *e-gov* attempts in the U.K. and U.S. seem to bear out some of the general criticism leveled at the sweb visions concerning issues of actual usage.

The question is simply how much citizens interact with government Web sites. Some news reports during 2004 suggested that doubts were growing about whether such efforts were worthwhile.

A report by the *Pew Internet & American Life Project* think-tank (*www.pewinternet.org*), 'How Americans Get in Touch With Government' (May 2004), found that U.S. citizens often prefer to pick up the phone when dealing with officials. They want a 'real-time' conversation over the phone so they can get immediate feedback on what to do next. Face-to-face interactions are also deemed preferable for certain kinds of problems.

Subtitled 'Internet users benefit from the efficiency of e-government, but multiple channels are still needed for citizens to reach agencies and solve problems', this report is overall less critical than summaries suggested. The main benefits involve expanded information flows between governments and citizens, and how the Internet assists contacts with respective authorities.

The limits have to do with people's technological assets, preferences, and the wide range of problems people bring to government. It was specifically noted that about one third of adult citizens did not then have Internet access, which clearly skewed both expectations and results.

The report concluded: 'In sum, *e-gov* is a helpful tool among several options for reaching out to government, but it is by no means the *killer app* among them.' It should, however, be noted in context that the primary comparison to personal contacts was with e-mail, not any interactive sweb technology as discussed in this book. Overwhelmingly, *e-gov* webs were static collections of HTML content and PDF-forms for download. E-mail, in this day and age of massive volumes of junk e-mail, is hardly a reliable or preferred medium.

On the other hand, a later report ('The Internet and Democratic Debate', October 2004) noted that wired citizens increasingly go online for political news and commentary, and the Web contributed to a wider awareness of political views during the 2004 campaign season.

This finding is seen as significant because prominent commentators had expressed concern that growing use of the Internet would be harmful to democratic deliberation. They worried that citizens would only seek information that reinforces their political preferences and avoid material that challenges their views. Instead, surveys found users were exposed to more varied political arguments than non-users, contrary to the expectations.

Again, the report results have little direct bearing on usage of sweb technology. For example, more intelligent and adaptive agent filtering to prepare information summaries for users could easily go in the opposite direction by simply trying to be 'more efficient'.

Bit 10.6   Information filters should never be allowed to become completely efficient

Some modicum of unsolicited ('random') information is needed in the flow. Even the biological 'proof-reading' mechanisms that exclude defects during DNA replication were found to be capable of several orders of magnitude better filtering than is observed. Selectively allowed random noise (generating mutations) is evidently of great value in natural systems.

Compare this concern with the issue of aggressive e-mail filtering as against the need for some unsolicited contact to succeed, as mentioned in the earlier section about FOAF. In both cases, a certain level of 'random' or unfiltered content seems desired, even if ultimately under the control of user-set policy rules.

Doubts also exist in the U.K. about the *e-gov* in general, and the recent portal consolidation under *DirectGov* (*ukonline.direct.gov.uk*). As reported by the BBC, a survey by newsletter *E-Government Bulletin* (*www.headstar.com/egb/*) found that nearly two-thirds of people working in *e-gov* were sceptical that the new initiative would attract any more visitors than its predecessor.

In other European countries with high penetration of broadband access, the *e-gov* situation is less debated and more assimilated. In part, this deployment and acceptance is promoted both from above (national governments and the EU), where authorities see online public access to government and its various agencies as a necessary fact of life, and from the citizens as meeting expectations due to their greater familiarity with basic Web and e-mail.

With a decade or so of ingrained use habits with online banking, information search, and form requisitioning, the general population therefore expects easy online *e-gov* access and streamlined functionality for applications, tax declarations and payments, and perusal of applicable regulations. Other differences between Europe and the U.S., due to social and infrastructure factors that affect sweb adoption, are discussed in the section 'Europe versus the U.S.' on page 7.

Also, in Europe, there is a greater acceptance of 'stated facts', a belief in authorities, as opposed to the overall tendency in the U.S. for citizens to want to argue their personal cases with officials. Europeans seem to view the impersonal nature of WS-based government as more 'fair' and impartial than personal contacts with bureaucrats – hence an easier acceptance of *e-gov*.

It is also far easier for government to meet the requirements of a multilingual citizenry with multilingual WS, than to attempt to provide multilingual staff in all departments and at all levels. This factor weighs heavily in the EU and its individual member countries, who by law must provide essential information to everyone in their native languages, (easily ten or twenty major ones in any given context).

Small wonder, therefore, that in much of the EU, where both Internet access and multiple languages are common, *e-gov* is embraced as a natural and necessary step forward.

Proposed Rebuttal to Meta-critique

Earlier chapters often make the point that any single construct is insufficient for a complete Semantic Web. Neither metadata, nor markup, nor ontology, nor any other single component is enough. Meaningful solutions must build on an intricate balance and synergy between all the parts.

More importantly, the solution must clearly be a dynamic and flexible one, and ultimately one that grows with our use and understanding of it.

In this context, we must dismiss the hype: the Web, semantic or otherwise, is *not* for everyone – nor is it likely to ever be so. Perhaps, it is time for the computing industry to stop trying to make the technology be everything to everyone. Most things are not.

Bit 10.7   Memo to vendors: Don't sell the hype; sell me the product

Useful infrastructures are inherently product-driven – essential ones are demand-driven. Either way, adapt to what people want to do and the products they want to use, or fail.

To be sure, machine-augmented Web activities are attractive and exciting to contemplate, and well worth the developmental effort for those who are interested. However, at the end of the day, enough people (or companies) have to be interested in using the technology, or it remains just a research curiosity.

If solutions later become free or commercial products/services that catch a broader public fancy, you can be sure they will quickly become part of the default infrastructure, thanks to demand.

Bit 10.8   However, do users really want to empower their software?

Knowledge is power; so who wants to share power with the machines? Organizations especially might seem unlikely to rush in to the new world of autonomous agents. In general, the answer to that question ultimately depends on the perceived trade-off against benefits.

Whether a given user wishes to delegate responsibility and initiative to software agents depends on many factors, not just the perceived benefits. For companies, a similar question often involves the issue of empowering the user-customer to manage ordering process and traditional phone-in support on the Web instead. The bottom-line benefits are quickly realized due to lower costs.

In government and its authorities, the issues are similar. What are the benefits and savings in letting citizens browse self-help, order forms, and submit required input online? Administrations are less sensitive to human-resource costs (though even there harsh cutback demands do appear), but control and follow-up benefits are quickly realized in many areas when automated self-help is successfully deployed.

Europe versus the U.S.

We can note rather large differences in attitude to all these issues, both across companies and across countries. Some are cultural, reflecting a greater trust in authority and hence a greater willingness on the part of Europeans to delegate to services and agents, and to trust the information received.

A significant difference affecting sweb implementations has to do with existing infrastructure. The relative larger proportion of Europeans enjoying 'fat' broadband access (cable or DSL at 1 to 8 Mbit/s, or better) provides a very different backdrop to both pervasive Web usage and viable services deployment, compared to most areas of the U.S. that are still dominated by 'thin pipes' (500 Kbit/s or less) and slow dial-up over modem.

Retrofitting copper phone lines for high bandwidth is a much more expensive proposition in most of the U.S. due to market fragmentation among many regional operators, and it is further constrained by greater physical distances. The situation is reflected in the wide deployment of 'ADSL-lite' (64 and 128 Kbit/s) access in this market, not much different from modem access (56 Kbit/s) except for cost and the always-on aspect.

Such 'lite' bandwidth was IDSN-telephony standard a decade ago, mostly superseded by the affordable Mbit/s broadband prevalent in large parts of Europe today – typically up to 8 Mbit/s ADSL in most urban areas for less than USD 50 per month. Population densities in Europe mean that 'urban' tends to translate into 'vast majority' of its domestic user base.

Therefore, even though bare statistics over broadband penetration may look broadly similar, these figures obscure the functional difference due to real bandwidth differences of one or two orders of magnitude. It is a difference that impacts users regardless of the efficiency of the services and underlying networks.

The 'bandwidth gap' to U.S. domestic users continues to widen in two ways:

- Backbone bandwidth is increasing rapidly to terabit capacity, which benefits mainly the large companies and institutions, along with overseas users with better domestic bandwidth.
- European 3G cellular coverage and *Wi-Fi* hotspots promise constant-on mobile devices for a potentially large user base, and developers are considering the potential for distributed services in such networks.

Operators are aware that compelling reasons must be made to persuade users to migrate quickly in order to recoup the investments, whether in land-line or wireless access.

The XML-based (and by extension sweb-enabled) automatic adaptation of Web content to browsing devices with widely varying capabilities, therefore, is seen as more of a necessity in Europe. Such different infrastructures provide different constraints on wide sweb deployment.

Bit 10.9   Technology deployment depends critically on existing infrastructures
Sweb deployment will surely play differently in Europe and in the U.S., if only because of the significant differences in infrastructure and public expectations. Differences in relevant legislation must also be considered, plus the varying political will to make it happen.

In the overviews of development platforms in previous chapters, it must be noted that European developers also work mainly on Java platforms for distributed agents. A major reason for this focus is the unified GSM mobile phone infrastructure, where it is now common for subscribers to roam freely throughout Europe using their own cellular phones and subscriber numbers.

Cellular access is more common than fixed-line access now, especially in the line-impoverished former East-block countries. The latest device generations provide enhanced functionality (embedded Java), more services, and a growing demand for automation.

Don't Fix the Web?

As noted, most of the Web remains unprepared for sweb metadata. Addressing the issue of wholesale conversion of existing Web documents does seem impractical even if it could be automated.

More promise lies in the approaches outlined in Chapter 5 and to some extent implemented in projects described in subsequent chapters, where a combination of KBS and KBM systems is used to identify, formalize, and apply the implicit semantic rules that human readers use natively to make sense of the text documents they read. On-demand semantic conversion to augment existing syntactic search may thus suffice for many interesting sweb services.

Success in this field does not have to be perfect processing, just good enough to be useful. Automatic translation between human languages is a practical example where the results are still awful by any literary standards, but nonetheless usable in a great variety of contexts, especially within well-constrained knowledge domains.

A similar level of translation from human-readable to machine-readable could prove just as useful within specific contexts, and probably provide better agent readability of existing Web content than the 'translate this page' option for human readers of foreign-language Web pages.

- Incidentally, this limited success would enable a significant improvement in the quality of human-language translation because it could leverage the semantic inferences made possible by the generated metadata.

As for the reliability (or lack thereof) in metadata generated (or not) by people, the critique does have relevance for any 'hard' system where all metadata is weighted equally. However, the overly pessimistic view appears unwarranted just by looking to the Web as it exists today.

Search engines prove that the averaged weighting of backlinks, despite being about as primitive a metric as can be imagined for reputable content, does in fact give a surprisingly useful ranking of search hits. On the average, therefore, the majority of people putting out links to other sites appear to do a decent job of providing an indirect assessment of the target site. (Attempts by a few to manipulate ranking are in context mere blips, soon corrected.)

Bit 10.10   People tend to be truthful in neutral, non-threatening environments

A good 'grassroots metadata system' would leverage individual input in contexts where the user perceives that the aim of the system is to benefit directly the user. Self-interest would contribute to a greater good. The Internet and Web as a whole, along with its directories, repositories, services, and Open Source activity, stand testimony to the power of such leveraging.

A related 'good enough' critique concerning the incomplete way even such a large database as a major search engine indexes the Web (many billions of pages, yet perhaps only a few percent of all published content) suggests that most users do not *need* a deep and complete search, providing even more hits. Existing results can already overwhelm the user unless ranked in a reasonable way.

The reasoning goes that the data required by a user will be at most a click or two away from the closest indexed match provided by the search. This argument assumes that the relevancy ranking method is capable of floating such a parent page towards the top.

Trusting the Data

From this observation on relevancy, we can segue to the broader issue of trust and note that the issue of source reliability and trustworthiness has long been the subject of study in peer-to-peer and proposed e-commerce networks. Awareness of trust matters have increased in the past few years, but not equally in all fields.

Models of the Semantic Web have as yet been fragmented and restricted to the technological aspect in the foreground for the particular project. This focus has tended to neglect the trust component implicit in building metadata structure and ontologies. On occasion, it has been suggested that 'market forces' might cause the 'best' (that is, most trustworthy) systems to become *de facto* standards.

Well, yes, that is the theory of the free market, but practical studies show that such simple and ideal *Darwinian* selection is never actually the case in the real world. Anyway, 'real natural selection' would appear to necessitate a minimum plurality of species for enough potentially 'good' solutions to exist in the diversity, which could then continue to evolve and adapt incrementally. Single dominance invariably stagnates and leads to extinction.

Bit 10.11   Natural market selection needs, at minimum, diversity to start with
A corollary would be that even the most driven 'survival of the fittest' selection process does not eliminate the competition entirely.

Getting back to trust, in a plurality of solutions, one also finds a plurality of trust networks based on individual assessments. Therefore, mechanisms to weigh and aggregate individual trust pointers should be part of the metadata structure and processing design.

In this context, we shall not discuss in detail the perceived merits or dangers of centralized authentication and trust, as opposed to decentralized and probably fragmentary chains of trust that grow out of diverse application areas (such as banking, general e-commerce, academic institutions, p2p networks, digital signing chains, and member vouching in Web communities). However, we can note that people on the whole trust an identified community a lot more than remote centralized institutions. In fact, people seem to trust averaged anonymous weighting of peer opinion (such as search or supplier rankings) far more than either of the preceding.

Despite the fact that such 'peer opinion' is rarely based on anyone's true peers, and is instead formulated by a more active and opinionated minority, it is still a useful metric. Smaller systems that incorporate visitor voting on usefulness, relevancy, and similar qualities generally give enhanced utility of the system, including a greater perceived trustworthiness by the majority of the users/visitors.

In its extreme form, we have the peer-moderated content in the open co-authoring communities that arise around public Wiki implementations. While these systems are essentially text-only relational databases, with one side-effect metric being the degree of referencing of any given page, one can discern certain intriguing possibilities in similar environments that would also include the capability to co-author and edit metadata.

Bit 10.12   Sweb-generated indicators of general trust or authoritative source are likely to be quickly adopted once available

Many contexts could benefit from agent-mediated reputation systems, weighted by input from both community and user, to which various clients can refer.

Trusting What, Who, Where, and Why

It is a common fact that we do not trust everyone equally in all situations – trust is highly contextual. Why should it be different on the Web? As it is now, however, a particular site, service, or downloaded component is either 'trustworthy' through a CA-signed certificate, or not, without other distinctions. This puts your online bank's SSL connection on par with some unknown company's digitally signed software component, to take one absurd example.

It is not the concept of digitally signing as such that is questionable here, it is instead the fallacious implication that the user automatically can trust a certificate signed by *Thawte, VeriSign*, or other self-appointed commercial CA. Anyone can self-sign a certificate, and many Web sites use such to set up SSL servers – the encryption is just as secure as with a CA-signed key. It is just that most Web clients show an alert that the certificate is initially unknown/untrusted.

So what dignifies CA-vouched so-called trust? Actually, nothing more than that the certificate owner can afford the expensive annual renewal fees. Interestingly, this condition favors the normally untrustworthy but wealthy shell-company that in fact produces malicious software over, for example, an open-source programmer. Economy is one good reason why most open-source developers or small businesses, who might otherwise seem to have an interest in certificate-branding their deliverables, do nothing of the kind.

But there is another, and better reason why centralized CA-certificate trees are ultimately a bad idea that should be made obsolete. Trust in emphatically *not* a commodity, and neither is it an absolute. The trading of certificates has misleadingly given it the appearance of being so. However, the only thing anyone really knows about a 'trusted' certificate today is that the owner paid for a stamp of approval, based on some arbitrary minimum requirement.

One might instead envision a system more in tune with how trust and reputation function in the real world. Consider a loose federation of trust 'authorities' that issue and rate digital certificates, an analog to other 'open' foundations – *'OpenTrust'*, as it were. Unlike the simple binary metric of current Web certificates, the open local authority would issue more useful multi-valued ratings with context. Furthermore, here, the certificate rating is not static, but is allowed to reflect the weighted reputation rating given by others in good standing. Presumably, the latter are also rated, which affects the weighting factor for their future ratings.

Trust is a network as well, not just value but interacting and propagating vectors; word-of-mouth in one form or another. This network is easy to understand in the social context and contacts of the individual. People have numerous informal (and formal) 'protocols' to query and assess relative trust for one another in different situations, well proven over a long time. It would be remiss not to try and emulate some of this functionality.

With corporations, as with some associations, any trust network is highly simplified, codified into a formal 'membership' requirement. Insuring trustworthy status on a case-by-case basis on application is not really done, although proven extreme transgressors may after the fact be excluded. Another trust mechanism is a contractual agreement, ultimately relying on external legal systems. It is easy to see how this simplified caricature of trust evolved into the commercial CA system.

Where Are We Now?

After the vision, and all the critique, you might be wondering where the reality lies. You are not alone in that because the view is somewhat fragmentary here and now at ground level.

The accepted 'Web standards' category has advanced to include XML, RDF, RDF-S, and OWL. Next in turn appear to be *rules/query* and *crypto* specifications. The reasoning and dependencies behind developing these recommendations, in-all-but-name 'standards', is discussed in previous chapters. A major motivation for their adoption has become the new royalty-free policy that ensures safe implementation and easy acceptance.

Starting at the higher level, OWL adds the ability to indicate when two classes or properties are identical, which provides a mechanism for linking together information in different structures (or schemas). It also enables declarations to provide additional information so that RDF-structured data can be subjected to automated rule-checking and theorem-proving.

OWL is still a work in progress, albeit now a W3C recommendation specification. Fortunately, the OWL-level of representation is not required for many practical applications. The lower level of RDF schema, longer established, is gradually becoming more important because of the flexible way URI referencing in the models can be made to use extensible models.

RDF, on the other hand, is a practical core structure, widely implemented and well supported by toolsets and environments. Yet one hears relatively little direct reference to it outside the specialist discussions.

Bit 10.13   RDF is a mature core technology that deserves wider recognition
Still, it is hardly mentioned in general Web development and publishing – compared to XML, for instance.

Several perceived problems might explain this silence:

- RDF is harder to learn than lower-level and older Web technology.
- Envisioned use often sounds too futuristic for those unfamiliar with it.
- Business sees little reason to publish RDF.

Plus RDF just does not make a catchy Web page (at least not directly).

And yes, although the development tools for RDF are out there, not all RDF features are supported in them yet. Core specifications are fragmented and hard to read, not especially conducive to practical development (though the online primer at *www.w3.org/TR/rdf-primer/* is considered good), and the higher level interfaces in RDF software are still on the drawing board.

If you are publishing information, the advice is to consider publishing an RDF version in addition to (X)HTML. Sweb-enabled applications are neat, if you can think of one that links two different sets of published data – not hard to implement given RDF-published data to begin with.

People do publish RDF content, however; it is just not always visible. For the most part, RDF is an enabler for services that might equally have been implemented using other technologies. The difference arises when these services start communicating with each other and with more intelligent agent software to deliver results not possible with the other technologies.

Bit 10.14   Initial size is never an indicator of importance

The original Web also started small. Adoption is based on many factors, including a threshold effect dependent on degree of adoption overall.

What People Do

Not everything is about Web content, something discussions about underlying protocols can obscure. It also concerns the users and how they benefit from the technology.

A number of surveys periodically report how often people are online and what they do during this time, highlighting trends and demographics of Web usage. We can examine and comment on selected data from a recent published statistical sample based on the roughly 63% of Americans adults who went online in 2003/2004 (*www.pewinternet.org/trends/ Internet_Activities_ 4.23.04.htm*).

- 93% read and sent e-mail, which is not surprising as this activity is the killer-app of the Internet as we know it.

The high number may seem surprising in light of the increasing complaints that e-mail has for many become virtually useless due to the volume of junk e-mail and malicious attachments. Evidently, people are still struggling and setting hopes on better filters. Clearly, e-mail remains an important technology for almost all people with Internet access.

- 84% used a search engine to find information, which confirms points made earlier in this book about the value of search on the Web as a starting point.

Equally frequent are subcategories such as finding maps or driving directions, seeking answers to specific questions, researching products or services before buying, or catering to hobbies or special interests. Other common tasks include checking the weather and news, looking for health/medical or government information, collecting travel information, or researching education-related topics.

- 67% surfed the Web 'for fun', as a form of entertainment.

This amount may also seem a bit high today considering the prevalence of Web annoyances such as intrusive advertising, inconsistent and poor navigation design, and outright browser hijacking. The value is further evidence that despite the considerably annoyances and drawbacks of the Web today, it remains attractive enough to be considered fun.

- 65% bought products or services online.

E-commerce to the masses is undoubtedly established, regardless of the fact that the relevant sites often show poor and confusing design, and that it can be difficult to complete transactions successfully. Sites such as *Jakob Nielsen's Alertbox* (*www.useit.com*) provide eloquent testimony on the continued inadequacies of consumer-oriented Web sites.

Participating in online auctions, such as *eBay*, with payments usually mediated by *PayPal* or similar services, shows a strong rise at 23%, but has suffered considerably from numerous scams and e-mail exploits (*'phishing'*) to try and obtain account or credit card information. Security is low in current implementations, and most transactions take place on blind trust.

Buying and selling stocks, bonds, or mutual funds online received 12%, as did grocery shopping.

- 55% bought or made reservations for travel.

Considering the post-2001 inconveniences of traveling and consequent drop in travel overall, this value is still respectable, perhaps even surprising.

- 54% looked up a telephone number or address.

Online directories are now an established standard over most of the world, much more up-to-date and cheaper to maintain than previous printed counterparts. Especially interesting is that the Web gives instant access to any directory in any region or country that has published it in this way.

- 52% watched a video clip or listened to an audio file.

Multimedia resources online are clearly important, so the ability of sweb technology to better find and correlate them is important to at least half the users already. The greater the bandwidth, the greater the probability that multimedia content will surpass text content in importance. It is well-documented that much of this 'consumption' is through p2p file downloads, even though an increasing number also purchase legitimate rights from the online stores that now exist.

We may also note growing, if still relatively minority, support of 'Internet radio', a technology which despite several setbacks concerning 'broadcasting' rights and heavy fees for music content, is evolving into an increasingly interactive relationship with its listeners. 'Internet TV' is assuredly not far off (see later section), though it too faces similar severe hurdles from the content-owning studios and television networks. Entertainment activities in general range from about 40% downwards.

- 42% engaged in *Instant Message* conversations.

IM technology is a precursor to some sweb features, such as pervasive yet (mostly) anonymous identity, presence indication, filters, direct interaction, and free mix of various media types. Some clients implement primitive automation and agent functionality.

Chat (IRC), the older interaction form, showed only 25% usage. Adult users typically find chat-room or online discussions too chaotic for their tastes, and the commands too obscure.

- 40% downloaded 'other' files, such as games, videos, or pictures.

The survey categories have some overlap/fragmentation due to formulation. In this case, 'other' appears to be in contrast to music files, a special category that came in at 20%, while 'sharing files' snowed 23%. However interpreted, collecting and sharing files is a significant activity. Videos as a special group received 17%.

- 34% bank online (while 44% got financial information online).

This is a low figure compared to Europe, but note also that online banking is designed with better security in some of these countries. Where a U.S. (and for that matter U.K.) bank customer relies on simple password authentication to do business, a European user may require digital certificates and special one-use codes to gain account access.

Sweb technology can provide similar high security levels and chains of trust on a network-wide scale, not dependent on particular vendors, and potentially less inconvenient to use.

- 19% created content for the Web.

Presumably, this category mainly means creating and updating Web site or weblog. It correlates well with the 17% who say they read someone else's weblog. Most active content creation by individuals today occurs in weblogs, not in traditional homepages, although the specific category creating a weblog received only 7% in the survey. Weblogs are very often RSS-enabled, which puts them part way into sweb-space.

- 10% studied college courses online for official credit.

The *MIT ESpace* project, noted in Chapter 8, is probably an indicator that online studies at advanced levels will become vastly more commonplace, both for credit and for personal gratification, as more free resources of quality become available.

- 7% made Internet phone calls (*VoIP*).

Although telephony is predicted to be subsumed by *VoIP* technology, the latter is still under the critical threshold where enough users are subscribed and available for it to become commonplace, which is partly a consequence of the relatively low access figure overall. Operators are also reluctant to move away from a known income source into an area that so far provides telephony for free.

- 4% played lottery or gambled online.

On the subject of vice, 15% admitted to visiting adult websites.

In short, most of the time, people tend to do what they did before, only over the Internet when the suitable services are available online. An important consideration is that these services are perceived, affordable, convenient, and as safe (or safer) than the corresponding real-world services.

Therefore, the caution should perhaps be that it is unreasonable to expect people in any large number to adopt new technology to do novel things any time soon. Then again, such predictions have often enough been proven wrong when the right 'thing' comes along – no matter what the intended purpose might have been for the innovation in question.

Comments on Usage Trends from the European Perspective

Although surveys in different parts of the world are difficult to compare directly, certain observations can be made in relation to the previous overview of current U.S. usage patterns. As noted earlier, these differences also have significant consequences for sweb development trends.

First, the percentage of online users is typically higher in Europe, especially in the well-connected countries in central and northwestern Europe. Internet access can approach and surpass 90% of all households in these countries, and a large number of these have broadband. The average connectivity for the entire EU 25-state region is 45% of the entire population (September 2004), which corresponds to a quarter of the world's total Internet users.

Such an infrastructure affects both usage patterns and services offered. In countries such as Sweden, world online leader today, it is difficult for many to reach the few remaining bank offices physically – Internet banking totally dominates the picture. Some recent bank start-ups do not even have brick-and-mortar counter services at all.

In Sweden, the post office, another long-standing institution, is completely marginalized, in part through its own inept attempts over the years to become more cost-effective. People no longer send letters and cards, they send e-mail and SMS. Over-the-counter service is outsourced to local convenience shops. The PO survives only on a combination of state-legislated public service and on special deals with businesses for delivery of mass advertising.

The broad user base with broadband has also promoted a shift away from traditional broadcast media to an over-cable/Internet custom delivery of pay-per-download digital content and services. Public air broadcast of television in Sweden is set to cease entirely in 2006, and that even after a delay from an earlier date. Almost all urban buildings are wired and ready for this new model of information distribution – 'three holes in the wall' as a recurring advertisement expresses it (telephony, tv, and broadband, already merging into single-supplier multi-service subscriptions).

Telephony operators are also heavily committed to offering broadband access and related services at competitive rates to complement or replace the traditional offerings. Therefore, *VoIP* telephony is a simple add-on to basic subscriber plans, as are commercial movies at affordable pay-per-view rates, try-and-buy games, and an ever expanding marketplace for goods and services. Customer care Web sites may at times seem to have minimal connection with classic telecom operations – reminiscent of how the classic *American Drugstore* evolved into a mini-mall.

It is worth dwelling a moment on the subject of mobile telephony, as the several rapid revolutions in popular telephony caused by the cellular are in many ways a precursor and indicator of the kind of usage revolutions we might expect for widespread sweb deployment. As noted earlier, mobile devices with always-on connectivity can play important roles in a sweb infrastructure.

Deployment of third-generation mobile telephony, however, promising much-hyped Internet capability, has lagged badly the past few years. Infrastructure and services are delayed, and consumer interest remarkably low in the new 3G handsets.

Although many GSM subscribers easily and regularly change handsets and operators, they seem unwilling to take the plunge into uncharted new services such as streaming music anywhere and television in a matchbox-sized screen – despite the subsidized entry and cut-throat introductory pricing. For most users, the so-called 2.5G (GPRS) seems sufficient for their mobile Internet use, and they have really no inkling of the kinds of services a sweb-enhanced 3G could provide.

- The one group to take to 3G and unmetered always-on call time quickly has been the deaf. One can see hand-signing people everywhere, who for the first time can really use telephony by signing through the video connection.

Table 10.1 indicates the potential for mobile access based on subscribers with Internet-enabled handsets (2.5G or 3G) according to interpreted statistics from many sources.

Actual usage of the corresponding services is much lower than the penetration figures for the technology would suggest, although it is difficult to get meaningful figures on it. Some providers figure 'Internet usage' including SMS traffic and ring-signal downloads, for example.

Strictly speaking, only 3G usage should really be significant. If we examine the 47% figure for the EU, the breakdown is roughly 3% (3G) and 44% (2.5G) subscribers. Based on typical call rates and unit capabilities, only the former group would be inclined to use even the basic Internet services to any significant degree. The 2.5G group have high per-minute/per-KB rates and are unable to use the more advanced services for 3G.

Table 10.1 Market penetration of Internet-enabled mobile telephony world-wide. Percentage of 2.5/3G is relative market share, that of all mobile is by total population, statistics for around mid-2004

Region	2.5/3G(%)	All mobile(%)	Comments
North America (U.S. + Canada)	37	28	Mostly CDMA standard. Only some 30% of mobiles here were GSM, though it is expected to reach over 70% after conversions.
Europe (EU)	47	85	GSM throughout. Some countries, like the Netherlands, show over 90% market share.
Japan	79	64	Mainly i-mod standard moving towards 3G and 4G.
Asia (excluding Japan)	54	No data	Taiwan has had more mobiles than people since 2002.
Brazil	37	No data	Leader in South America
Africa	low	6	Fastest growing mobile market
World (total)	low	25	(No data for India and China)

- Take one of the most connected populations, the Japanese, who have long had access to i-mod with many Internet-like services. Although over 80% of mobile subscribers are signed up for mobile Internet, other indicators suggest that only about half of them are in fact active users. Worse, a significant number of Japanese subscribers may nominally have the services plan even though they lack the Internet-enabled handset to use it.

It seems that the adult mobile phone user just wants to talk, and the juvenile one to send SMS or MMS, or use the built-in camera. And, in boring moments, the popular pastime is not to play the built-in games but to stand in a ring and play toss-the-handset (the previous unit, not the current one) in an approximation of *boule*. Strangely, they seldom break . . ..

Bit 10.15   Should user conservatism worry the advocate for sweb technologies?

Probably not. As noted earlier, the technology is not for everyone and all contexts. The overriding concern should be to provide the capability for those who wish to use it, and as flexibly as possible to cater for all manner of usage that cannot be predicted *a priori*.

- Here is my personal reflection as a user who recently upgraded to 2.5G mobile service: I found to my surprise that what I appreciated most was the ability to upload the built-in camera's photos as MMS attachments to e-mail – for immediate Web inclusion, family list distribution, or to my own account (more convenient than the local *IrDA* link). It was not a usage I would have predicted before being in that situation. I now expect sweb-3G usage to be full of surprises.

Creative Artists Online

One group of users has enthusiastically embraced the potential of the Web, both as it is now and as it may become with sweb-enhanced functionality: musicians and other creative artists. Not everyone, of course, but a significant majority according to the surveys made.

They see the Web as a tool that helps them create, promote, and sell their work. However, they remain divided about the importance and consequences of free file-sharing and other copyright issues. Some of these issues can likely be solved with sweb technology, but not until broad deployment and acceptance occurs. Payment (micropayment) capability is high on the list.

Other important online activities mentioned include browsing to gain inspiration, building a community with fans and fellow artists, collaborative efforts, and pursuing new commercial activities. Scheduling and promotion of performances or showings benefit from an online presence, and online distribution of free samples are often mentioned. Both explicit and implicit reference is made to various aspects of *easier communication* – with fans, customers, resellers, organizers, and friends and family when on the road.

Overall, one can here see an example of a transition in business model that seems characteristic for e-business in general, and is particularly apt for creative artists of all kinds. The most important aspect of this model is active social networking, with both peers and customers, marked by the absence of distancing intermediaries.

Bit 10.16   In e-business, part of the 'new economy', the emphasis is on the client–customer relationship rather than on any actual product

Digital products are incredibly cheap to mass-produce (that is, copy) and distribute. Transaction costs approach zero. What the customer is willing to pay for is less the actual product than the right to enter into an ongoing and personal relationship with the creator-producer. Traditional distributors seem not to grasp this essential point but instead focus on maximizing return on units sold.

Creative artists as a group would greatly benefit from a wider deployment of sweb core technologies, mainly RDF to describe and make their works more accessible on the Web. Such technology would also make it easier to manage and enhance their professional relationships with both colleagues and fans.

Despite the lack of a purposely media-friendly infrastructure so far, digital artists have clearly thrived regardless. The basic technology of the Web allowed partial realization at least.

It is interesting that the vast majority of artists in recent surveys do *not* see online file-sharing as a significant threat to their creative industries. Instead, they say that the technology has made it possible for them to make more money from their work and more easily reach their customers. They rarely feel it is harder to protect their work from unlicensed copying or unlawful use, a threat that has always existed, long before the Internet (for example, bootleg recordings and remixes).

Note that these artists are the people most directly affected by technologies that allow their works to be digitized and sold online. Therefore, they should be the most concerned about technologies for easy copying and free sharing of those digitized files, yet only a very small percentage of those interviewed find this issue a problem for their own livelihood.

In fact, a majority of all artists and musicians in U.S. surveys say that although they firmly believe current copyright regulations are needed to protect the IP-rights of the original creator, application of these rights generally benefits the purveyors of creative work more. They are also split on the details of what constitutes 'fair use' when copying or making derivative work, even though they seem satisfied that their own extensive borrowing in the creative process is legitimate and transmutative enough not to require prior creator consent.

This grassroots opinion stands in stark contrast to the almost daily diatribes by the dominant entertainment and distribution industries (music, cinema, and gaming) who over the past few years have been on a self-proclaimed crusade on behalf of the creative artists against decoding and file-sharing technology. Their efforts also actively discourage many of the very tactics adopted by the artists to promote their own works, in defiance of the minimal chance to be distributed in the established channels.

Modern artists are unquestionably early adopters of technologies to 'publish' their work on the Web, whether performances, songs, paintings, videos, sculptures, photos, or creative writing. Therefore, their passions and often their livelihoods critically depend on public policies that may either encourage or discourage creativity, distribution, and the associated rewards.

In particular, the ability to *exchange ideas freely* and *access published material* is central to their usage of the Web.

- For example, over half say they get ideas and inspiration for their work from searching online. Ever more restrictive legislation on and application of 'copyright' affects not only their resulting works, but increasingly also their ability to create these works without running afoul of IP-claims from those people or corporations whose work provided the inspiration.

On the other hand, the Web also allows creative artists to search more actively for inspirational works that are expressly public domain or free to use regardless, and to research the status and possible licensing of works that are not.

The complex issue of 'digital rights' deserves a more detailed discussion.

Intellectual Property Issues

In 2003, the entire issue of management of digital intellectual property 'claims' became even more contentious and infected, and two years on it shows little sign of improving soon. (The use of the term 'rights' is becoming just as questionable as the use of the term 'piracy' in this same context.)

Although the public conflict is largely confined to the areas of unlicensed copying and trading of music and film, thus infringing on current copyright, the overall encroachment of traditional unfettered access to information on the Web (and elsewhere) is now a serious threat to much of the vision of the Web, Semantic or not.

Even free exchange of formerly open research is seriously under attack from several directions, such as more aggressive assertion by companies of copyright and patents on ideas and concepts, terrorist concerns, and tensions between countries. The stemming of scientific exchange would make meaningless the entire concept of scientific peer review, already aggravated by the merged and ever more expensive scientific publications where papers are usually published.

Sweb solutions might be made to a great extent irrelevant if new information is no longer published for sharing. Limited deployments would perhaps still be possible in the closed intranets of the corporations, but the global vision would quickly dim in the face of the legal and authoritarian measures to control and restrict access. True exchange would be relegated to anonymous and encrypted networks, driven 'underground' and for the most part considered illegal.

Finding a Workable Solution

Various countermeasures are, however, being tried, such as a consensus-based licensing to make information explicitly shareable and thus avoid the default trap of copyright lifetime-plus-70-years. Publishers and others, who still wish to profit from their work, embrace the limited-term *Creative Commons* copyright or similar constructions, which release the works into the public domain within a far shorter time than current copyright.

Others experiment with combining commodity selling of hard copy and concurrent free access over the Web, reasoning that the free exposure drives enough extra users/readers to buy physical copies to more than compensate for those who prefer to stay with the free digital versions. Open Source Vendors use a similar strategy by selling support and customizing services to complement the free releases on the Web. For the most part, these open strategies seem to work well.

Even in the contentious commercial music market, a few have dared go against the norms. Some artists sell their works directly on the Web, offering free downloads. Distributors may offer select tracks for free, and most catalog items for modest prices.

In late 2004, a certain shift in stance was evident among even the large media distributors, as several scrambled to set up legal file download/share systems attractive enough to make consumers willing to pay. Different strategies are deployed to limit content spread and use.

The incentive was to act before an entire new generation of music and movie consumers was lost to unlicensed file sharing, though arguably it might well have been too little, too late. Time-limited playability, for example, does not sit well with the consumer. Neither do constraints on ability to copy across different playback devices.

In the field of sweb implementations, commercial and free versions of agents, services, and content might well coexist if rational strategies are chosen. The shareware/freeware software market has used this approach for some time, with varying but overall rewarding results for both developers and users. Other content media are experimenting with variations of the theme.

Sweb technologies, though in themselves committed to remaining free for the greater public good, are not *a priori* in opposition to commercial solutions for content. As noted in Chapter 5, RDF vocabularies could become a saleable commodity, yet co-exist with other free versions. Ontologies would fit the same model.

Free Infrastructure

Nevertheless, the basic infrastructure as such needs to be open and unfettered, and in realization of this fact, the W3C asserted the policy of not recommending any technology for inclusion in the infrastructure standards unless eventual licensing claims are waived.

Subsequent levied costs for usage of applications deployed on the infrastructure and for metered use of resources can be accepted – the 'free as in beer' approach of general affordability. This kind of rates charging is a different issue, and non-critical.

Sweb technology can even assist in several key aspects of IP management:

- Global registration and retrieval of provenance records for created content.
- Easy evaluation of access and use (license) status for any given document or media-file component.
- Per-usage tracking of licensed content for reporting and payment services.
- Trusted distributed regulation of usage based on policy rules.

Online Protection Registration

Returning to the subject of copyright, the ability of creators to register formally their works for easier protection in the U.S. is being streamlined and made into a WS. Though copyright is 'automatic' on creation in most countries, it is clearly easier to defend the claim on the basis of a formal registration with a verifiable date.

CORDS (*Copyright Office Electronic Registration, Recordation, and Deposit System, www.copyright.gov/cords*), run by the U.S. Copyright Office, is a system to accept online filings for literary texts, serials, and musical works, currently in HTML, PDF, ASCII-text, or MP3 format. Although delayed and as yet only available to a select few copyright holders, it is expected to be fully operational sometime in 2005. The service should level the playing field somewhat so that not only the large corporations register their claims.

Other countries vary in their application of copyright registration, and in some cases a niche exists that is filled by corporate registration services for those willing to pay. In other cases, a kind of registration exists in the context of number assignment for unique identification, such as ISBNs for books, where copies of each work are accepted and stored in national libraries or other repositories. National allocations of URIs allow a truly global way to establish work identity and associate it with required metadata, typically in Dublin Core format.

A more intelligent, sweb-based management of IP-registration data, with capability for worldwide exchange of national registrations, would assuredly simplify matters for many people. Determining the copyright state of much material is extremely difficult today given the fragmented way such status is recorded or not.

The Road Goes Ever On

What about the future goals of ontology representations, since so much of high-level development and reasonable implementation appears to focus on devising workable ontologies?

The guiding concept seems to be 'expand what can be expressed' and push the declarative representation beyond the current constraints. To be more specific, the following areas constitute research areas:

- *Uncertain knowledge*, which requires the incorporation of probability and failure modes into the hitherto simple assertion models.
- *Multiple perspectives*, to allow for the human-world experience of seeing and interpreting the same data from several points of view, depending on context and intended purpose.
- *Relationships among ontologies*, to deal with alternative representations. Relevant concepts: abstracts, alternatives, approximations.
- *Capability for formal annotations of ontologies*, to provide ongoing feedback for development. Relevant concepts: annotations about made assumptions and competency questions.

Apart from research into regions that as yet exist largely in theoretical speculation, pencil-marks beyond the contour lines on the map, there are issues of usability.

It is necessary to make ontologies easier to build, test, use, and maintain. In the long term, complete ontologies can become commodities like specialist vocabularies, but for this to happen, several design and construction aspects need improving:

- Better support for collaborative development, automatic extraction from existing data sources, and finding applicable libraries for re-use.
- Comparison strategies (and tools) to choose between alternative ontologies.

- Capability for testing ontologies by inferring properties from example situation models.
- Ability to quantify and process probabilities in a reasonable way.
- Task-specific agreements between agents, better to define the framework for their cooperation in the context of referencing common ontologies.
- Schema translation into application-specific representations.

Ultimately, of course, before serious adoption occurs the developers must demonstrate convincing examples of ontology use in practical situations. Such examples must be either of great public utility (such as library access) or of great personal benefit (such as PIM, agenda, and e-mail management). The assessment can only come with actual usage.

Bit 10.17   Massive adoption is more likely to come from self-interest
The rule of the Internet/Web has always been that people adopt the technologies that let them do what they want to do – irrespective of external regulation, default-bundled client constraints, or formal legal issues (if not accepted by common opinion).

Reusable Ontologies

Declarative knowledge is seen as a critical enabler for many systems, including large scale 'intelligent' systems, interoperative networks, and effective query systems (such as required within e-commerce and customer support). The trouble is that encoding such knowledge in KBS is very expensive and time consuming, requiring intense involvement by experts in the field.

Representational ontologies, both to mine knowledge and to manage it for online access, are thus extremely attractive propositions despite the initial investments to develop workable ones with sufficiently generic scope. It also makes the initial KBS investment more attractive, knowing that the data can be reused and re-purposed, and perhaps merged with other collections.

Given working ontologies and suitable tool sets, a great number of published data resources can be automatically analyzed, converted, and repackaged to satisfy semantic-based queries.

Bit 10.18   Ontologies can become the great enablers to access knowledge
The Web makes it possible to publish information so that anyone can read it. Ontologies enable publishing knowledge so that anyone can use it.

We will now explore some of these early KBS conversion studies.

Public Data Conversions

As discussed in Chapter 9, and elsewhere, much Web-published information must be either augmented with metadata or converted to agent-friendly forms. Possibly we can use 'scraping' technologies to collect and convert on demand, or leverage annotation technologies to use third-party metadata descriptions.

The *CIA World Fact Book* (WFB, *www.cia.gov/cia/publications/factbook*) can strike the casual visitor as an unlikely beast in many ways, yet the online version has quickly become a standard reference for anyone needing a quick and up-to-date political and statistical overview of a country. The WFB content is in the public domain, and therefore its free use raises no intellectual property issues, which makes it an interesting candidate source for automated data mining.

The proposed use case demonstrates adding semantic structure to a widely referenced data source, including proper handling of temporal and uncertain (estimated) values. Extra resource value can be possible by extending the ontology to allow reasoning based on geographic region, industry categories, etc. The basic WFB representation is in key-value pairs, making an RDF representation or the HTML source (or CD distribution SGML source) fairly straightforward.

- Early on, the PARKA project on very large databases demonstrated the capability to extract the published data and build a database from it using a Web robot (see *www.cs.umd.edu/projects/plus/Parka/parka-kbs.html*). Several versions of the database are downloadable.
- A *World Fact Book Ontology* capable of representing the data structures is developed and described at Stanford University's Knowledge Systems Laboratory (see *www-ksl-svc. stanford.edu:5915/doc/wfb/index.html and www.daml.org/ontologies/129*, and other *Ontolingua* projects at KSL covered in Chapter 9).

The WFB ontology can be browsed on the Ontology Server. The aim for the ontology is to represent the facts in the WFB, allowing for easy browsing in an OKBC-based environment, and providing machinery for reasoning in a first-order logic theorem prover like ATP.

Published scientific data are also good candidates for conversion into RDF databases. Many repositories of scientific data already exist, and numerous others are being developed (increasingly in the open RDF structures that promote ontology descriptions and reuse).

An example already discussed in Chapter 7 is the *Gene Ontology* (GO) project, which has the goal of providing an annotation ontology framework for research data on molecular functions, biological processes, and cellular components, as attributes of mapped gene products. It is thought that the synergy benefits will be considerable.

One vision is that instead of copying and recopying data in different contexts, it will often be more convenient to reference a common source using a URI. This directness is clearly a good thing for data that change over time. Clever clients (or servers) could inline the most recent values on the fly. The concept is nothing new, but implementations of the feature have been largely custom coded for each context and available data source.

Commodity Data

What remains to be solved is how to deal with the vast volumes of commercialized data that exist. For example, online dictionaries and encyclopaedias do exist as WS implementations, but to a large extent in a split deployment:

- a *free* service, comprising simple queries or 'sampler' articles (abstracts);
- a *subscriber* service, with access to the full work and extended search functionality.

The split is artificial in the sense that large segments of the public do expect knowledge repositories of this nature effectively to belong to the public domain – a view engendered by the free accessibility of the corresponding paper-published volumes at public libraries.

Increasingly, this emphasis on subscriber access is true not just of corporate resources and archives, but even of resources originally funded by the public taxpayer yet now expected to return a profit as an online commodity.

Bit 10.19 Users expect resources already available for free offline to remain free online

Few users are willing to pay out-of-pocket monthly subscription fees to access the same material online. Most users who do pay are in fact having their employers pay.

A similar split is observed in other established commercially published resources, where the full online resources are reserved for paying subscribers, while others may sample for free only a limited and feature-constrained selection.

As a rule, such resources are rarely accessible to search. Even if they are, the subscriber paradigm breaks a fundamental rule of the Web in that the links from the search hits are broken. A few sites do implement workable redirect-to-log-in-then-resume-fetch, but many fail, and the impediment to access is significant even when deferred access works.

Device Independence

One issue yet to be fully resolved in the Semantic Web is how devices should communicate their capabilities in a standardized way. The ideal is that any user or agent software should be able to query any connected device and receive some formal profile – that is, a description of device capabilities and user preferences that can be used to guide the adaptation of content presented to or retrieved from that device.

Proposed standards to resolve HTTP requests containing device capability information are CC/PP (Composite Capabilities / Preferences Profile, the proposed recommendation by the W3C, see *www.w3.org/Mobile/CCPP/*) and *UAProf* (User Agent Profile, proposed by the WAP Forum). The involvement of WAP (Wireless Application Protocol) should come as no surprise as the new generation of mobile phones comprise an important category of connected devices in this context.

RDF is used to create the profiles that describe user agent and proxy capabilities and preferences. However, vendor implementations of the proposed standards have so far tended to use a simpler XML representation of the RDF metadata, which seriously constrains their capabilities and interoperability. Work continues to develop specifications and implementation guidelines at the RDF level.

The protocols are new and few browsers support them. A weakness of CC/PP is that it does not resolve two key requirements concerning device independence:

- *Standard vocabulary*, which enables clients to communicate their capabilities to servers.
- *Type descriptions*, which specify the transformations and customizations that servers are expected to perform on behalf of devices, based on their communicated capabilities.

These problems are beyond the scope of the CC/PP working group (though perhaps not of its successor, the *Device Independence WG*), but they must be addressed in order for the protocol to be of practical use.

A strength of *UAProf is* that it defines five different categories of device capability: *software, hardware, browser, network*, and *WAP*. However, the weakness is that it does not resolve how servers and proxies should use the information provided by clients.

The nominal aspects of device plug-and-play are complicated by issues of trust in a roaming context on a global network. Presumably true device independence will not be completely realized until the sweb infrastructures for proof and trust are finalized as recommendations.

11

Extending the Concept

As with much of the material in this book, it might seem that the Semantic Web is mainly about software on the Web, specifications, or protocols. All very intangible to the average user. However, it is really as much a symbiosis between humans, devices, and the software – interacting in our real world.

From this view, it is not enough to develop the virtual network; in fundamental ways, the network must be made manifest in our physical world. It must have sensors everywhere, to be able to relate to the world we inhabit, and have real-world actuators, to manipulate physical objects.

Meaning comes from this interaction – it is after all how humans learn about meaning, and language, by interacting with the environment and with the people in it. It is likely that meaning, as we understand it, is not possible without this interaction.

Anyway, the point of this chapter is mainly to examine the importance of network extension into the physical world to achieve the kind of functionality that the goal of *anywhere-anytime* interactive computing requires. The idea is to push more and smaller devices deeper into the daily environment, ready to respond to our requests, as a new physical infrastructure. It will be a functional layer on top of the basic power and communications utilities, and the extension of the network in this way will profoundly change the way we do things.

Chapter 11 at a Glance

This chapter explores some directions in which future functionality might develop. *Externalizing from Virtual to Physical* examines the requirements that arise from pervasive connectivity and ubiquitous computing.

- *The Personal Touch* notes that a sweb-endowed world will be an animated one, with our environment actively responding to our wishes.
- *Pervasive Connectivity* suggests that always-on, everywhere access to the Web will be the norm.
- *User Interaction* explores the new ways that users and agents will interact.

The Semantic Web: Crafting Infrastructure for Agency   Bo Leuf
© 2006 John Wiley & Sons, Ltd

- *Engineering Automation Adaptability* explores the reasons for automatic device adaptability when interacting and how this function might work in a deep-networking
  environment.

  Whither the Web? speculates on possible consequences of a deployed Semantic Web.

- *Evolving Human Knowledge* explores the idea of publishing knowledge instead of simply
  publishing information.
- *Towards an Intelligent Web* discusses the entire issue of 'intelligent' behavior in the
  context of machines, revisits 'AI' and reflects on our problem of ever knowing if
  something is sentient.
- *The Global Brain* mentions a general resource on cybernetics appropriate to the
  philosophical context of intelligent machines and sentience in general.

  Conclusions brings us to the end of this journey, and this section is in part a reasoned
  motivation why the Semantic Web is 'a good thing' as long as we care enough to determine
  the practical applications by the way we use it.

- *Standards and Compliance* considers the current state of interoperability and what
  remains to be done in implementing support for the new sweb infrastructure. It is not
  just about technical specifications, but also about 'social' ones.
- *We can choose the benefits* concludes with the observation that individuals matter. We can
  control how new technology is implemented, if at all, and ultimately shape it by how we
  use it.
- *The Case for the Semantic Web* sets out the social aspects that can benefit from sweb
  technology.

  This chapter ends the narrative and presentational text: the appendices and some other
  support matter follow.

Externalizing from Virtual to Physical

The suggestion was made that important aspects of the Semantic Web dealt with its relations
with and reactions to the external world, the physical world we ourselves inhabit and which
forms our semantic notions. To act in this world, sweb technology must have a ubiquitous
presence in it and be able to refer reliably to the objects in it. What, then, can a ubiquitous
Semantic Web do?

A better question is perhaps: *How far do we want it to go?*

Bit 11.1   URIs can point to anything, including physical entities
As a consequence, the RDF language (based on URI-notation) can be used to describe physical devices such as cell phones, TVs, or anything else. And if these devices are then endowed with connectivity....

With access to schemas describing the external world, and the described edge devices able to converse with distributed software and each other, we have the foundation to do something quite remarkable with our environment. *We can animate it*, and make the physical world respond to requests from the software, anywhere, and by extension the user acting through the software.

In effect, we can externalize the virtual model of the world, built up in the Semantic Web using stored data, RDF schema, ontologies, inference engines, agent software, and device drivers.

The model then gets to push the objects around. . ..

The Personal Touch

The best way to predict the future is to invent it.   Alan Kay

Fiction has long described some of the possible situations, be it legend and myth, or speculative fiction in the modern sense. What the best of these stories have been maddeningly vague about is the underlying mechanisms, the technology – the responsive environment is in effect just one or more plot personalities, animated as if by magic.

Actually, this aspect is probably why the best stories are good because, by necessity, their focus is on the manifest functionality and people's interaction with the technology. The worst stories plod along trying to provide hopelessly confused technical detail, often blatantly impossible even according to the elementary laws of physics.

Yet now we stand on the threshold of actually being able to deploy such constructs, albeit with very primitive response patterns, as yet not deserving the description 'personalities' in any true sense. We may presume some framework of rules to guide the logic, and autonomous entities following such guidelines may well seem 'intelligent' to the casual observer.

People commonly anthropomorphize anything that even begins to seem as if it 'has a mind of its own' – think of cars and computers, for example, and how people talk to and think of these inanimate and largely unresponsive things.

Like Magic

Any sufficiently advanced technology is indistinguishable from magic.   Arthur C. Clarke

The impact of truly responsive, seemingly autonomous entities all around us on a routine basis will surely be profound to an extent we cannot even begin to imagine, despite the best efforts of fiction and cinema. Much of the change will be in how we view and interact with information in the networks, and the empowered and individually responsive environment.

We need not postulate entities with which we can hold intelligent conversations to get this kind of perception revolution. Intelligent-seeming actions are enough. Early prototyping of 'intelligent' environments has invariably been described by participants as a remarkable experience – when mere movement and gestures cause doors to open, lights to respond, and other appropriate actions to be performed by various devices in a room.

It's like magic! Indeed.

More sophisticated and adaptive systems able to read mood, infer probable intentions from context, and network with all available devices would take the experience to a whole new level. Include speech recognition and synthesis, and we begin to approach real conversations between user and software agent, albeit 'natural' perhaps only in the narrow context of a task at hand rather than in any abstract sense of idea exchange.

The list of requirements to realize this vision is long, and perhaps not fully specified. Mainly, it must support highly dynamic and varied human activities, transparently. A short-list of agent-service properties might look as follows:

- **Adaptable**, not just in the sense of flexible, but also in the autonomous capability to show initiative and quickly respond to changing requirements and conditions.
- **Empowered**, in that it can adaptively seek out and allocate new resources to meet new demands on the system caused by user tasks.
- **Embedded**, in that it is a participant in our real world, and is capable of also sensing and affecting the material surroundings.
- **Intentional**, in that it can infer user intent from context and interpret user-centric (or 'relative') identification/location of other devices the way people normally refer to objects in their environment (including gestures).
- **Pervasive**, in that the capability is available everywhere, with the same access profile, much in the same way that roaming lets a mobile phone user connect from any location that has compatible cell coverage.
- **Nomadic**, in that both user and data/software environment can be unrestricted in locality, and not materially changing as the user moves about.
- **Eternal**, in that it is always available, irrespective of local errors, faults, upgrades, and other disturbances to infrastructure components.

Ambient intelligence technologies addressing these requirements have the focus on human needs.

Pervasive Connectivity

When not only our desktop and notebook computers achieve constant connectivity, but also most devices imbued with enough software to communicate with the Web, the resulting level of functionality will probably astound even the experts.

Pervasive and ubiquitous connectivity to most devices will in any case profoundly change our everyday environment – working and leisure. The related term 'ubiquitous computing' tends to mean the same thing, but with a focus mainly on the actual devices rather than the connectivity.

Developing this next generation Web requires the compatibility, integration, and synergy of the following five technology areas:

- *User interface*, where the most common current human interface of screen, keyboard, and mouse must be broadened significantly to include (or be replaced by) human speech and gestures. The 'human–computer' interface paradigm should be replaced with new 'human–information', 'human–agent', and 'human–human' models supported by smart devices.

Many devices may lack support for a keyboard or pointing device. Pragmatics suggest that the system should be able to use any connected proximate device as an impromptu interface for simple interaction, adapting to the constraints, or including new devices when required.

A multitude of flat screens in a room may function as decorative items when not in use. A variety of adaptable embedded projection/sensor devices might create functional illusions of more complex ones when needed (such as an optical keyboard). Furthermore, the system (whether local or distributed) must be capable enough to infer user intent within defined contexts, yet subtle and adaptive enough to avoid becoming intrusive.

- *Network access*, where today's predominantly hard-wired access is complemented and perhaps largely replaced by a wireless infrastructure – IR and more recently *Bluetooth* connecting local peripherals.

The cellular phone is also evolving into a network portal device, only the current bandwidth constraints and per-KB rates keeping it from broader use as a roving access point. At the very least, a combined wired/wireless access must be integrated and function seamlessly for the locally or globally mobile user/device.

The rapid deployment of *Wi-Fi* hotspots seen since 2002, combined with the default inclusion of the technology integrated into notebook and hand-held computers, shows the way. Hard-wired LANs are in many cases being replaced by wireless versions, even in the home. *WiMAX* aims to broaden this to wireless broadband connectivity at even higher data rates.

- *Protocols*, which means further deployment of higher, application-level layers to handle the complex messages that distributed software and agents will use to implement intelligent functionality.

We should also see a transition from location-based URL identities to location-independent URI identities more uniquely tied to the resource itself. This shift can be a subtle one, and may go largely unnoticed, given that URL notation can be subject to various server-side rewrites and redirections to hide the actual and variable instance location of a public virtual resource.

- *Web architecture*, where the static, lexical-associated content in today's Web provided by stateless servers is gradually replaced by a dynamic, services-oriented architecture that is continually adapting to user requests and requirements.

This aspect is related to the previous comment about dislocation. Typically, a dynamic architecture is also distributed, visible only as a virtual single source. Content will in many cases increasingly be an agent-compiled presentation derived from many sources.

- *Personal space*, which can be seen as another kind of user interface defined by the network access points and enabled devices reachable by and affecting the user.

This space increasingly includes hand-held and wireless connected devices, but can also include virtual spheres of influence where some aspect of the network has a monitoring and actuating presence through a multitude of embedded and roaming devices.

An important change affects the latter, in that personal spaces in this sense have hitherto been defined by location, requiring mobile users moving between physical workspaces to log in manually from new machines or access points, perhaps even with different identities in each location. Shares and working environments might vary unpredictably, and access critically denied from some locations.

The new mobile paradigm requires the system to track the user and seamlessly migrate connectivity and information context to appropriate proximate devices. In fact, carried devices (cellular phones) can often be used to accurately pinpoint user location – useful information for either user or network.

The future Web will increasingly enable *virtual spaces defined by identity* – mobile users would in effect carry mobile spaces with them, with a constant and consistent presence on the Web.

Pervasive Personal Access

In analogy with telephony, where formerly location-bound phones (home, office, or public) defined a person's access (and accessibility), the focus has changed. This infrastructure was rapidly complemented and sometimes replaced by the almost ubiquitous mobile phone, and a person's access (and accessibility) became more personal and consistent.

With land-line phones, you generally knew *where* you were phoning, but not necessarily who would answer. With mobile phones, you are almost always confident of *who* will answer, but you know nothing of the roaming location – and probably no longer much care. The subscriber number belongs to the hand-held device, independent of location but usually personally tied to the user.

Now extend this concept to Web presence (as access and accessibility). A person's Web identity will become more consistent, just like the phone identity. The effects can be far-reaching, and the user's access immediate (or continuous) and intuitive. The concept can be implemented in various ways, but it is intimately associated with online identity.

In the context of wireless devices, this kind of personal space is implemented as a network infrastructure called a *Wireless Personal Area Network* (WPAN).

A typical application of a WPAN is in the office workspace, where initially it just frees the user from the tyranny of physical cords. In the broader perspective, however, components of this office context can become virtualized, decoupled from particular physical devices. If the user physically moves to another location, such contexts (including user profiles and current workspace) can then be transferred to and manifested on any set of conveniently located devices.

Any set of compliant devices can in this view function as a 'screen' for projecting the workspace. The capability can include 'scaling' the display and functionality gracefully to the capabilities of the current devices. The need to carry physical paraphernalia around is reduced.

Bit 11.2   Personal space can be seen as roaming sphere of access and actuation

Ideally, such a virtual space is globally portable and provides transparent yet secure access to both personal-profile workspaces and personal information spaces.

The vision is ultimately of a pervasive and transparently accessed network that provides *anytime, anywhere* access to *arbitrary* information resources. With efficient user interfaces, and applications that learn and adapt, the Web could provide useful services in any user context – at home, at work, or on the move. It is probable that with good design such on-demand functionality in any context can become as natural to the user as sensor-driven doors and lighting is today, and as convenient as always-on Internet on the desktop with the Web only a click away.

The most profound technologies are those that disappear. They weave themselves into the fabric of everyday life until they are indistinguishable from it.   Mark Weiser, Xerox PARC (ca. 1991)

An in-depth investigation and expansive vision of the wireless aspect is found in the book *The Intelligent Wireless Web*, by H. Peter Alesso and Craig F. Smith (Addison Wesley, 2002) and the associated Web site (The Intelligent Wireless Web at *www.web-iq.com*).

One of the early aims of the site was to stimulate discussion and solicit input on developing a Web Performance Index (amusing called the Web IQ) for evaluating the relative performance of intelligent applications over the Web. The site currently provides considerable overview and background material on the subject.

A more popularized approach to the subject is found in the book *Brave New Unwired World: The Digital Big Bang and the Infinite Internet*, by Alex Lightman (Wiley, 2002). The subject grows inevitably 'hotter' as more intelligent devices proliferate.

Tracking the Physical

With pervasive connectivity supporting networks and embedded devices, many contexts can benefit from more intelligent tracking solutions. We have long had real-time tracking of vehicles and shipments – bar codes were invented to track rolling stock on railway lines, after all, long before Web interfaces enabled any customer to track the status and location of that latest online order.

These simple bar code labels and badges are in many contexts being replaced by tags with embedded storage and functionality, often as microchips powered by induction for readout or configuration at strategic locations. Smart transponder tags (RFID) seem likely to fill a central role linking the physical and the virtual, as a physical aspect of an entire distributed application.

Bit 11.3   To function intelligently in the real world, the network must identify and locate real objects, uniquely, and reliably

In conjunction with IPv6 and the vastly more unique Internet addresses, RFID tags provide the 'glue' to connect model and physical world through transponder identification and location.

Part of the functionality is embedding local information about the object that is tagged – whether device or building, permanent or updated. It need not be very much information that is stored in the tag itself, although one of the attractions of RFID in many contexts is the flexible ability to store (and update) more data than corresponding 13-digit bar codes.

A metadata structure with identity assertion and URI pointers to other resources on the network is, however, often preferable from many perspectives. More capable tags might be active and have sensors to monitor their environment (or bearer) for diagnostic, integrity, or safety reasons, and only periodically upload stored information to the network. Their presence might also selectively enable proximate devices, or present personal data (such as bearer-implant credit cards).

In summary, remote-readable tags are at the very least a technology that enables the network application to map the physical world, monitor its changing status, and track any 'significant' objects (including people) moving in it. Given this capability, the network can then move and modify corresponding virtual objects and contexts in appropriate ways.

Other smart tags (with displays, for instance) can provide more direct human-readable information, perhaps mediated through the network between locations: such as warnings for expired-date, unfit-for-consumption content, unsuitable storage or operating environment, inappropriate location or orientation, inappropriate use, and so on.

A plethora of 'nag-tags' everywhere might be more annoying than really useful, but it is not difficult to see application areas where such functionality would provide real benefits. The health-care community, for instance, is highly motivated to promote research into these areas, since many forms of remote monitoring both improve quality of life for patients and reduce overall costs.

User Interaction

A sometimes neglected aspect is how users interact with computers. It is easy to talk and speculate blithely about automatic agents and intelligent services on the Web, yet still take for granted that the user usually sits in front of a desktop screen, and uses mouse and keyboard.

Even when discussing speech and gesture recognition, hand-held devices, and user roaming, it is easy to spot the underlying assumption held by the speaker/writer that the 'normal' environment for the user is in front of the desktop GUI.

Bit 11.4   What is normal for me, now, is normal for everyone, always – Not true

Mindset assumptions of this nature are unwarranted, but far too common. The 'always' is insidious because of what it implies, even when not said, that we never really believe that our lives will change very much – despite ample personal historic evidence to the contrary.

The assumed future is somehow always just a simple extension of today. This is why true visionaries, who can see the potential for change, are so rare; or for that matter rarely listened to when one does speak out about some new concept.

We tend to forget that, not so very long ago, things looked very different. It is almost unimaginable (very much so for our younger children) that only a decade or so ago, we did not have pervasive access to the Web. A comparatively short time earlier, computing in the 1960s and 1970s was relegated to carefully guarded computer centers, where access was through rooms of terminals or batch card readers. User interaction was . . . *different*.

A New Paradigm

Anyway, given the vision of pervasive device connectivity, the current 'human–computer' paradigm, of one-computer-to-one-user interaction, that underlies so much of how user interfaces and application models work, is assuredly inappropriate. Most people must already now deal with computer at work and computer at home, sometimes several in each location. Progressively more people also use portable notebook and hand-held devices.

Synchronization of locally stored user data between multiple machines is a serious issue. Just consider the problems people already face in connection with their e-mail. Originally, it was seen as a convenience to download mail from a server mailbox to the local machine. This view is still the prevailing one for the design of e-mail software, despite it no longer being convenient in the multiple-machine context. Only the gradual adoption of webmail/ IMAP and server-based storage begins to loosen the artificial constraint that ties a user's e-mail archive with a particular physical machine's local storage.

As we move towards hundreds, and even thousands, of computers interacting with each individual, the current one-on-one interaction paradigm becomes unmanageable. We simply can no longer consciously interact with each device, configuring it to our needs.

Bit 11.5   Sweb computers must autonomously adapt to the individual's context
The change in focus necessitates that the computers respond pro-actively to inferred user needs, instead of demanding that users focus on them, individually. They must anticipate our needs and sometimes initiate actions on our behalf.

No doubt we will continue to interact with a few of our computers in more traditional ways, but the vast majority of the computers we encounter will be embedded within our physical environment. These devices will capture data and may act without direct human intervention, guided only by a combination of general and individual policy (or delegation) rules.

Think about it. By analogy, precursors to this kind of pro-active and seamless environment have been implemented in several ways already:

- In cities we hardly ever open doors in public spaces any more; they are all automatic, activated by motion detectors – primitive pro-active mechanics.
- Roaming mobile telephone users take for granted that successive relay transmitters can seamlessly track location and maintain connectivity, even across different operator networks, with no user intervention and no change to the subscriber number.
- The process of connecting to peripherals or networks has already been changed from laborious manual configuration to 'plug-and-play' – self-negotiating configuration, with some degree of auto-detection.
- Changing networks can be as easy as plugging in and waiting for DHCP to lease the machine its new network identity. In future, of course, the identity would never change.

What if all computer devices near you were always prepared to notice and react to your presence, ready to accept your requests? Pro-active computing will in fact soon be a reality, not just in the labs; virtual doors will open as we come near.

Numerous research and prototype projects are looking into new interface models and the devices to support them: interactive wall displays, tabletop screens, communicating chairs, ambient displays, intelligent rooms, and so on. Novel interaction models include gesture interpretation to allow users to 'throw' and 'shuffle' displayed objects into new arrangements and locations in a shared workspace, various forms of awareness and notification mechanisms to facilitate informal encounters between people, and different ways to enhance the public-space noticeboard concept.

Intel is one of the major players involved in research to achieve this goal (*www.intel.com/ research/*), and has formulated a series of relevant 'challenges' that must be addressed along the way:

- *Getting physical*, which means connecting computers with the physical world, both to acquire data directly and anywhere about it and about us (using sensors and biometrics) and to interact with physical objects (with *microelectromechanical* systems or MEMS).
- *Deep networking*, which means interconnecting all these sensors and devices in useful and adaptive ways. The 802.11 wireless standard now appearing as 'hotspots' for notebook and hand-held connectivity in public places is perhaps an early indicator of the trend. The future may mean embedding wireless connectivity at the microchip level, everywhere.
- *Macroprocessing*, or distributed processing at very deep levels of the system, which enables processes to work coherently in a roaming mode of their own, independent of underlying devices, to follow a mobile user.
- *Dealing with uncertainty*, which is necessary in the context of the 'messy' physical world filled with 'unpredictable' events and people.
- *Anticipation*, which is another way of expressing the capability of inference about probable events and requirements in changing contexts (such as with human actions) and being prepared to respond or prompt appropriately.
- *Closing the loop*, which is to be interpreted as the ability of computers to act appropriately in the circumstances, whether taking initiative or responding to specific requests. As the loop implies, it is a feedback system, so control and stability issues are vital when humans are no longer directly in the interaction loop to mediate the possibly deleterious effects.
- *Make it personal*, which is the long-term goal, and includes appropriate adaptation to user preferences, language and locality, cultural aspects, privacy and security concerns, and a host of other 'people' issues.

The autonomous functionality of this new infrastructure is fundamental.

All these objects will be communicating; therefore, we need to develop ways to have them autonomously configure and maintain themselves. People have more important things to do.
                                                                    Gaetano Borriello, *Intel* researcher

In this context, it was early on identified that large-scale embedded environments require extremely easy-to-use interaction methods with simple modes that support multiple users accessing common resources. They also require fast 'embedded' ways to identify (authenticate) users reliably.

Privacy in the New Paradigm

The thing is, will we be comfortable in this new information environment? Caught in a 24/7/365 web of the Web extended into our every surrounding, perhaps even into our very body; constantly connected, monitored, anticipated. . .. It may seem kind of ominous. . ..

Privacy is a cultural and relative concept. We only need look at its variations within human culture to see that privacy/integrity can mean very different things in different contexts. So-called western civilization has come to have a fairly strict, individualistic view of personal privacy, but it has not always been so, nor is it so everywhere even today.

Nor will it continue to be so, if we read the indicators correctly.

Bit 11.6   Privacy on the Web will have a different context from what we are used to now

It is already the case that many things that were considered 'private', previously unutterable outside intimate contexts, are today casually stated loudly in conversation over a mobile phone in earshot of complete strangers, or published on a homepage for all the world to see.

Nor is 'privacy' an especially easy semantic term to pin down, as it has numerous dimensions: physical and non-physical. Ultimately, it is a *perception*, not a thing, one that furthermore is projected into the social context. Some people manage to remain exceedingly 'private' in situations where most of us would agonize at being so exposed, for example.

Variations in acceptable and unacceptable 'infringement' of the privacy sphere can be remarkable even between, for example, European countries and the U.S. Some historic and more exotic cultures deprecated the concept of individuality (and hence our understanding of privacy) in ways that are totally unacceptable to the average westerner, yet which may have similarities to the way the concept is evolving on the Web.

The fiction of treating legal entities, such as businesses, as if they have the same (or more) rights compared to the individual (in terms of ownership, for example) is a problematic aspect of modern society, which is increasingly in opposition to the basic tenant of information access on the Web.

Western corporate culture is often in opposition to individual privacy, while at the same time promoting the notion of complete privacy for the organization as a kind of super-individual with rights transcending those of any real person. Companies routinely gather and correlate information from customers and casual visitors, building 'personal profiles' with an intimacy and a level of detail that might easily shock the unaware subject.

So, perhaps our perception will adapt to the new reality? It is a trade-off: benefits against costs. Available technologies also provide some innovative new twists to the concept of 'identity'. The following examples illustrate some issues:

- Users must expect to be 'recognized' by devices and systems, automatically. This ability may be based on biometric sensors, digital signature keys, or agent negotiation between a hand-held device and the surrounding bubble of responding devices.

- Roaming users will expect 'personal' profiles to roam with them, in whole or in part, over a wide variety of devices. Key features are to be recognized anywhere, and to experience seamless connectivity at all times.
- Roaming users may expect their working environments and data to be virtually transportable. Personal contexts should be capable of being 'served' across arbitrary devices near them.

However, note that *recognition* (that is, a verifiable and consistent identity) need not necessarily translate into the user's normal everyday identity.

Bit 11.7   Privacy, in some meaningful sense, can still be maintained despite ubiquitous recognition

A consistently recognized public identity can be an anonymous alias, thanks to digital signatures, yet still securely tied only to the user. Public-key cryptography techniques can safeguard the data, irrespective of location.

On the other hand, if device recognition is coupled to say biometric sensing, and from there (through authentication) to national identity registries, roaming anonymity is no longer an option. Therefore, the implementation decision is more political than technical, but persuasive argument can be made that it should not be allowed to just 'default' to central identity.

Engineering Automation Adaptability

Automation today is generally based on a limited repertoire of pre-programmed responses in specified devices. In the ubiquitous 'deep networking' view, all devices in the environment are 'plug and play' (or to adapt the metaphor, 'unplug and still play') in the broadest sense with respect to the local network. They connect in proximity, negotiate, and self-organize to complete tasks.

Advertising device functionality and hence openness for control is something that is gradually being deployed in all new designs that incorporate networking capability. With increasing logical sophistication also comes the capability for reprogramming – software upgrades on the fly from the network. *Debian-Linux* users may have some feel for what this feature can mean.

Now add the virtual model maintained within the network, the RDF schemas that describe the devices and their capabilities, and input the location and status of each device within the environment at any given time.

It starts to get interesting, no?

What then if we add software agents capable of mediating communications and actions between devices, the people in their proximity, and their owners (irrespective of location)?

Automation Cooperation

Take it all one step further. The key technology of true *ad hoc* networks is *service discovery*.

Devices can automatically seek out and employ services and other devices for added information or functionality, provided they can communicate meaningfully with each other to discover and adapt functionality to each other's capabilities. Several of the fundamental sweb protocols and specifications discussed in the core chapters of Part II provide the mechanisms.

Such behind-the-scenes functionality, by which device 'services' (actually, any functions offered by various mundane devices) can be described, advertised, and discovered by others, is being developed in a semantic context.

- This new system must replace the current service-discovery implementations within computing (such as *Sun*'s JINI, *Microsoft*'s, PnP and UPnP). The latter are based on representation schemes that rely heavily on *a priori* standardization of known and detail-specified characteristics for the devices. Such schemes cannot cope in the semantic cooperation context: ontology-based schemas can.

Bit 11.8   Sweb connectivity aims to be 'use anything, anywhere, automatically'

The goal of ubiquitous connectivity is 'serendipitous interoperability' under 'unchoreographed' dynamic conditions, without direct supervision of humans. In practice, access control logic would dictate whether a given person is granted specific access in a given context.

Devices that are not necessarily designed to work together should under appropriate circumstances still be able to discover each other, exchange functionality capabilities, and adapt to use appropriate ones in the course of ongoing task fulfillment. Therefore, in some measure devices must be able to 'understand' one another and reason about functionality. Standardizing usage scenarios in advance among perhaps hundreds of different device types is an unmanageable task.

Further autonomous tasks in this negotiation process might involve contracting and compensation issues (payment for services provided), and meeting required parameters concerning trust, security, privacy, preference, and any user-specified rules on either side (the devices may belong to different users).

The visual image in the discovery context might well be the little maintenance robot from the Star Wars films, *R2D2*, who showed the ability to plug into and scan or manipulate virtually any system it encountered. This image is very crude, however, in comparison with the true potential envisioned – but it does, after all, derive from the mainframe plug-in mindset of the late 1970s, in a galaxy far, far away.

An important aspect of device negotiation in the corporate or organization setting is that security policies can be expressed in a form that is independent of application. It must be possible to express the representation of security constraints consistently and reliably, across an arbitrary and heterogeneous mix of enforcing devices and services.

Automation and Context Semantics

Today, devices have little understanding of their real-world context, and as a result they often make what we consider to be stupid mistakes. Just providing automatic discovery and ubiquitous connectivity does not solve this problem, it only magnifies the potential for havoc.

Sweb ontology addresses the issue by providing the structures that might enable applications to make sense out of particular contexts and avoid the literal-minded pitfalls so easy to encounter when dealing with human situations.

Take a simple application such as an agenda/calendar/planner, where automated logic functions can often run into intractable problems that human users routinely solve with 'a little common sense'.

- One proposed early solution, *SensiCal*, was presented as 'A calendar with common sense' in *Proceedings of the 2000 International Conference on Intelligent User Interfaces* (pp. 198–201, ACM, see *portal.acm.org/citation.cfm?doid=325737.325842*). The article discusses the pieces of common sense important in calendar management and presents methods for extracting relevant information from calendar items.

Contextual reasoning around calendar planning typically involves weighting and a host of implied priorities. Many of these guidelines are difficult to encode into formal rules and must instead be inferred from the semantic contexts. User commands and intent are ideally passed on in natural idiom, yet such human imprecise formulations are very hard to process reliably by software.

An intelligent agent must cope with such ambiguity, integrate input with its model of the world, and additionally be able to make intelligent suggestions to the user based on inferred relations.

Whither the Web?

What is the long-term perspective for the Web? We do not really know, but can of course speculate. Few could at the time foresee the consequences and development of other technologies that have subsequently come to be accessible to everyone. And even when the importance was seen, the detail of the vision was invariably off – sometimes seriously wrong.

It is not just about the technology either. Often far more important and decisive are issues such as social/cultural acceptance of a new technology, preconceptions that determine design and acceptable use, and unexpected interactions with other technologies (whether existing or as yet unknown). Not to mention seemingly random events that turn out to be critical, and political, legislative or corporate decisions that directly or indirectly determine a technology's future.

Take the telephone, our recurring example, and compare the predictions of early visionaries and the criticisms of the detractors, with the complex communications infrastructure that eventually evolved out of telephony, becoming essential to our civilization – manifest in the handsets in practically everyone's pocket.

In the process, telephony indirectly spawned the Internet and the Web, which arose out of interaction with another new technology, the computer. Quite unforeseen, that, yet the result has utterly transformed the day-to-day life in western societies in only decades, and is soon doing so for an even larger portion of the world.

On the other hand, take the *Zeppelin*, which but for some quirks of history and politics that got in the way, might have become the dominant form of air travel in the latter half of the 20th century. Never happened, except in the realms of fantasy and alternate worlds.

The recent trend of measures and changes occasioned by heightened 'security' after 2001 might perhaps be seen by future historians as part of an overall shift that also marked the end of 'free information' in the sense used earlier. The Semantic Web vision might therefore become an impossibility due to social and political factors, or at least fail to achieve global deployment.

Possible. Unfortunate in many ways, but most assuredly a possible outcome.

The point is, even the best visionary cannot see the complex and unpredictable path a given technology will take in the future. The best that can be accomplished is only to suggest what might be, and perhaps add a few caveats and warnings along the way for guidance.

Evolving Human Knowledge

The main issue in broader application of sweb technology and promoting the development of the Semantic Web is undoubtedly *addressing*, specifically the deployment and use of URIs in all aspects of implementation.

In naming every concept using a published URI, the Semantic Web lets anyone express with minimal effort new concepts that they invent. Not only does this URI notation make the concepts freely available to others, but also an increasingly rich web of referencing links can leverage other resources.

URI addressing is the basis of a unified WSW (Web Services Web). It will determine security, WS orchestration, and a host of other essential features. Its importance cannot be overestimated.

Many services today still use proprietary, internal ways to reference resources of interest to external queries. These identifiers are readable by other services, but they require further interfaces ('glue'). RPC coding is the prevalent technology to implement such inter-service glue.

Expressed as and accessible through URI notation, however, these pointers would be easily reusable throughout the Web by any application without requiring extra coding. Existing, published data may thus be leveraged by remote services to create new and innovative structures.

The new, ontology-based layers of representing the knowledge in abstract structures, amenable to both human and machine processing of concepts and assertions, add further to this universal referencing.

As noted in the previous chapter, the name of the game is changing from publishing information on the Web so that it is *accessible* to anyone, to publishing knowledge (information plus meaning and relationships) on the Web so that it can be *used by* anyone – including machines.

This shift is comparable to what happened when libraries were invented: it became possible to collect and store vast repositories of written knowledge, for anyone to read. Now these repositories, and new previously undreamed of resources, are being moved into a virtual space where the combined knowledge will be available to all.

Or so we hope.

Towards an Intelligent Web

Perhaps we are also approaching something more than just a vast globally-accessible knowledge repository, There is the potential to develop devices, services, and software agents that we might converse with as if they were fully sentient and intelligent beings themselves. Here lies a promise of real Artificial Intelligence.

I have in this book intentionally avoided the casual (or at least unqualified) use of the term 'AI' in several places. This seeming lapse leads some reviewers to remark that in their understanding the Semantic Web and AI are intimately interlinked, since much of sweb technology in the realms of structure, KR, and reasoning builds extensively on earlier and contemporary AI research.

My response is that contrary to the impression perhaps given, I am not denying this sweb-AI connection. However, I found it necessary to at least qualify my usage into 'lesser AI' and 'greater AI' technology. Though I have noted a variant reading of AI as '*Artifactual* Intelligence', this latter interpretation seemed only to obfuscate the point I wanted to make.

I introduced the lesser or greater distinction of AI early on in Chapter 1, mainly to distance the subsequent discussions from the original reach of AI research, a topic reserved for the speculative visions explored here. When 'intelligent' Web Services are mentioned in earlier discussions, the meaning is generally constrained to mean self-guiding, task-oriented systems that can reason from rules, make reasonable inferences, and arrive at reasonable results.

Even so, the potential of the simple implementations is remarkable, and may become hard to distinguish from real intelligence. Should we care? It is not clear that we could tell the difference.

Bit 11.9   Assessment of intelligence is generally based on test results of performance

For most people, 'intelligence' simply means to arrive at sensible conclusions in a given context, and to learn from mistakes. By such criteria, even fairly simple agent software might pass. And not a few real people would fail, miserably.

Humans show an innate tendency to anthropomorphize, to 'animate' the inanimate and read 'human' traits, even intention, into even the most recalcitrant mechanical device. How could we not then relate to hidden software that gives the semblance of a real personality by showing independent, goal-oriented action?

Add some programmed 'politeness' and many users will interact with the software agent as with a real person; perhaps even prefer their company to real people.

Even the earliest '*Eliza*' programs, simulating intelligent conversation by echoing fragments of user text response recast into prompting questions, managed to fool many people into thinking a real person was typing the program-generated half of the 'conversation' held. Intelligent agents with responsive interfaces will seem that much more alive as they fly to our bidding.

So, yes, in the pragmatic sense of the word, the Web will evolve into an entire ecology of 'intelligent' systems. By some metrics, selected agents might reasonably be classed as more intelligent than many of the users that summon them, and thus qualify for the AI moniker!

How Intelligent?

Perhaps we are creating the prerequisites for a *cyber-Gaia*?

Undoubtedly, new fields of study will be spawned, to study and determine the precise degree of 'intelligence' of the system, in time applying a considered definition of actual sentience. I remain unconvinced that a human might conclusively pass such a test while a highly functional but still non-sentient system would fail. Nevertheless, the study is sure to keep a cadre of researchers and philosophers busy for a very long time to come.

Pragmatics and consensus might instead simply decide that 'human is as human does' and hope that any emerging sentience in the machine will share our best human values. Any other result might prove too frightening to contemplate, a theme much science fiction in the 1950s and 1960s explored, echoing many earlier speculations on the same theme but with magical overtones.

Creating artificial life and suffering the consequences has been a hubris theme for as long as humans have had stories to tell. The concept of created intelligence is seemingly as old as our own intelligence capacity to think the thought. Legend and myth recognizes the conceit, as it were, by warning of the dangers inherent in creating non-human forms of intelligence.

Exploring the speculation that the Semantic Web might be the catalytic medium for developing true artificial intelligence of one form or another is beyond not only the scope of this book, but also beyond the scope of current knowledge.

Still, an interesting speculation has it that the only way for us to create, or engender the creation of an artificial intelligence that we can understand – and which has the potential to understand us – is to ensure that it can perceive and affect the physical world in ways that are compatible with our own senses and actions. Judging from some of the functional speculation, such a move might be prudent for both sides to remain on friendly, collaborative terms.

The thrust to provide sensors and actuators to deal with the real world on the same terms as humans do is probably a sensible move in this light. Human concepts and values are shaped by our own *sensorimotoric* experiences, so AI shaped from similar real-world interactions and human-shaped ontologies should be compatible, and safe. . ..

The Global Brain

The *Principia Cybernetica Project* (PCP, see *pespmcl.vub.ac.be*, or mirror site in the U.S. *pcp.lanl.gov*) is an international organization that aims to develop collaboratively a complete philosophy based on the principles of evolutionary cybernetics. Fittingly enough, the effort is supported by collaborative computer technologies.

It is is one of the oldest hypertext repositories on the Web, registered in July 1993. Regarding the associated claim that it is the best organized, and largest, fully connected hypertexts on the Web is harder to credit of late. Other repositories based on newer technologies have led to the rapid growth of far larger webs of knowledge, such as *Wikipedia*, to name perhaps the most well-known.

Nevertheless, the PCP site has over 2000 'pages' of highly interesting material to browse that often relates to AI-issues discussed in the context of sweb technologies.

The *Theory* section of the site collects epistemology, metaphysics, ethics, concepts, principles, *memetics*, and the history and future of evolution. The *R&D* section collects algorithms, experiments, and applications to develop distributed, self-organizing, knowledge networks, inspired by the 'Global Brain' metaphor.

The *Reference* section of the site offers volumes of background material, including an electronic library with free books, a web dictionary, related Web sites, and info about cybernetics and systems theory, such as bibliographies, associations, and journals.

PCP contains far more history and philosophy than technology, but this, too, is needed in the context of the Semantic Web. As noted in KR and Ontologies, the structures used have deep roots in classical logic and philosophy. The context is required reading, and for thinking about in light of possible future developments.

Possible Evolutions

As the computer networks and devices participating in the Web become more 'intelligent', it will all start to look more like a global brain or super-brain, a coherent entity in its own right – or possibly an incoherent one. Its capabilities would far surpass those of individual people, and a defining moment would be when it formulates its own agenda for future autonomous actions.

This line of development is sometimes called 'IA', *intelligence amplification*, rather than AI, the creation of intelligence. The starting point is the enormous amount of information and resources available in the Web and using interactive, self-organizing systems to leverage it. The IA concept pre-dates but echoes some of the thoughts behind the transition from the traditional Web to the Semantic Web.

Such an evolutionary transition to a higher level of complexity and 'collective thought' certainly seems possible. The question is whether it would eventually lead to the integration of the whole of humanity, producing a single human 'super-being', or evolve in opposition to the collective will of humans. It could alternatively enhance the capabilities of only a selection of individuals over the majority, for good or bad.

Success or flaws in the infrastructure and in the coupling between physical and virtual can then become decisive factors in determining the outcome.

Conclusions

Retreating from the misty visions of future potential, it is perhaps best to take stock of where we are at the moment and summarize.

Standards and Compliance

As noted in several places in the text, some standards and specifications are in place, while others are still in the draft stage. Many applications are early prototypes and highly experimental. The infrastructure is not quite 'there' yet, but significant parts of it are being deployed, even though the results may look different on different continents and in different countries.

However, the foundations are solidly in place, even if the higher edifice is not yet visible in all its details. The core specifications for it exist and are being refined: they represent a consensus commitment to the Semantic Web.

When you connect a cat-5 ethernet cable to your computer, you effectively commit to taking part, with your computer, in a very special system. It is a system in which the meaning of messages is determined, in advance, by specifications.   Tim Berners-Lee

The cat-5 is these days often replaced by wireless access, just as the network cable earlier replaced the modem and phone line, but the point is valid whatever the connectivity.

Interoperability Agreements

The Internet works because of interoperability between different computers and an *agreement* to use a specific set of protocols. Despite different hardware, operating systems, local language context, and software supplier, it all works together to do the bidding of the user. Increasingly, this bidding is not simply for the information, but for advanced functionality.

Developers and vendors of the equipment and implementations explicitly agree to use the open specifications. If they refuse and opt for proprietary solutions, they only constrain usability to proprietary technology with a far smaller application area. They play in their own sandbox.

Providers who seek the widest audiences deploy their systems using the widest-adopted specifications. Users of the Web implicitly agree to use the specified languages when they use the Internet, if only by the choice of system and client software. If either lets them down, they will eventually migrate to the systems that fulfill the expectations.

Without this interlocking agreement, nothing comes down the pipe.

Open Ensures Diversity

The fact that the defining specifications are open ensures diversity in application. Anyone can create compliant implementations that will work with the network and its resources. Anyone can build on the specifications to leverage existing resources in new contexts.

Previously, such diversity has served the Internet and Web well, allowing adaptation to contexts and scales unimagined by the framers of the original Web specifications, yet requiring but minimal changes to them. Open diversity should serve the new Semantic Web just as well.

Social Specifications

A virtual interoperability also exists in the social fabric that governs how people use the technology. This implicit 'specification' manifests on several levels:

- *formal* (legislative constraints);
- *normative* (vendor constraints, whether hardware or licence);
- *informal* (consensus on acceptable use).

As happens in times of rapid change, as now, one or more of these may diverge sharply from the others. The resulting tension and conflict may take years to resolve.

The old hierarchies are losing their attraction as new forms of 'peer' networking between individuals and between businesses create richer possibilities and greater value. Old-new hierarchies will not give up without a fierce struggle, but they can only prevail if the majority accepts (if only by inaction default) the imposed laws and norms without question. If consistently opposed, they *will* adapt, or fall.

The Case for the Semantic Web

What are the arguments for the Semantic Web revolution? And why should we care?

Knowledge Empowers

The motivation that knowledge brings power to the individual is ultimately why repressive regimes everywhere have always tried to assert control over and limit the spread of knowledge, often by crippling or outlawing the technology of the medium. Book burning, fax banning, censorship, or central Web filters; the form it takes matters not. What matters is how resilient and resistive the infrastructure is to the threats.

The fear of empowered individuals is also evident in the ranks of the larger corporations, since informed customers cannot be manipulated as easily. In fact, they might just go to the competition. But information is not easily suppressed once released. Better then to just outlaw the competition, forming an alliance between normative and legislative; such is the common response.

On the other hand, the more powerful and pervasive the technology is to disseminate the information, the more difficult it is to keep the individual ignorant. It was far easier in past centuries for the tyrants to rule absolutely by choking the information flow, since it was then merely a trickle.

Today, the hard choice for presumptive despots is whether to impose absolute rule by cutting off society completely from the global information exchange, and thus irreparably cripple the economy, or to allow the flow and accept the inevitable looser control over citizens it means. North Korea's leaders committed to the former. China's leaders reluctantly chose the latter. A decade down the line, the results are plain to see.

The examples should be a warning for those traditionally open societies that are now contemplating some of the new technologies to rule more absolutely than ever before. This is where popular consensus comes in, and it is why we should care.

- Consensus can change legislation to align with the will of the people.
- Consensus can bring down megalomaniac rulers and overly greedy corporations alike.
- Consensus can make people leave some technology untouched, despite wide deployment, and it can encourage people to use and spread other implementations in novel and empowering ways.
- Consensus can realize a future radically different from any envisioned by formal leaders.

History provides several examples, some recent, of popular will effecting dramatic changes against all odds. Unfortunately, history also provides even more examples of when public passivity or contention led to disastrous results.

Scale of Information Management

As our information repositories increase, and the desire grows to use them in concert to derive maximum synergy, we run into the problem that the 'non-intelligent' solutions do not scale very well. Centralized management can no longer meet the requirements.

The agent-based and semantic-structure technologies hold the best promise for making sense out of the networked information. Structured abstractions of the data are the only hope to deal with the vast amounts of information and allow widespread re-use. Bioscience research has already seen the light in this respect, other disciplines are sure to follow.

Metadata structures are necessary, if only because text and unannotated media formats on their own are not sufficient to categorize and retrieve properly. Ontologies lend clarity to both design and representation, a clarity required ever more as the system is scaled up.

In addition, interactions with the real world involve important concepts (such as events, processes, intentions, relations, and goals) that are not easily or at all expressible as flat text so that they can be processed by machines. Semantic structures support implementation of validation and 'proof' mechanisms, and capture the concepts of user perspective, user intentions, and user preferences in a general way.

Independent agents capable of reasoning and inference are a necessary part of that vision, although the exact parameters of deployment and policies of use are ultimately subject to the decisions of all three levels of social specifications.

Economics and Advantages

One of the greater driving forces for change in the modern world is economics.

In a theoretical sense, a 'free market' of information and resources, of services and products, seems both possible and ideal on the Web. It would be composed of a collective system of agents making direct transactions that can match supply and demand without complex intermediary infrastructures.

With transaction costs approaching zero, global reach for niche markets, and incentives to fulfill diversity, the immediate expectation is of growth and less business 'friction'. We might see less waste as production can quicker adapt to demand.

This idealized picture makes the transition seem inevitable if only because of its economic advantages for all parties. Businesses adopting the new model would seem to gain such competitive benefits that they would outstrip competitors who did not.

Real evolution towards a global virtual economy is messier, and the situation is vastly more complicated due to social, cultural, and political factors. The intermediaries do not want to be made superfluous. Governments do not want to lose control or tax incomes. Bureaucrats find new ways to continue expanding and issuing permits. Large companies are disinclined to rethink their business process unless forced to do so. Few at any level really *want* change.

For these reasons, change is driven mainly by the actions of individuals and small businesses adopting the new technologies *en masse*, for basically very trivial and self-serving reasons, never imagining that collectively are they driving a watershed event. Such change also comes about in spite of studied non-interest or excessive meddling of the intermediaries or authorities.

Failing the determined efforts of individuals wanting to use the technology for their own purposes, the visions will falter and even if the technology is deployed, the result become something completely different – monolithic, stagnating, and serving vested interests.

We can Choose the Benefits

Empowered, and with empowered systems to manage information for us, we *can* choose in a new way the things that interest us, that enlighten us, and that enrich us. We can choose the people we communicate with, that can reach us, or those for whom we are prepared to vouch.

Bit 11.10   Focus on the positive, the visible benefits, and advocate for them
Let the overall scheme of things manifest through the system's capacity for self-organizing.

For the immediate future, however, just expect semantic software to proliferate as target-and-forget, fetch-my-data, summarize-these-sources, organize-my-day, field-the-distractions, brief-me-quickly, find-something-interesting, contact-this-person, etc.

We can Choose the Restrictions

Not everything 'doable' is necessarily desirable, hence design restrictions or regulation to make certain things less easy to do. Intentional impediments to functionality come in several flavors:

- *Precautionary*, like the prompt 'Do you really want to do this?' to make sure the user is aware of the consequences.
- *Regulatory*, in the sense that a user is not *allowed* to proceed, or is required to authenticate and prove the right to access information or functionality.
- *Registrations*, which means a user may proceed, but only at the 'cost' of leaving a unique audit trail and perhaps be blocked in future if the present action is later deemed inappropriate.
- *Filtered*, in that the user may be kept entirely unaware of certain options if these are deemed not appropriate for the context.

None of this is new.

What may happen in the sweb context, however, is that online activities may be more intensely scrutinized by agent software. Functionality may then be regulated in great detail by logic, applying policy rules to inferred contexts. Such automated regulation may become either more intrusive, or more transparent and flexible, depending on how it is implemented, by whom, and why.

Choose Diversity

One of the success-characteristics in the Web has always been *consensus but diversity*, innovations and recommendations free for anyone to follow but few hard regulations and requirements.

It is true that for such a system to be effective, the people participating in it need to agree about a set of common standards or rules to facilitate communication and cooperation. In the Web, these common rules are compliance to the core technologies, such as URI, HTTP, and TCP/IP, and basic rules of conduct. The latter suggest policy restrictions on exploits and intrusion attempts, or ways to combat the spread of computer viruses and worms. However, well-chosen rules increase rather than decrease freedom.

In actual fact, the bottom-line hard requirement is simply that whatever you implement must be gracefully compliant with existing infrastructure. Your overlaid protocols and applications can be as strange as you want, as long as they do not break the transport or interfere with the expected capabilities of others.

Bit 11.11   Infrastructure compliance is a self-correcting requirement and environment

There is no need for a tyranny of regulation and restriction as long as functionality is in everyone's self-interests. It is only against purely destructive efforts that blocking measures are required.

Become too distant from the consensus way of working and you will likely lose connectivity or common functionality. Become too intrusive and you bring down the ire of the community, who might first flame you, then filter you out of it. Become too strange and obscure in your innovation and nobody will adopt it – you may then continue to play in splendid isolation. The usual balance is mildly chaotic diversity, constantly evolving in various directions.

It is a state we often see in nature.

- Consider ants. If ants would always follow the paths laid down by their fellow ants, and never diverge to create paths of their own, then the colony would starve as soon as food sources on the existing paths became exhausted. So they have evolved to meander slightly, leaving the strongest scent trail, with occasional individuals striking out boldly where no ant has gone before. Some of these will perish, a few return. Sometimes the action of the one becomes the path for many, returning with new-found food, and then ultimately the majority path shifts.

Bit 11.12   Successful collective problem solving relies on a diversity in the individual approaches and different paths

Significant advances may then attract consensus attention, the chosen divergent path become a dominant one in future, but it never becomes the only path.

Since the same rules democratically apply to everyone, the net result is that otherwise dominant organizations, governments, or corporations have less power to censor or impose their rules on the people who use the Web. The individual gains freedom.

Who Controls It?

A distributed and partially autonomous system like the proposed Semantic Web, and like the Web before it, is ultimately controlled by the people who make themselves part of it and use it.

Bit 11.13   The Internet is functionally a collective; a complex, self-organizing system
It is a direct result of many autonomous entities acting in a combination of self-interest and advocacy for the common good. This collective is guided by informed but independent bodies.

If people stop using the network, then effectively it will cease to exist. Then it no longer matters or has any relevance to the people, simply because it no longer connects to their lives.

This is not the same as saying a network, controlled by a central authority, with extensions into and controlling our physical environment, would not matter. Some people, a very much smaller collective, are then still using the system when all the others have opted out and relinquished their distributed and moderating control over it.

The choice is ours, collectively – yet any one individual action can be pivotal.

Part IV

Appendix Material

Appendix A

Technical Terms and References

This appendix provides a glossary of technical terms used in the book, along with the occasional technical references or listings that do not fit into the flow of the body text.

At a Glance

- *Glossary* of some of the highlighted terms in the text.
- *RDF* includes:

  - *RDF Schema Example Listing* gives the entire Dublin Core schema defining book-related properties.
  - RDF How-to, a simple example of how to 'join the Semantic Web' by adding RDF metadata to existing HTML Web pages.

Glossary

The following terms, often abbreviations or acronyms, are highlighted bold in their first occurrence in the book text. See the index for location. This glossary aims to provide a convenient summary and place to refer when encountering a term in subsequent and unexpanded contexts.

Agent, in the sweb context, is some piece of software that runs without direct human control or constant supervision to accomplish goals provided by a user. Agents may work together with other agents to collect, filter, and process information found on the Web.

Agency is the functionality expressed by agents, enabling for example automation and delegation on behalf of a user.

API (*Application Programming Interface*) is a set of definitions of the ways in which one piece of computer software communicates with another – protocols, procedures, functions, variables, etc. Using an appropriate API abstraction level, applications can reuse standardized code and access or manipulate data in consistent ways.

Architecture, a design map or model of a particular system, showing significant conceptual features.

Authentication, a procedure to determine that a user is entitled to use a particular identity, commonly using login and password but might be tied much tighter to location, digital signatures or pass-code devices, or hard-to-spoof personal properties using various analytic methods.

Bandwidth, a measure of the capacity a given connection has to transmit data, typically in some power of bits per second or bytes per second. Extra framing bits mean that the relationship between the two is around a factor 10 rather than 8.

Broker, a component (with business logic) that can negotiate for instance procurement and sales of network resources.

Canonical form is the usual or standard state or manner of something, and in this book it is used in the computer language sense of a standard way of expressing.

ccTLD (*country-code Top Level Domain*) designates the Internet domains registered with each country and administered by that country's NIC database. The country codes are based on the ISO3166 standard, but the list is maintained by IANA along with information about applicable registrar – for example, *.uk* for the United Kingdom, *.se* for Sweden, and *.us* for U.S.A. Also see *gTLD*.

CGI (*Common Gateway Interface*) is in essence an agreement between HTTP server implementers about how to integrate gateway scripts and programs to access existing bodies of documents or existing database applications. A CGI program is executed in real-time when invoked by an external client request, and it can output dynamic information, unlike traditional static Web page content.

Client-Server, the traditional division between simpler user applications and central functionality or content providers, sometimes written server-client – a seen variant is '*cC-S*' for centralized client-server, though '*cS-C*' would strictly speaking have been more logical to avoid thinking the clients are centralized.

Content classification system is a formal way to index content by subject to make it easier to find related content. Examples mentioned in the metadata context of this book are DDC (Dewey Decimal Classification Number, for U.S. libraries), LCC (Library of Congress Classification Number), LCSH (Library of Congress Subject Heading), and MESH (Medical Subject Headings). Also see *identifier*.

CSS (*Cascading Style Sheets*) is a systematic approach to designing (HTML) Web pages, where visual (or any device-specific) markup is specified separately from the content's structural markup. Although applicable to XML as well, the corresponding and extended concept there is XSL.

DAV or **WebDAV** (*Distributed Authoring and Versioning*), a proposed new Internet protocol that includes built-in functionality to facilitate remote collaboration and content management. Current, similar functionality is provided only by add-on server or client applications.

Dereferencing is the process required to access something referenced by a pointer – that is, to follow the pointer. In the Web, for example, the URL is the pointer, and HTTP is a dereferencing protocol that uses DNS to convert the protocol into a usable IP address to a physical server hosting the referenced resource.

DHCP (*Dynamic Host Configuration Protocol*) is a method of automatically assigning IP numbers to machines that join a server-administrated network.

Directory or **Index services** translate between abstraction names and actual location.

DNS (*Domain Name Service*) is a directory service for translating Internet domain names to actual IP addresses. It is based on 13 root servers and a hierarchy of caching nameservers emanating from registrar databases that can respond to client queries.

DOM (*Document Object Model*) is a model in which the document or Web page contains objects (elements, links, etc.) that can be manipulated. It provides a tree-like structure in which to find and change defined elements, or their class-defined subsets. The DOM API provides a standardized, versatile view of a document's contents that can be accessed by any application.

DTD (*Document Type Definition*) is a declaration in an SGML or XML document that specifies constraints on the structure of an SGML or XML document, usually in terms of allowable elements and their attributes. It is written in a discrete ascii-text file. Defining a DTD specifies the syntax of a language such as HTML, XHTML, or XSL.

DS (Distributed Service) is when a Web Service is implemented as across many different physical servers working together.

End-user, the person who actually uses an implementation.

Encryption, opaquely encoding information so that only someone with a secret key can decrypt and read or use it. In some cases, nobody can decrypt it, only confirm correct input by the fact it gives the same encrypted result (used for password management in Unix/Linux, for example).

Gateway (also see proxy), a computer system that acts as bridge between different networks, usually a local subnet and an external network. It can also be a computer that functions as a portal between a physical network and a virtual one on the same physical machine that use different protocols.

gTLD (*generic* or *global Top Level Domain*) designates the Internet domains that were originally intended not to be reserved for any single country – for example, the international and well-known *.com, .org, .net*. Also see *ccTLD*.

Governance is the control of data and resources and who wields this control.

Hash, a mathematical method for creating a numeric signature based on content; these days, often unique and based on public key encryption technology.

HTML (*HypeText Markup Language*) is the language used to encode the logical structure of Web content. Especially in older versions, it also specifies visual formatting and visual features now deprecated and consigned to stylesheet markup. HTML uses standardized 'tags' whose meaning and interpretation is set by the W3C.

HTTP (*HyperText Transfer Protocol*) is the common protocol for communication between Web server and browser client. The current implementation is v 1.1.

HTTPS (*HTTP over SSL*) is a secure Web protocol that is based on transaction-generated public keys exchanged between client and server and used to encrypt the messages. The method is commonly used in e-commerce (credit card information) and whenever Web pages require secure identity and password login.

Hyperlink is a special kind of pointer defined as an embedded key-value pair that enables a simple point-and-click transition to a new location or document in a reader client. It is the core enabling technology for Web browsing, defined in HTTP-space as a markup tag.

IANA (*Internet Assigned Numbers Authority, www.iana.org*) maintains central registries of assigned IP number groups and other assigned-number or code lists. Domain country codes, protocols, schemas, and MIME type lists are included, although many earlier responsibilities have been transferred to ICANN (whose motto is 'Dedicated to preserving the central coordinating functions of the global Internet for the public good').

ICANN (*The Internet Corporation for Assigned Names and Numbers, www.icann.org*) was formed as an international NGO to assume responsibility for the IP address space allocation, protocol parameter assignment, domain name system management, and root server system management functions previously performed under U.S. Government contract by IANA and other entities.

Identifier, generally refers in metadata context to some formal identification system for published content. Examples of standard systems mentioned in the text are *govdoc* (Government document number), ISBN (International Standard Book Number), ISSN (International Standard Serial Number), SICI (Serial Item and Contribution Identifier), and ISMN (International Standard Music Number).

IETF (*Internet Engineering Task Force, www.ietf.org*) is the body that oversees work on technical specifications (such as the RFC).

Implementation, a practical construction that realizes a particular design.

IP (*Internet Protocol*) is the basis for current Internet addressing, using allocated IP numbers (such as 18.29.0.27), usually dereferenced with more human-readable domain names (in this example, *w3c.org*).

IP (*Intellectual Property*) is a catch-all term for legal claims of ownership associated with any creative or derivative work, whether distributed in physical form (such as book or CD) or as electronic files, or as a published description of some component or system of technology. The former is legally protected by copyright laws, the latter by patent laws. Related claims for names and symbols are covered by trademark registration laws.

Living document means a dynamic presentation that adapts on-the-fly to varying and unforeseen requirements by both producer and consumer of the raw data.

MARC (*MAchine-Readable Cataloging*) project defines a data format which emerged from an initiative begun in the 1970s, led by the U.S. Library of Congress. MARC became USMARC in the 1980s and MARC 21 in the late 1990s. It provides the mechanism by which computers exchange, use and interpret bibliographic information and its data elements make up the foundation of most library catalogs used today.

Message, a higher logical unit of data, comprising one or more network packets, and defined by the implementation protocol.

Metadata is additional information that describes the data with which it is associated.

Middleware, a third-party layer between applications and infrastructure.

MIME (*Multipurpose Internet Mail Extensions*) extends the format of Internet mail to allow non-US-ASCII textual messages, non-textual messages, multi-part message bodies, and non-US-ASCII information in message headers. MIME is also widely used in Web contexts to content-declare client-server exchanges and similarly extend the capability of what was also originally ASCII-only. MIME is specified in RFC 2045 through 2049 (replacing 1521 and 1522).

Namespace is the abstract set of all names defined by a particular naming scheme – for example, all the possible names in a defined top level Internet domain, as constrained by allowable characters and name length.

NIC (*Network Information Center*) is the common term used in connection with a domain name database owner or primary registrar – for example, the original *InterNIC* (a registered service mark of the U.S. Department of Commerce, licensed to ICANN, which operates the general information Web site *www.internic.net*), a particular *gTLD* database owner (such as *www.nic.info*), or a national *ccTLD* administrator (such as NIC-SE, *www.nic-se.se,* for Sweden).

NIC (*Network Interface Card*) is a common abbreviation for the ethernet adapter card that connects a computer or device with the network cable on the hardware level.

Ontology is a collection of statements (written in a semantic language such as RDF) that define the relations between concepts and specify logical rules for reasoning about them. Computers can thus 'understand' the meaning of semantic data in Web content by following links to the specified ontologies.

Open protocol, the specifications are published and can be used by anyone.

Open source, opposite of proprietary 'closed' source. 'Open' means that the source code to applications and the related documentation is public and freely available. Often, runnable software itself is readily available for free.

OSI reference model (*Open Systems Interconnect protocol layers*), see Figure A.1, with reference to the OSI diagrams in Chapter 1 and 2, and to the native implementation examples. (.NET usually runs at the Application layer.)

OWL is the W3C recommendation for Sweb ontology work.

OSI Reference Model

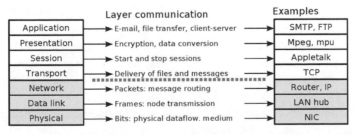

Figure A.1 An indication of what kind of communication occurs at particular levels in the OSI model, and some examples of relevant technologies that function at the respective levels. The top four are 'message based'

p2p *(peer-to-peer)* designates an architecture where nodes function as equals, showing both server and client functionality, depending on context. The Internet was originally p2p in design, and it is increasingly becoming so again.

P3P *(Platform for Privacy Preferences)* is a W3C recommendation for managing Web site human-policy issues (usually user privacy preferences).

Packet, a smallest logical unit of data transported by a network, which includes extra header information that identifies its place in a larger stream managed by a higher protocol level.

Persistency, the property of stored data remaining available and accessible indefinitely or at least for a very long time, in some contexts despite active efforts to remove it.

PIM, Personal Information Manager.

Platform, shorthand for a specific mix of hardware, software, and possibly environment that determines which software can run. In this sense, even the Internet as a whole is a 'platform' for the (possibly distributed) applications and services that run there.

Protocol, specifies how various components in a system interact in a standardized way. Each implementation is defined by both model (as a static design) and protocol (as a specified dynamic behavior). A protocol typically defines the acceptable states, the possible outcomes, their causal relations, their meaning, and so on.

Provenance is the audit trail of knowing where data originate, and who owns them.

Proxy (also see gateway), an entity acting on behalf of another, often a server acting as a local gateway from a LAN to the Internet.

PURL *(Persistent Uniform Resource Locator)* is a temporary workaround to transition from existing location-bound URL notation to the more general URI superset.

Push, a Web (or any) technology that effectively broadcasts or streams content, as distinct from 'pull' that responds only to discrete, specific user requests.

QoS (Quality of Service) is a metric for quantifying desired or delivered degree of service reliability, priority, and other measures of interest for its quality.

RDF *(Resource Description Framework)* is a model for defining information on the Web, by expressing the meaning of terms and concepts in a form that computers can readily process. RDF can use XML for its syntax and URIs to specify entities, concepts, properties, and relations.

RDFS *(RDF Schema)* is a language for defining a conceptual map of RDF vocabularies, which also specifies how to handle and label the elements.

Reliable and **unreliable packet transport** methods are distinguished by the fact that reliable transport requires that each and every message/packet is acknowledged when received; otherwise, it will be re-sent until it is acknowledged, or a time-out value or termination condition is reached.

Representational, when some abstraction is used for indirect reference instead of the actual thing – a name, for example.

Reputability is a metric of trust, a measure of known history (reputation).

Resource is Web jargon for any entity or collection of information, and includes Web pages, parts of a Web page, devices, people and more.

RFC (*Request For Comment*) in the Internet context designates a body of technical specifications overseen by the IETF which encompasses both proposals and consensus standards that define protocols and other aspects of the functional network.

RPC, remote procedure call, a protocol extension that enables remote software to directly invoke a host's local API (application program interface) functionality.

RSS is a common term for several related protocols for summary syndication of content. It is a simple way for clients to 'subscribe' to change notification.

Schema is a term widely used to designate a kind of relationship table between terms and their meanings in a given context. Such tables can be used to map the meaning of particular terms to corresponding terms in different logical structures and different application contexts.

Semantic Web (sweb) is the proper name for the 'third-generation' Web effort of embedding meaning (semantics) in Web functionality.

Service discovery is the term for the process of locating an agent or automated Web-based service that will perform a required function. Semantics enable agents to describe to one another precisely what function they carry out and what input data are needed.

SGML (*Standard Generalized Markup Language*) is an ISO standard (ISO 8879:1986) which defines tag-building rules for description encoding of text. HTML, XML, and most other markup languages are subsets of SGML.

SSL (*Secure Socket Layer*) is a protocol for securely encrypting a connection using exchanged public keys between the endpoints, usually seen in but not limited to the HTTPS Web document request.

Swarm distribution, when peers adaptively source downloaded content to other peers requesting the same material. Random offsets ensure quick fulfillment. Swarm services in general are network services implemented by cooperating nodes, often self-organizing in adaptive ways.

Swarm storage, when content is fragmented and distributed (with redundancy) to many different nodes. On retrieval, swarms adaptively cooperate to source.

Sweb (Semantic Web, SW) is a common abbreviation used to qualify technologies associated with the Semantic Web effort.

SWS (Semantic Web Service) is to Web Service what the Semantic Web is to the Web.

TLD (*Top Level Domain*) is the root abstraction for HTTP namespaces, dereferenced by Internet DNS. Also see *gTLD* and *ccTLD*.

Triple is a *subject-predicate-object* expression of three terms that underlies RDF.

UDDI (*Universal Description, Discovery and Integration*) is a specification that enables businesses to find and transact dynamically with one another. UDDI encompasses describing a business and its services, discovering other businesses that offer desired services, and integrating with them.

URI (*Uniform Resource Identifier*) is a complete and unique scheme for identifying arbitrary entities, defined in RFC 2396 *(www.ietf.org/rfc/rfc2396.txt)*.

URI persistence is the desired characteristic that URI addresses remain valid indefinitely. Its opposite is 'link-rot' expressed as resource-not-found.

URL (*Uniform Resource Locator*) is a standard way to specify the location of a resource available electronically, as a representation of its primary access mechanism – it is the addressing notation we are used to from Web and other Internet clients (including e-mail). URLs are a subset of the URI model and are defined in RFC 1738.

URN (*Uniform Resource Name*) is another subset of URI, and refers to resource specifiers that are required to remain globally unique and persistent even when the resource ceases to exist or becomes unavailable. It is thus a representation based on resource name, instead of location as in the familiar URL. It is defined in RFC 2141.

W3C (*World Wide Web Consortium, www.w3c.org*) was created in October 1994 to develop interoperable technologies (specifications, guidelines, software, and tools) to lead the Web to its full potential. W3C is a forum for information, commerce, communication, and collective understanding, and is the custodian of numerous open protocols and APIs.

WebDAV, see DAV

Web Services (WS) is a common name applied to functionality accessed through any URI, as opposed to static data in stored documents.

WSDL (*Web Services Description Language*), is a modular interface specification to Web Services.

WUM (Web Usage Mining) describes technologies to profile how users utilize the Web and its different resources.

XML (*eXtensible Markup Language*) is a markup language intended to supplant HTML, transitionally by way of an intermediate markup called XHTML (which is HTML 4.2 expressed in XML syntax). XML lets individuals define and use their own tags. It has no built-in mechanism to convey the meaning of the user's new tags to other users.

XMLP (*eXtensible Markup Language Protocol*) defines an XML-message-based message protocol to encapsulate XML data for transfer in an interoperable manner between distributed services.

XLink (*XML Linking Language*) defines constructs that may be inserted into XML resources to describe links between objects, similar to but more powerful than hyperlinks. XLink also uses XPath.

XPath (*XML Path Language*) is an expression language used by XSLT and XLink to access or refer to internal parts of an XML document.

XPointer (*XML Pointer Language*), is based on the XML Path Language (XPath), and supports addressing into the internal structures of XML documents. It allows for traversals of a document tree and choice of its internal parts based on various properties, such as element types, attribute values, character content, and relative position.

XSL (*Extensible Stylesheet Language*) is a language, or more properly a family of W3C recommendations, for expressing stylesheets in XML (see CSS). It consists of three parts: XSL Transformations (XSLT, a language for transforming XML documents); XPath; and XSL Formatting Objects (XSLFO, an XML vocabulary for specifying formatting semantics). An XSL stylesheet specifies the presentation of a class of XML documents by describing how an instance of the class is transformed into an XML document that uses the formatting vocabulary.

RDF

The following sections complement the Chapter 5 descriptions of RDF and RDF-Schema.

RDF Schema Example Listing

An RDF schema for defining book-related properties is referenced in Chapter 5, apropos RDF schema chaining. The chosen example is from a collection of Dublin Core draft base schemas (at *www.ukoln.ac.uk/metadata/dcmi/dcxml/examples.html*). It spreads over several pages in this book, so it is not suitable for inclusion in the body text. The example schema is referenced by name as *dc.xsd.*

```
<xs: schema
xmlns:xs="http://www.w3.org/2001/XMLSchema"
xmlns:x="http://www.w3.org/XML/1998/namespace"
xmlns:xlink="http://www.w3.org/1999/xlink"
xmlns="http://purl.org/dc/elements/1.1/"
targetNamespace="http://purl.org/dc/elements/1.1/"
elementFormDefault="qualified"
attributeFormDefault="qualified">
<xs:annotation>
<xs: documentation xml:lang="en">
XML Schema 2001-12-18 by Pete Johnston
Based on Andy Powell,
Guidelines for Implementing Dublin Core in XML, 9th draft.
This XML Schema is for information only
</xs:documentation>
</xs:annotation)
<xs: import namespace="http://www.w3.org/XML/1998/namespace"
  schemaLocation=
  "http://www.ukoln.ac.uk/metadata/dcmi/dcxml/xmls/xml.xsd">
</xs:import>
<! —
  <xs:import namespace="http://www.w3.org/XML/1998/namespace"
  schemaLocation="http://www.w3.org/2001/xml.xsd">
  </xs:import>
  —>
<xs:import namespace="http://www.w3.org/1999/xlink"
  schemaLocation=
  "http://www.ukoln.ac.uk/metadata/dcmi/dcxml/xmls/xlink.
        xsd">
</xs:import>
<xs:complexType name="elementType">
  <xs:simpleContent>
  <xs:extension base="xs:string">
  <xs:attribute ref="x:lang"/>
```

```
   </xs:extension>
   </xs:simpleContent>
 </xs:complexType>
 <xs:complexType name="titleType">
  <xs:simpleContent>
  <xs:extension base="elementType">
  </xs:extension>
  </xs:simpleContent>
 </xs:complexType>
 <xs:element name="title" type="titleType"/>
 <xs:complexType name="agentType">
  <xs:simpleContent>
  <xs:extension base="elementType">
  <xs:attributeGroup ref="xlink:metadataLink"/>
  </xs:extension>
  </xs:simpleContent>
 </xs:complexType>
 <xs:element name="creator" type="agentType"/>
 <xs:complexType name="subjectType">
  <xs:simpleContent>
  <xs:extension base="elementType">
  </xs:extension>
  </xs:simpleContent>
 </xs:complexType>
 <xs:element name="subject" type="subjectType"/>
 <xs:complexType name="descriptionType">
  <xs:simpleContent>
  <xs:extension base="elementType">
  </xs:extension>
  </xs:simpleContent>
 </xs:complexType>
 <xs:element name="description" type="descriptionType"/>
 <xs:element name="publisher" type="agentType"/>
 <xs:element name="contributor" type="agentType"/>
 <xs:complexType name="dateType">
  <xs:simpleContent>
  <xs:extension base="elementType">
  </xs:extension>
  </xs:simpleContent>
 </xs:complexType>
 <xs:element name="date" type="dateType"/>
 <xs:complexType name="typeType">
  <xs:simpleContent>
  <xs:extension base="elementType">
  </xs:extension>
  </xs:simpleContent>
```

```
</xs:complexType>
<xs:element name="type" type="typeType"/>
<xs:complexType name="formatType">
 <xs:simpleContent>
 <xs:extension base="elementType">
 </xs:extension>
 </xs:simpleContent>
</xs:complexType>
<xs:element name="format" type="formatType"/>
<xs:complexType name="identifierType">
 <xs:simpleContent>
 <xs:extension base="elementType">
 </xs:extension>
 </xs:simpleContent>
</xs:complexType>
<xs:element name="identifier" type="identifierType"/>
<xs:complexType name="sourceType">
 <xs:simpleContent>
 <xs:extension base="elementType">
 </xs:extension>
 </xs:simpleContent>
</xs:complexType>
<xs:element name="source" type="sourceType"/>
<xs:complexType name="languageType">
 <xs:simpleContent>
 <xs:extension base="elementType">
 </xs:extension>
 </xs:simpleContent>
</xs:complexType>
<xs:element name="language" type="languageType"/>
<xs:complexType name="relationType">
 <xs:simpleContent>
 <xs:extension base="elementType">
 <xs:attributeGroup ref="xlink:metadataLink"/>
 </xs:extension>
 </xs:simpleContent>
</xs:complexType>
<xs:element name="relation" type="relationType"/>
<xs:complexType name="coverageType">
 <xs:simpleContent>
 <xs:extension base="elementType">
 </xs:extension>
 </xs:simpleContent>
</xs:complexType>
<xs:element name="coverage" type="coverageType"/>
<xs:complexType name="rightsType">
```

```
   <xs:simpleContent>
   <xs:extension base="elementType">
   </xs:extension>
   </xs:simpleContent>
 </xs:complexType>
<xs:element name=" rights" type="rightsType"/>
<xs:group name="elementsGroup">
   <xs:choice>
   <xs:element ref="title" />
   <xs:element ref="creator" />
   <xs:element ref="subject" />
   <xs:element ref="description" />
   <xs:element ref="publisher" />
   <xs:element ref="contributor" />
   <xs:element ref="date" />
   <xs:element ref="type" />
   <xs:element ref="format" />
   <xs:element ref="identifier" />
   <xs:element ref="source" />
   <xs:element ref="language" />
   <xs:element ref="relation" />
   <xs:element ref="coverage"/>
   <xs:element ref="rights" />
   </xs:choice>
 </xs:group>
 </xs:schema>
```

RDF How-to

The following example shows how to 'sign' any published Web page so that it provides RDF meta-data. The method shown here is a quick-start way to 'join the Semantic Web' without having to change existing content or website structure.

The process requires at least two components:

- an RDF description file of you as the content creator;
- a metadata block on a relevant HTML or XML page.

The Creator Description File

The RDF description is stored as a flat-text file on your server, in any suitable public-web location that can be accessed using a normal URL (URI). We use a W3C-defmed 'example' domain.

```
<rdf:RDF
  xmlns:rdf="http://www.w3.org/1999/02/22-rdf-syntax-ns#"
  xmlns:rdfs=http://www.w3.org/2000/01/rdf-schema#"
  xmlns:wn="http://xmlns.com/wordnet/1.6/"
```

```
  xmlns:dc="http://purl.org/dc/elements/1.1/"
  xmlns:="http://xmlns.com/foaf/0.1/">
<wn:Person rdf:ID="sw">
  <name>Sem Webb</name>
  <mbox rdf:resource="mailto:sem@example.com" />
  <homepage rdf:resource="http://example.com/~sem/"/>
  <pubkeyAddress rdf:resource="http://example.com/~sem/mypubkey.asc" />
</wn:Person>
</rdf.RDF>
```

You would of course edit the variable values and pointers (highlighted bold) to suit your particular situation. Note the optional reference to a public PGP key.

Locating this file as *http://example.com/~sem/about.xrdf* (typical for a user homepage root) provides you, as referenced person, with a fully qualified **sweb address**. In this example, it is:

```
http://example.com/~sem/about.xrdf#sw
```

The Metadata Block

Metadata blocks are inserted in the head-block of normal HTML pages, describing content metadata in Dublin Core terms. In this example, we place a description with reference to the RDF description file just created:

```
<rdf:RDF  xmlns:rdf="http://www.w3.org/1999/02/22-rdf-syntax-
     ns#"
        xmlns:dc="http://purl.org/dc/elements/1.1/"
        xmlns:wot="http://xmlns.com/wot/0.1/">
  <rdf:Description rdf:about=""
       dc:title            = "My document"
       dc:description      = "Experiments with sweb and rdf."
       dc:date             = "2003-10-12" >
     <dc:creator rdf:resource="http://example.com/~sem/about.
           xrdf#sw" />
     <wot:assurance rdf:resource="http://example.com/~sem/page1.
           asc" />
  </rdf:Description>
</rdf:RDF>
```

Again, replace highlighted variable values with your own.

At this point, your page is sweb-compliant in terms of RDF. Obviously, further metadata may be added to better describe the content than in this minimal example.

The optional web-of-trust (*wot*) 'assurance' entry refers to a digital signature created of the completed page (source) using your private key – for example using the *GnuPG* command '`gpg -ba page1.html`' if that is the program you use. A user client can thus validate that the received document is identical to the page you created and signed.

Appendix B

Semantic Web Resources

This appendix collects references for further reading, which in some cases goes beyond what is mentioned in the text. For the many URI links, you might prefer to visit the book's Web site where these links are published in active form.

At a Glance

Further Reading summarizes resources of interest.

- *Book Resources* lists some current sweb resources in print, sorted into overview, technical and other/AI groups.
- *Other Publications* notes significant periodical publications.
- *Web Resources* summarizes important online sweb resources, grouped by main technology or focus area.
- *Implementation Resources* has pointers mainly to How-to tutorials for core technologies.
- *Miscellaneous* collects supplementary or slightly off-topic resources.

Further Reading

It is in the nature of hot new subjects that most resource material is in the form of scattered documents and resources on the Web. Much of this material is both very specific and narrow, dealing with only one or another implementation. This fact was one motivation to write this book, to try and collect useful information in one place for people who are looking for a concise technology overview.

Book Resources

When this book was started in 2002, few books were published specifically about Semantic Web technologies or how they function. This situation improved over the following two years, and relevant titles that seem worth pursuing are listed here.

The Semantic Web: Crafting Infrastructure for Agency   Bo Leuf
© 2006 John Wiley & Sons, Ltd

Overview

These titles provide an overview, at least, within one or more core sweb technology areas.

Spinning the Semantic Web: Bringing the World Wide Web to Its Full Potential, edited by Dieter Fensel, James A. Hendler, Henry Lieberman, and Wolfgang Wahlster, MIT Press, 2002 (Foreword by Tim Berners-Lee)
The Semantic Web: A Guide to the Future of XML, Web Services, and Knowledge Management, by Michael C. Daconta, Leo J. Obrst, and Kevin T. Smith, John Wiley & Sons, 2003
A Semantic Web Primer, by Grigoris Antoniou and Frank van Harmelen, MIT Press, 2004
Explorer's Guide to the Semantic Web, by Thomas B. Passin, Manning Publications Company, 2004
Towards the Semantic Web: Ontology-Driven Knowledge Management, by John Davies, Dieter Fensel, Frank van Harmelen, and Frank van Harmelen, John Wiley & Sons, 2003
Semantic Web: A Field Guide, by Thomas B. Passin, Manning Publications Company, 2003
Introduction to the Semantic Web and RDF, by A.M. Kuchling, PyCon, 2003
Service-Oriented Computing: Semantics, Processes, Agents, by Munindar P. Singh and Michael N. Huhns, John Wiley & Sons, 2005

Technical

The following books take more technical approaches to specific areas.

XML Databases and the Semantic Web, by Bhavani Thuraisingham and Bhavani Thuraisingha, CRC Press, 2002
Definitive XML Application Development, by Lars Marius Garshol, Prentice-Hall, 2002
Creating the Semantic Web with RDF: Professional Developer's Guide, by Johan Hjelm, John Wiley & Sons, 2001
Practical RDF, by Shelley Powers, O'Reilly & Associates, 2003
Ontologies: A Silver Bullet for Knowledge Management and Electronic Commerce, by Dieter Fensel, Springer Verlag, 2001
Ontological Engineering with examples from the areas of Knowledge Management, e-Commerce and the Semantic Web, by Asunción Gómez-Pérez, Mariano Fernández-López, and Oscar Corcho, Springer Verlag, 2002
Developing Semantic Web Services, by H. Peter Alesso and Craig F. Smith, AK Peters, Ltd, 2004

Other sweb /AI-related

Internet Based Workflow Management: Towards a Semantic Web, by Dan C. Marinescu, John Wiley & Sons, 2002
Web Intelligence, by Ning Zhong, Jiming Liu, and Yiyu Yao, Springer Verlag, 2003
Visualizing the Semantic Web, by Vladimir Geroimenko and Chaomei Chen, Springer Verlag, 2003
The Description Logic Handbook: Theory, Implementation and Applications, edited by Franz Baader, Cambridge University Press, 2003

Other Publications

Some longer descriptions of various sweb technologies have been published in various periodicals over the past few years.

'The Semantic Web', *Scientific American* special issue, May, 2001, by Tim Berners-Lee, James Hendler and Ora Lassila (also see SciAm Web site: *www.sciam.com/2001/0501issue/0501berners-lee.html*). The seminal article for greater developer, and to some extent public, awareness.

'Ontologies and the Semantic Web for E-learning', *IFETS Journal* Special Issue, **7**(4), October 2004 (*ifets.info*). The journal addresses education issues.

'Ontology specification languages for the Semantic Web', Asunción Gómez-Pérez, and Oscar Corcho, *IEEE Intelligent Systems*, Jan/Feb 2002 *(www.computer.org/intelligent/)*. Most issues of this periodical have one or more articles relating to intelligent agents and other aspects of sweb technology.

'Intelligent Agents Meet the Semantic Web in Smart Spaces', *IEEE Internet Computing*, **8**(6), Nov/Dec 2004. This periodical has occasional sweb-specific articles.

Just before the publication process stopped further updates, I ran across the following highly interesting article about a study in how easy sweb implementation can be:

'The Semantic Web in One Day', York Sure, Pascal Hitzler, Andreas Eberhart, and Rudi Studer, in *IEEE Intelligent Systems*, May/June 2005. To determine just how far Semantic Web technologies have come, the authors created a snapshot of what you could do by applying and assembling existing Semantic Web technologies – in one day. In the summary section, the authors note: '. . . we were surprised to see that the systems that emerged after 24 hours were much more sophisticated and functional than we expected.'

Web Resources

Some of the more active and comprehensive Web resources that deal with the Semantic Web and related technologies are found either on the W3C Web site or are linked from it.

Significant URLs include:

- *W3C Org* (*www.w3.org/2001/sw/*): World Wide Web Consortium Semantic Web Initiative.
- *Semantic Web Org* (*www.semanticweb.org*): Portal of the Semantic Web Community. Projects, tools and ongoing events.
- *Ontology Org* (*www.ontology.org*): Ontology Org was formed in May of 1998 to highlight the need for ontology in Internet commerce.

Some relevant Internet search categories can be found in the *Google Directory Listings* – for example:

- *directory.google.com/Top/Computers/Artificial_Intelligence/Knowledge_Representation/ Semantic_Web/*
- *directory.google.com/Top/Reference/Libraries/Library_and_Information_Science/Technical Services/Cata loguing/Metadata/RDF/Technical_Articles_and_TechNotes/*

Internet Interoperability and Recommendations

Both these sites have many further links to specific issues:

- W3C, World Wide Web Consortium, *w3.org*
- IETF, Internet Engineering Task Force, *www.ietf.org*

Primary Technical References

This section summarizes the primary online technical references for the core sweb technologies.

A full list of published W3C recommendations is at the TR-root *(w3.org/TR/)*, along with candidates and draft proposals. Note that the given W3C links *(www.* prefix optional) are the preferred ones, which are when applicable automatically repointed at the Web site to the most recent (date-qualified) version of the respective document.

XML

- W3C recommendation: *w3.org/TR/REC-xml*
- Extensible Markup Language (XML) 1.1: *w3.org/TR/xml11*
- *Namespaces* in XML: *w3.org/TR/REC-xml-names/*
- XML events: *w3.org/TR/xml-events*
- *XForms* 1.0: *w3.org/TR/xforms/*
- DOM Level 3 Validation Specification: *w3.org/TR/DOM-Level-3-Val*

RDF

- RDF/XML Syntax Specification: *w3.org/TR/rdf-syntax-grammar/*
- Concepts and Abstract Syntax: *w3.org/TR/rdf-concepts/*
- RDF Semantics: *w3.org/TR/rdf-mt/*
- RDF Vocabulary Description Language 1.0: RDF Schema: *w3.org/TR/rdf-schema/*

OWL

- Technical reference: *w3.org/TR/owl-ref/*
- Features overview: *w3.org/TR/owl-features/*
- Language Guide: *w3.org/TR/owl-guide/*
- Language semantics: *w3.org/TR/owl-semantics/*

Web architecture

- Architecture of the WWW (Vol 1, December 2004): *w3.org/TR/webarch/*

Implementation Resources

Sweb implementations must at this stage be seen as advanced prototypes at best, subject to rapid change or alternatively abandonment. As noted in the text, some are more responses to specific problem statements than any attempt to be a generic or compliant SW implementation.

How-to references

- An introduction to ontologies: *www.SemanticWeb.org/knowmarkup.html*
- Simple HTML Ontology Extensions Frequently Asked Questions (SHOE FAQ): *www.cs.umd.edu/projects/plus/SHOE/faq.html*
- How to sign your pages with RDF and join the Semantic Web: *logicerror.com/signYourPage*

- RDF Primer: *w3.org/TR/rdf-primer/*
- RDF/OWL Tutori al: *www.w3schools.com/rdf/default.asp*
- Other free 'Web building' tutorials online (XML, XSL, XPath, XQuery, XML Schema, SOAP, WSDL, etc.): *www.w3schools.com*
- XML Schema Primer: *w3.org/TR/xmlschema-0/* (part 1 of 3)

Miscellaneous

These are assorted references that are only implied in the text, or are slightly off-topic.

- Sharon Hopkins: 'Camels and Needles: Computer Poetry Meets the Perl Programming Language': *www.wall.org/~sharon/plpaper.ps*
- Obtaining and using *dmoz ODP* RDF data: *dmoz.org/help/getdata.html*
- IEEE Transactions on Evolutionary Computation: *www.ieee-nns.org/pubs/tec/*
- IEEE Transactions on Fuzzy Systems: *ieee-cis.org/pubs/tfs/*
- IEEE Transactions on Knowledge & Data Engineering: *www.computer.org/tkde/*
- IEEE Pervasive Computing: *www.computer.org/pervasive/*
- IEEE Systems, Man and Cybernetics Society (3 publications): *www.ieee-smc.org/*
- IEEE Technology and Society Magazine: *www.njcc.com/~techsoc/*

Appendix C

Lists

This appendix collects quick reference lists of included page elements: Bits, Tables, and Figures.

Bits

Figures

Tables

Index   355

Index